BOTTLED ENERGY

ELECTRICAL ENGINEERING AND THE EVOLUTION OF CHEMICAL ENERGY STORAGE

BOTTLED ENERGY:

ELECTRICAL ENGINEERING AND THE EVOLUTION OF CHEMICAL ENERGY STORAGE

RICHARD H. SCHALLENBERG

AMERICAN PHILOSOPHICAL SOCIETY
Independence Square • Philadelphia

Cover illustrations: top figure taken from
E.J. Wade, *Secondary Batteries*, p. 105;
bottom figure taken from Gaston Planté,
The Storage of Electrical Energy, p. 30.

Printed in the United States of America by
George H. Buchanan Co., Inc.
Philadelphia, Pennsylvania 19146

Library of Congress Catalog Card No.: 80-68493
International Standard Book Number: 0-87169-148-5
US ISSN 0065-9738

Contents

Foreword: On the Author

Richard Schallenberg completed work on *Bottled Energy*, his history of the electric storage battery, only a few months before a diabetic seizure took his life. He was only 36 years old, had just been approved by the university for tenure and promotion, and seemed to be well started on a productive and useful career. The promise of spring 1980, however, was not to be fulfilled.

Schallenberg took his first degree, in engineering, in 1965, and worked as a research engineer for a year before going into the Army. He was awarded the Bronze Star for service in Vietnam. In 1969 he entered graduate school at Yale in the history of science, receiving his M.S. in 1970 and his Ph.D. in 1973. He was an assistant professor in the history of science and technology at Virginia Tech from 1973 until the time of his death. His classes were not only heavily enrolled but also generally praised. He made a lasting impression on students and colleagues alike as a man of boundless energy, enthusiasm, and insights.

The present work developed out of his doctoral dissertation and was completed with a very generous grant from the National Science Foundation. In describing his project before entering on the major part of his work, Dr. Schallenberg noted that he was "fascinated by studying the process by which modern industry decides to innovate." His purpose was to examine the ways in which industrialists and engineers have opted for change in the design and construction of batteries. He planned to use a "comparison of the need for batteries to perform certain functions and the actions of inventors to provide such special purpose batteries . . . as a case study of the motivation for industrial research and development." True to the historical side of his profession, Schallenberg was convinced that "an historical study of past battery innovation should be required for all new battery engineers, since it is common for new men . . . to underestimate the difficulties involved in battery design and to repeat old mistakes."

It seems certain that, had he lived, Schallenberg would have dedicated this volume to his parents, Mr. and Mrs. J. E. Schallenberg. The family relationship was very close, and both mother and father had been completely supportive through the years of education, army service, graduate work, and establishment. In the circumstances Richard's colleagues would like to augment that dedication—to Richard's mother and father for their endless love and encouragement, and to the memory of the author, Richard H. Schallenberg.

WILLIAM E. MACKIE
Chairman, History Department
Virginia Polytechnic Institute &
State University

*   *   *   *   *

Richard H. Schallenberg was born in Chicago in June, 1943, and got his B.S. in chemical engineering from Brooklyn Polytechnic Institute in 1965. In 1968 he came back from the War in Vietnam and started reading economic history at Columbia, then continuing at Brooklyn Polytechnic Institute with Rom Sviedrys, who whetted his interest in the history of technology and led him to do his early pioneer work on the charcoal iron industry. He came to Yale in 1970, still markedly affected by his war experience and by a brittle diabetic condition. From the beginning he was a model student; I have never known another who habitually turned in all papers a week early and his thesis six months ahead of deadline. He saw at the outset that the crucial areas in history of technology were its interfaces with business and economic history on the one side, and history of science on the other. He set out to make himself well read and competent in both areas, and he succeeded, and exhibited mastery of such consolidation of internal and external histories of science and technology in his thesis on "The Electric Battery 1800-1930, a case study in technological innovation." On appointment to Virginia Polytechnic Institute and State University he rapidly became an excellent teacher, full of new plans for the substantive development of our field. Professionally he gave tremendous service within our societies, most notably by doing all the work that enabled us to have a series of Guides to Graduate Study and estimates of manpower and employment at a time when such documentation was crucial for all institutions in the fields. Just as he had won his tenure and associate professorship and was in the midst of publishing increasingly important papers, Dick died suddenly at his home in April, 1980. He had achieved so much already, and we are in his debt; but the loss of one of the few such competent persons is a professional as well as a personal matter of heavy grief.

DEREK DE SOLLA PRICE
Avalon Professor of the History
of Science
Yale University

Acknowledgments

The following are the names of individuals, companies, and institutions gleaned from the author's files and tapes who should be acknowledged as being of assistance and contributing to the content of this book:

Richard G. Acton, M. Barak, James E. Brittain, Leah Burt, Francis Celoria, George Dubpernell, Ralph Fries, Arthur J. Hedges, Nels E. Hehner, Thomas Parke Hughes, Ray Jardin, W. James King, Martin J. Klein, Robert E. Kohler, Philip E. Krouse, D.E. Lighton, David Lindberg, Edgar Oldham, Nathan Reinberg, George Rosen, and Elliot Sivowitch.

Bell Telephone, Chloride Group Ltd., Electric Storage Battery Company of Philadelphia, Pa., Exide Corporation, Gould-National Batteries Inc., McGraw Edison, Oldham & Son, Ltd., Ray-O-Vac, Saft Company, Union Carbide Corporation, Varta Batterie AG.

Eleutherian Mills Historical Library, Greenville, Delaware; Federal Records Center, Suitland, Maryland; Engineering Societies Library, New York City; National Archives and Smithsonian Institution Archives, Washington, D.C.; Mr. C.G. Povey, Telecommunications Museum, U.K.; Mrs. I.M. McCabe, Royal Institution, U.K.

Dr. Derek J. de Solla Price; the Virginia Polytechnic Institute and State University's Photo Laboratory staff; the Carol M. Newman Library staff; and the staff and faculty of the History Department at Virginia Polytechnic Institute and State University.

Introduction

The Evolution of a Technology

Storage batteries have been made and sold for more than a century. To the modern reader, the storage battery may seem to be little more than the device which starts his car, but during the hundred years of its commercial existence, the battery has filled many different roles. These roles have changed with the years, the technology of the storage battery adapting itself to these changes. This book is a study of the evolution and interaction between changes in the needs for batteries and the response of battery designers to these needs.

The choice of the words "evolution" and "adaptation" to describe the development of storage battery technology is done to draw the analogy with the mechanisms and dynamics of biological change. Contrary to the opinions of some recent critics, technology is not an exogenous force in society, developing in terms of its own kinetics, and leading rather than being led by society. The technological structure of society is the sum of all the individual technologies, large and small, of that society. These technologies can be heuristically compared to varieties, species, genera in the biological world, and their evolution can and should be analyzed in terms of the changing environment of the society in which they evolve.

No attempt has been made to develop an elaborate analog model of technological development using the image of biological evolution. Rather, the analogy has been used loosely as an overall framework for understanding the evolution, adaptation, and century-long survival of the one particular "species" of modern technology under study.

The storage battery has been a very successful species. This success comes from the readiness with which the technology has been adapted to new needs. This adaptability was essential for the survival of the storage battery, since, up until the third decade of the twentieth century, the purposes for which batteries were used continually changed, with old uses constantly disappearing, thereby destroying old technological environments in which the batteries thrived.

Storage batteries emerged out of the environment of early nineteenth-century science. Electrochemistry was the glamour field of physical science of its day; like sub-atomic physics today, it promised to unlock the basic secrets of matter. The nineteenth century also saw the birth of the concepts of conservation and interconvertibility of energy. Rechargeable batteries are an excellent physical illustration of these concepts, since they convert electrical energy into chemical energy and vice versa.

By the middle of the last century, the growing field of telegraphy started to create a favorable technical environment for storage battery innovation. As the technology of telegraphy grew in complexity, it developed a need for different kinds of power sources; the simple primary batteries used in the 1830s and '40s were insufficient for the lengthening lines and complicated instruments. During the 1850s, '60s and '70s, a series of inventors tried to adapt the unique properties of rechargeable cells to the needs of telegraphy. They failed, but their failure succeeded in laying the foundation for the successful storage battery industry of the 1880s.

The electrical revolution of the 1880s greatly expanded the innovative environment for electrical technology. The 1880s were a technically chaotic period; standard solutions to specific technical problems had not yet been developed. The new environment of electric light and power was filled by engineers from the older environments—electroplaters, telegraphic engineers, chemists, and mechanical engineers. Each of these different groups developed solutions for the new technologies which were abstracted from each group's prior technical methods. By the 1890s, a set of generally accepted electrical engineering methods and standards started to take shape as textbook writers, technical society journals, and engineering college professors began the codification of the technology. In the 1880s, however, the technological environment was too new for a separation to have occurred between techniques of different relative values.

One result of this rapid mushrooming of the environment and its subsequent confusion was that the storage battery was proposed for a wide diversity of new applications—as a voltage and current controller; as a voltage divider; as a transformer; as an emergency power source; as a dynamo-excitation power source; as a powerhouse load-equalization device; as a drive for self-propelled streetcars; as a means for dispensing with labor in power stations; as a substitute for primary batteries in telegraphy and telephony; and more. This sudden expansion in the environment for storage batteries led to the creation of many storage battery manufacturing companies in the 1880s. Most of this environment soon proved to be illusory, as batteries were found to be unsuitable for many of these roles, or, as better solutions were found, most of the companies soon disappeared. But enough of the environment did survive the 1880s to establish the storage battery industry on a small but firm base by the 1890s.

Two significant storage battery applications created in the 1880s survived into the 1890s: as an emergency power source and as a load-equalizer in both electric lighting and trolley power stations. The environment created by these applications stimulated the creation of the earliest commercially-successful storage batteries. These batteries were massively-built devices designed to store great amounts of electrical energy. At the same time, however, the rapid growth of the electric streetcar industry stimulated inventors to develop small, light-weight cells for vehicular drive.

Unfortunately, the self-propelled streetcar proved to be a failure, but by the time this had been recognized in the mid 1890s, the electric automobile had emerged as a new and more important environment for the evolution of vehicular batteries. Enthusiasm for the electric car only lasted about ten years, but by the time this enthusiasm had run its course, its stimulating influence was taken over by the new technology of the self-starter for gasoline cars. Since the 1920s, the gasoline automobile battery has remained the chief mainstay of the storage battery industry.

The chapters of this book separate the different environmental periods of the storage battery. Since the standard lead-acid battery has comprised the vast bulk of all rechargeable battery manufacture, the first four chapters deal solely with it. The last chapter is devoted to nickel-iron and nickel-cadmium batteries; these cells comprise almost all of the market not covered by lead-acid batteries. The history of other, more exotic forms of rechargeable batteries has been excluded, since none of these batteries has yet to achieve any sort of economic significance.

The first chapter, covering the years 1801-1879, concerns the period when the needs of science and telegraphy maintained a meager environment for the evolving storage battery. The second chapter, 1880-1889, deals with the fertile but continually changing environment of the first decade of electric light and power. The chaotic nature of the broader electrical technological environment of this decade is mirrored in the chaotic state of storage battery design and in the rapid formation and dissolution of battery companies. The third chapter, 1890 to 1900, covers the beginning of the mature phase of storage battery technology. During this period, electrical engineers identified a few restricted applications in powerhouse design for which storage batteries could be used efficiently. As the roles which storage batteries were called upon to play grew more restricted, so also did the designs of the batteries themselves grow more limited and standardized; the growth of techniques in the electrical engineering profession as a whole which stressed precision and the use of mathematics stimulated a similar response towards precision in the storage battery industry.

The fourth chapter covers the period from 1900 to the present, and stresses the dominant role of the automobile in battery design in this period.

The major portion of this study is restricted to a period of about thirty years at the beginning of the commercial history of the invention that is, 1880-1910, the reason being that this is a study of the interaction between the technology of chemical energy storage devices and the various technologies it was designed to serve. These technologies are considered as environments which directed the evolution of chemical energy storage technology. Therefore, the greatest emphasis has been placed upon those decades in which the environment changed most rapidly and when the greatest number of environments for the growth of chemical energy storage existed. This is not to imply that significant change in the nature of storage battery technology has not occurred during the past fifty years. However, although great changes have taken place in the more recent period, these changes have generally not been produced by changes in the uses to which batteries have been put, but have come about as the result of a more gradual evolutionary process of improvement stimulated from within the industry itself. In addition, the internal histories of storage battery companies have been omitted, except where they illustrate a historical point about the technological innovative process.

A number of general trends in the evolution of the storage battery industry can be made out against the background of the changing demand environment. One of these trends, which was most pronounced in the period 1880-1910, was the failure of attempts to manufacture batteries in the context of a larger industrial enterprise. Those companies which made batteries in addition to other items of electrical equipment were universally unsuccessful with their battery enterprise, regardless of how successful they may or may not have been with their other electrical goods. Only those firms which concentrated all their attention on storage batteries succeeded in developing economically viable batteries. Moreover, even those specialist firms which succeeded commercially, did so only when they were given external financial support by larger financial or industrial enterprises. The reasons for these very rigorous standards of economic survival will be examined in the second and third chapters.

Another general trend of the storage battery industry, which is partly related to the trends described above, was the tendency towards consolidation. In the early years (1880-1895) a large number of battery firms sprang up in Europe and the United States. Most of these firms were small, and their lifespans were roughly proportional to their size. Beginning in the mid 1890s, a single, large firm emerged in each of the major innovating countries—Britain, Germany, France and the United States. By the early years of the twentieth century, each of these firms was well on the way to creating a monopoly in its own country, and a worldwide cartel existed between the four major national firms. Successful but smaller-sized battery companies usually were amalgamated into the larger company in each country. This pattern of consolidation was not unlike that which Bright has described for the electric lamp industry.[1]

About the time of World War One, this process of consolidation was partly reversed. New firms emerged and existing ones grew stronger. The motivating factor here was the appearance of the gasoline car starting battery, which created a greatly expanded economic environment. The pre-1912 environment had been small and inflexible, whereas the new environment expanded with the explosive growth of the automobile industry. Moreover, by the 1920s, the technology of the storage battery had become sufficiently standardized so that new, smaller firms could enter the field successfully. A pool of professional storage battery engineers existed who could be hired away from the older companies to start up a new plant with a minimum of costly experimentation and false starts.

---

[1] Arthur A. Bright, *The Electric Lamp Industry*, Mac Millan Co., N.Y., 1949, p. 12.

Technological innovation itself has tended to follow this same general pattern of consolidation, although with no tendency towards greater diversity after 1912. Indeed, ever since 1880, the tendency has been for the various battery firms to develop a consensus technology for each application. Today, where small differences exist between the batteries made for the same purposes in different countries, these variations are usually caused by variations in the prices of alternative materials of construction.

I. A Solution in Search of a Problem

1. The Scientific Background

The origins of the storage battery idea lie in the history of early nineteenth-century science, where its birth is connected with the greatest triumph of nineteenth-century physical science—the concept of conservation of energy. The storage battery was the technological incarnation of that scientific principle. Conservation of energy grew out of the closely related idea of interconvertibility of forms of energy; the storage battery, by converting electrical to chemical energy and vice versa, was a physical embodiment of this idea.

The storage battery performs these functions by using a physical phenomenon known as electrochemical polarization. Polarization is a complex phenomenon which even today is incompletely understood. It is not so much a process as a symptom, and this symptom can be caused by a variety of physical-chemical processes, depending on the nature of the electrochemical systems which produce the symptom. This symptom is the decay of current output of a discharging battery, or, in an electrochemical system which receives current from an outside source, the increasing resistance of the system to current:

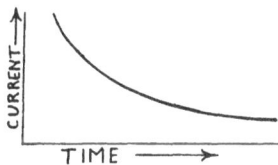

When polarization was first observed in the early nineteenth century, some scientists believed it was caused by the buildup of excess quantities of electricity in the electrodes. According to this model, the negative electrode attracts positively charged particles in the solution when the battery discharges. As the

1

positive particles give up their charges to the negative electrode, it becomes progressively positive. Consequently, the
initial polarity of the battery changes—the electrodes
"polarize" and the current diminishes. The technical term
"polarization" is still used today, although the modern scientific explanation has changed.

Battery current is produced by chemical reaction: reduction
at the cathode and oxidation at the anode. Polarization occurs
when something happens in the cell to slow down the rate of
oxidation or reduction. In most systems, it is the oxidation-
reduction reactions themselves which set up the conditions for
their own diminution, that is, polarization is a process of negative feedback.

The simplest case of this is gaseous polarization in a battery
with a zinc anode, a sulphuric acid electrolyte and a copper
cathode.[1] There is no problem at the anode; zinc oxidizes and
dissolves away into the solution. At the cathode, however, the
hydrogen ion is reduced and a microscopic film of hydrogen
builds up on the copper. This raises the electrical resistance of
the cathode, since hydrogen ions are prevented from penetrat-
ing to unite with an electron. Thus the current output of the
cell falls.

Polarization is also produced in passive systems; that is,
those which themselves cannot generate current, such as two
identical electrodes in a common electrolyte. If external current
is passed between the electrodes, chemical changes in either
the electrodes or the electrolyte or both take place. These chemical changes produce changes in the electrochemical potentials
of the electrodes. Unless the new chemicals are immediately
removed, a secondary EMF is created which opposes the
externally-supplied EMF, thus again "polarizing" the system.
If the external EMF is removed, the secondary EMF generates a
reverse current until the new chemical deposits are consumed.

Chemical energy storage is based on this kind of polarization. A storage battery consists of an electrochemically passive
system which has been deliberately polarized by passing external current through it. The chemicals generated on the electrodes by the current are called "active materials."

---

[1] Such a battery was usually called a Voltaic cell in the nineteenth century. The
original Volta battery consisted of a Zn anode, an Ag cathode, and a salt water electrolyte. The name Voltaic battery, however, was usually applied generically to any battery without a depolarizing system. The usual Voltaic battery had a Zn anode, a Cu
cathode and a sodium chloride, ammonium chloride or sulphuric acid.

The polarizing of a storage battery is called "charging." Charging places on the anode a chemical which is oxidized during discharge, and on the cathode, one that is reduced, that is, charging equips the cathode with a strong oxidizing agent. By contrast, in primary batteries (such as dry cells) the oxidizing and reducing agents are *mechanically* placed on the electrodes, rather than being placed there by *charging*. The cathodic oxidizer is called a "depolarizer," since it functions to oxidize positive ions which would otherwise build up around the cathode.

Most technologies have evolved independently of science. This has not been the case, however, for electrical technology. Almost all of the instrumentalities of electrical technology have undergone a three-phase evolutionary process: first, as the *objects* of scientific investigation; second, as *tools* used in scientific investigation; third, as tools for industrial or consumer use. For example, electromagnetic induction was first studied by Faraday and Oersted because it was an interesting physical phenomenon; then it was used in apparatus for the accurate, quantitative measurement of electrical phenomena; and finally, it was used in electric motors and the electromagnetic telegraph. The storage battery underwent the same three-phase pattern of evolution.

Phase one for the battery lasted from 1801 to the start of the 1850s. Phase two lasted from then until the start of the 1880s, although unsuccessful attempts were made during this period to use the battery as a practical engineering tool. Phase three—the successful engineering use of the storage battery—began in the 1880s—seventy-five years after its invention.

The reason for the late appearance of commercial storage batteries is that until the late 1870s only one practical source of electric current existed—the primary battery. The primary battery preempted the role of the storage battery since, in a manner of speaking, it is a storage battery; every battery stores electrical energy in the form of its chemical deposits. Therefore, until the late 1870s, the most valuable function of the storage battery—its ability to preserve energy for future use—was contained in the electrical *generator* itself. Only when dynamos replaced primary cells as the principal generators of current were the functions of *generation* and *preservation* separated, and the engineering role of storage batteries begun.

This chapter will deal with the first two phases of storage battery evolution. The first phase—in which the physical phenomena used in the battery were objects of theoretical study—can itself be broken down into three periods. The first period, from 1801 to the mid 1820s, saw the discovery of polarization and the invention of the first storage battery. Beyond this initial work, however, little was done—the work was not followed up by attempts to explain the phenomena, to measure them precisely or to expand upon the first discoveries.

In the second period, extending from the mid 1820s to the mid 30s, electrochemists made the first serious attempts to create theories of polarization and to measure its effects quantitatively. In the third period, the process of depolarization started to excite the interest of electrochemists, leading to the discovery of a great number of depolarizing agents for primary batteries in the next decades. In addition, the work of the second period was continued and expanded during the third period.

During the initial period, only two scientists made serious studies of electrochemical polarization—the French chemist Nicolas Gautherot and the German Johann Ritter. A few other scientists discussed polarization before the 1820s, but these discussions were little more than passing comments which did not arise from research undertaken by the scientists themselves.

Gautherot's work was an offshoot of that of Nicholson and Carlisle. In 1800, shortly after Volta announced his invention of the electric pile, the English scientists William Nicholson and Anthony Carlisle immersed the wire leads from a pile in a beaker of water and observed the decomposition of water as bubbles of hydrogen and oxygen formed on the wire ends. In 1801, Gautherot duplicated this work, and then expanded upon the earlier discovery:

> As I continued my experiments . . . I observed that the burning taste which one obtains, when he holds the two metal wires in the mouth, whose other ends are in the end cups of the apparatus[2]; I observed, I say, that when platinum or silver wires were used and I placed them in contact after removing them from the crown of cups, I again detected a slight galvanic taste,

---

[2] For most of his experiments, Gautherot used Volta's Crown of Cups battery, which consisted of a series of drinking glasses, each holding one copper and one zinc electrode. These cells were arranged in series.

which continued a short time, as long as the wires were kept in contact.

This taste is still more distinct if you place the two wires in a flask with saltwater and hold them in place by sticking them through the middle of a cork, so that they do not come into contact; if you now place the other ends of the wires in the end vessels of a crown of cups . . . and leave them there an instant until the water in the flask just begins to decompose, and then bring the two ends of the wires, which are still in contact with the apparatus (flask) in the mouth, the taste will be very strong; in some cases you will even feel a slight shock, and, in addition, the effect lasts long. I have also succeeded in decomposing water with this new apparatus.[3]

Gautherot then modified this experiment to disprove Volta's theory that the "galvanic taste" came from acids and bases generated on the electrodes by decomposition of water. He dipped the ends of the polarized wires into distilled water so that any adhering acid or base would be washed off. Nevertheless, if the washed wires were then allowed to discharge by the methods described above "you still get a very pronounced taste."[4] Reversing the usual experiment, Gautherot dipped the ends of the polarized platinum wires into nitric acid and a basic solution and then tried to detect the galvanic taste, but here there was nothing. The "galvanic taste" was due to the reionization of the layers of nascent oxygen and hydrogen on the platinum wires, thus generating a secondary current.

Nevertheless, despite this promising beginning, Gautherot proceeded no further with this study. He recommended that others should take up the work, however, since he believed that such studies might lay the basis for an understanding of the nature of electricity itself. This, in fact, was the same idea that Johann Wilhelm Ritter had.

Johann Ritter's interest in electrochemical polarization stemmed from that curious blend of science and romanticism called Naturphilosophie. The central concept of Natur-philosophic thinking was the *Weltseele*—a spiritual force which pervades the entire universe and animates all matter. All matter, therefore, must possess a form of life, since to be animated by a soul is to be alive. Moreover, since all matter is animated by the same World Soul, all forms of matter must participate in the same basic kinds of life-processes. The only

---

[3] Wilhelm Ostwald, *Elektrochemie-Ihre Geschichte und Lehre,* pp. 173-174.
[4] Ibid.

difference in these life-processes is that the simpler non-organic materials possess them in simpler form than more complicated organic materials and living beings. Thus, Friedrich Schelling, the originator of Naturphilosophie, says: "Auch in der bloss organisirten Materie sey *Leben:* nur ein Leben eingeschränkterer Art."[5]

The Weltseele manifests itself in matter by a few Lebensphänomene: in the case of organic matter, these are reproduction, irritability, and sensibility, while for inorganic matter, the analogous phenomena are chemical action, electricity, and magnetism. These second three are what Schelling meant by the "eingeschränkterer" forms of the organic three. Thus inorganic matter reproduces itself by chemical action—the inorganic analog of sex. The choice of chemical action, electricity, and magnetism as the analogs of life processes reveals the influence of contemporary physical science on Schelling's reasoning—chemistry and electricity were the most fashionable sciences of the day. Moreover, the two-sided polarity of these phenomena were particularly significant for Schelling and his followers, since such fundamental duality seemed to reveal the two basic processes which control all existence—attraction and repulsion.

For the Naturphilosopher, therefore, these three phenomena expressed the basic duality of Nature in the existence of inorganic matter. Schelling showed how this duality existed in all life forces, organic and inorganic—for example, plus-minus, north-south, acid-base, male-female, life-death. This duality is therefore the basic principle by which the *Weltseele* operates.

Here, then, is the root of Johann Ritter's interest in electrochemical polarization; it represented for him a particularly lucid example of the symmetrical and balanced structure of the forces governing all existence. According to Ritter, *all* substances, not merely those substances which form parts of a voltaic pile, are subject to the polarization of electricity. Volta's invention of the primary battery, however, only showed that a combination of zinc and silver or copper could produce electrical polarity. But the discovery of polarization showed that even identical plates of the same metal could participate in this universal process of duality.

---

[5] Walter D. Wetzels, "Johann Wilhelm Ritter; Physik im Wirkungsfeld der deutschen Romantik," p. 7; George Dubpernell & J.H. Westbrook, *Selected Topics in the History of Electrochemistry* (Princeton, N.J., Electrochemical Society, 1978) pp. 68-87.

Naturphilosophic reasoning was characterized by a freewheeling and undisciplined use of analog model building. Ritter's favorite analog device was the voltaic pile. To him, the world itself was one gigantic, infinitely complex pile in which all substances were positive or negative to all other substances. This was the source of the electrical manifestation of the World Soul, and the duality of the electrical poles governed the eternal oscillations between growth and decay. All earthly and cosmic phenomena could be shown to participate in these great, invariable oscillations.[6]

Ritter felt that the charge and discharge of electrochemically polarized systems was a particularly clear example of such eternal oscillations and he therefore set out to explore the phenomenon experimentally. Ritter began his polarization experiments at the same time as Gautherot, but independently of the French scientist. In the summer of 1801 Ritter connected two gold wires to the opposite ends of a voltaic pile and immersed their other ends in a saline solution. Upon disconnecting the voltaic pile, he observed the polarization current. Moreover, he noted that the ends of the gold wires produced a different taste on the tongue when they discharged than when they carried the current from the pile. What had earlier produced a sour taste, produced a bitter taste on discharge. Ritter correctly analyzed this change as the result of a reversal of current direction.[7]

Up to this point, Ritter had only duplicated the work of Gautherot. In December 1802, however, he began a series of experiments on polarization whose variety was not to be duplicated again until the 1850s and 1860s. Gautherot used only platinum or silver wires, but Ritter constructed secondary piles from all of the common metals, using either sodium chloride or ammonia solution as electrolytes. The significance of this is that Ritter obtained polarization currents whose action arose from the formation of *solid* reaction products on the electrodes, and not solely from the formation of hydrogen and oxygen gases.

Although Ritter used lead in some of his experiments, he did not recognize its value as a secondary battery material. Because he used sodium chloride electrolytes, Ritter obtained lead chloride as a polarization product on the cathode, whereas in

---

[6] Ritter expounded upon this in his famous treatise: "Das elektrische System der Körper," Ibid.

[7] Edmund Hoppe, *Die Akkumulatoren für Elektrizität* (Berlin, Springer, 1898) p. 77.

the modern storage battery, lead peroxide is the cathodic active material. Lead peroxide is a good electrical conductor, but the chloride is very poor. Consequently, Ritter observed no secondary current with lead, zinc, or tin.[8]

Ritter did recognize, however, that surface oxidation of the metals was the source of the polarization currents which he did receive from other metals, for example, copper. Ritter was also responsible for constructing the first device which can truly be called a storage battery. This was done by imitating the structure of the voltaic pile:[9]

> Fifty copper plates were prepared, each slightly larger than a dollar (Laubthaler) and about as thick as a card, along with the same number of salt-water impregnated cardboard discs about two inches in surface and a line (Linie) thick. These were assembled in the sequence: Copper, cardboard, copper, cardboard, copper, etc and ended with copper. . . .

The purpose of this construction was to prove that such batteries behave in precisely the same manner as the usual voltaic pile of alternate zinc and copper discs. Such proofs were designed to give support to his Naturphilosophic ideas. Thus, for example, Ritter showed that connecting such piles in series increased the voltage while paralleling them increased the current, just as in an ordinary pile.

Although Ritter was not interested in the utilitarian value of his storage piles, he studied their properties in some detail. He noted, for example, that his "Aufspeichurungssäule" (storage piles), tended to rapidly self-discharge. If a cell were charged and left on open-circuit for at least 30-45 minutes, it lost its charge automatically.[10] Ritter predicted that it should be possible to increase the storage capacity of his cells by proper choice of form and construction materials.

Between the time of Ritter's work and the mid 1820s no significant work on electrochemical polarization was done. This sterility may appear surprising, given the great scientific enthusiasm for electrochemistry after 1800. The problem, however, lay not in the lack of theoretical interest in electrochemical phenomena, but in the lack of proper instruments for measuring them.

---

[8] *Trans. Am. Electrochem. Soc.* 3 (1903): 161.

[9] Ostwald, *Elektrochemie*, p. 176.

[10] Edmund Hoppe, *Die Geschichtliche Entwicklung des Akkumulators*, p. 146.

Before the 1820s, scientists had no instruments for quantitatively measuring small electric currents, and gaseous polarization is a low intensity phenomenon producing only small currents. Gautherot's use of his tongue to taste the amperage was as sophisticated a technique as then existed for small currents. The frog's leg was the oldest standby—Ritter used the speed and intensity of reaction of frog's muscle to measure amperage. Small currents were also estimated by inserting the wire leads in tiny breaks in the skin and noting the degree of pain produced.

The quantitative measurement of higher amperages was somewhat easier during the early years of the century, although still very crude. Experimenters measured the rate at which thin wires were melted by the current, or determined the rate at which oxygen and hydrogen were generated from water by the current. For small, transient currents, however, such as those of polarization, these techniques were unsuitable. Consequently, during the first two decades of the nineteenth century the only definite conclusion which an experimenter could make about small currents was whether or not they existed; anything more specific or quantitative was beyond the limits of reproducibility. Ritter and Gautherot, therefore, had done just about all that could have been done with polarization studies up until the 1820s.

In 1820, however, Hans Christian Oersted laid the foundation for all future electrical instrumentation by moving a magnetized needle with an electric current. By the mid 1820s, Oersted's discovery was embodied in the electromagnetic galvanometer—a magnetized needle around which was wound a coil of wire for conducting the test current. Even the faint and transient currents of gaseous polarization could be measured with a galvanometer equipped with many turns of fine wire coil.

Consequently, a number of scientists undertook experimental studies of polarization in the mid 1820s. Most of this work was restricted to the study of *gaseous* polarization—that is, the production of secondary currents by formation of hydrogen and oxygen films on platinum wires. This was because much of this early work was not so much concerned with creating a theory of polarization *per se*, as it was part of the famous contemporary debate on the origin of electromotive force in the galvanic battery. These researchers, therefore, felt that the kind of polarization they produced was immaterial to the results. Nevertheless, a number of these scientists were concerned with polarization produced by solid deposits.

Since the nature of early nineteenth-century debates on the origin of chemical electromotive force is outside the scope of this study, it is sufficient to note that some chemists, such as Christian Schönbein, defended a chemical reaction model of its origin, while others, such as Auguste de la Rive, tried to find evidence for a mechanical model.[11] In 1825-1826 de la Rive published the results of a series of experiments designed to prove that the galvanic current is analogous in its operation to light or heat.[12] The immediate purpose of these experiments was to determine how the action of a galvanic battery is changed under different operating conditions. De la Rive indicates, however, that the nature of these particular experiments was prompted in part by recent advances in the instrumentation discussed above:[13]

> Les nouvelles données que les travaux scientifiques des dernières années ont fournies sur l'électricité; les moyens aussi exacts que délicats que quelques physiciens ont tirés des découvertes récentes pour s'assurer de la présence de cet agent et pour en mesurer l'intensité; tels sont les motifs qui m'ont engagé à m'occuper d'un sujet sur lequel on a beaucoup travaillé, mais que personne, a'ma connaissance, n'a repris dans son entier, en s'aidant des nouveaux secours que la science a mis entre les mains des expérimentateurs.

De la Rive used a galvanometer to measure the rate at which gaseous polarization diminishes the current output of a battery. The battery employed was a trough battery containing sixty pairs of zinc and copper plates. Trough batteries were a favorite for laboratory purposes in the early nineteenth century because of their portability and compactness:[14]

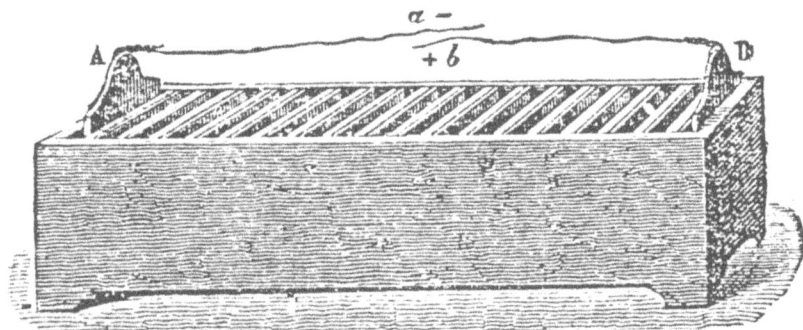

---

[11] Ostwald, *Elektrochemie, passim.*

[12] Ibid., p. 661.

[13] *Annales de Chimie et de Physique* (hereinafter: *Annales*) 28 (1825): 190-209.

[14] Alfred Niaudet, *Elementary Treatise on Electric Batteries*, p. 14.

De la Rive inserted a series of platinum plates into the trough. Since these passive plates were in series with the active zinc-copper pairs, de la Rive in essence reproduced the Gautherot experiment, with the platinum plates substituting for the platinum wires. The experiments were repeated for one, two, and then three interposed platinum plates. With each combination, de la Rive discharged the battery ten times, each time letting the battery discharge itself a little longer before inserting the plates. Consequently, the battery polarized itself a little more each time and produced a gradually diminishing initial current over the ten runs. In addition, the interposed plates were clearly seen to resist the flow of current.

From a theoretical standpoint, de la Rive's investigations did not contribute to a scientific understanding of polarization. Their importance lies in the fact that they were the first attempt to measure quantitatively the effects of polarization; they served as a source of stimulation for much future work in which de la Rive's methods would be imitated and improved.[15]

Another scientist who worked on quantitative measurement of polarization at the same time as de la Rive was Stefano Marianini, professor of physics and applied mathematics at Venice. Like de la Rive, Marianini developed a non-chemical model of galvanic action, in which he likened the flow of electric current to a beam of light. If current behaves analogously to light, then it can be made to undergo refraction.[16] This, says Marianini, is what causes the diminution in current strength called polarization. If a series of inactive plates are placed in a battery, as in de la Rive's experiment, the current is diminished because there occurs between these plates "une sorte de réfraction ou de réflexion électrique, analogue a' celle de la lumiere."[17] When electricity passes through alternate metallic and liquid conductors it must move at different speeds, which is analogous to light traveling between two ordinarily transparent media—Marianini chooses air and water as examples.

Now if the number of alternate passages between water and air are few, as when the layers are thick, the transparency of

---

[15] For example, de la Rive used the data from his 1825 experiments to prove that electrochemical systems polarize at a rate inverse to the strength of the primary current. In the 1840s Poggendorff conducted careful experiments to disprove this idea.

[16] Ibid., pp. 662-664.

[17] *Annales* 33 (1826): 119.

the air-water sandwich is little diminished, but if the numbers of layers are increased, and each layer made thinner, then the light will be retarded and the mixture appear opaque. Shifting his analogy to that of sound, Marianini concludes: [18]

> If we assume that the electric current is propagated in waves (propage par les ondulations), in the manner of sound, we find a simple explanation of the phenomenon (i.e. of the retardation of the electric current by the interposed inactive metal plates) in which we can compare its speed in metallic and bad conductors (electrolytes) to that of (the speed of) sound in solid and gaseous bodies.

Having presented this rather vague theoretical explanation, Marianini offered some experimental data in support. An eleven inch long trough battery was prepared, at one end of which was a six-inch chamber holding a zinc-copper pair, and at the other end were five, one-inch compartments, each holding a lead plate. In this apparatus, the zinc-copper pair was analogous to a light source, and the lead plates with their surrounding electrolytes represented the opacity-producing media which forced the electricity to continually shift its speed and thus be retarded.

If this model is correct, then as the number of shifts increases, the current will be increasingly diminished. That is, as more lead plates are inserted, the current should fall. This was what Marianini observed:[19]

Number of Lead Plates	Galvanometer Reading
0	3°
1	1°
2	.4°
3	.2°
4	.15°
5	.05°

Important as de la Rive's and Marianini's work in stimulating experimental studies of polarization was, it was a retrogression from the theoretical insights of Ritter. Ritter recognized that polarization was caused by the creation of a source of counter EMF in the electrochemical system. That is, just as he recognized that the primary source of EMF is an active

---

[18] Ibid.
[19] Ibid., p. 121.

agent, so also is the cause of the polarization effect *active*. Marianini and de la Rive, however, saw polarization as a *passive* resistance to the passage of current; neither of them attempted, for example, to measure any *reverse* currents produced by their polarized plates, since such currents were not to be expected from their theory. They measured only the diminution of the *primary* current.

Nevertheless, the anti-chemical arguments of de la Rive and Marianini did not remain unchallenged for long. In 1826 Humphry Davy published a lengthy defense of the chemical model of galvanism which was a rebuttal of de la Rive.[20] The latter part of the paper is devoted to Davy's chemical model of polarization. The experiments which Davy describes are little more than modifications of the quarter-century old work of Ritter. Davy's explanations of the results, however, are much clearer than those of the German chemist, and are similar to our modern interpretation.

Davy connected in series six pairs of platinum plates in nitric acid filled cups. These six neutral cells were then connected in series with a galvanometer and a fifty pair trough battery. After the current direction was observed, the trough battery was removed from the circuit and the galvanometer again read. Davy now found that the platinum pairs had themselves become sources of EMF, but of opposite direction to that of the trough battery. Davy repeated the experiment several times, substituting for the platinum plates pairs of zinc, tin, silver, copper and other metals in various saline solutions. He always obtained a reverse current of greater or lesser intensity. This he explained as follows:[21]

> As the chemical changes always tend to restore the electrical equilibrium destroyed by the contact of the metals with each other in the fluids,[22] it is evident that in cases in which arcs (i.e. pairs) primarily inactive (i.e. identical) are connected with those primarily active, the chemical changes produced by the electrical attractions must tend to produce in the primarily inactive parts of the combination an arrangement which must give it a power in direct opposition to that of the primarily active circles (pairs); so that when separated, their actions, if any, must be

---

[20] Humphry Davy, *The Works of Sir Humphry Davy*, Vol. 6-Misc. Papers & Researches, 1815-1828, (London, 1840).

[21] Ibid.

[22] In modern terminology, Davy is saying that the electrochemical potentials of the reaction products built up at the electrodes precisely balance the electrochemical potentials of the electrode metals themselves.

directly the reverse of the other. . . . These experiments, showing the nature of the chemical changes in combinations made
active by their connexion with voltaic batteries . . . offer a simple and adequate solution to the circumstances observed by M.
de la Rive. . . .

Davy then demonstrated the perfect reversibility of polarization currents inside the *primary* battery itself; that is, without
using separate, inactive plates. He offered this demonstration
as his most convincing proof that polarization is caused by the
formation of chemical deposits on the electrodes.[23] The experiment employed two zinc-copper trough batteries, one of fifty
pairs, and the other of six. First the batteries were connected in
series so that they would send current in the *same direction;*
that is, with the positive end of one connected to the negative
end of the other. After a ten-minute discharge, the troughs
were disconnected and the six-pair one immediately discharged through a galvanometer. The current was *less* than
would have been obtained if the battery had not been first
discharged with the fifty pair.

Now the experiment was repeated, except that the two
troughs now sent current in opposing directions. After ten
minutes, the troughs were disconnected and the six pair again
tested with the galvanometer. Now it deflected the galvanometer needle four times more than if it had not been connected to
the fifty pair. Davy attributed this correctly to formation of
copper oxide on the cathode, and, more dubiously, to the zinc
being given "a much higher positive state."[24] This explanation
agreed with the general theory of galvanic action as contained
in the first sentence of Davy's statement quoted above. These
experiments proved that polarization was caused by the creation of a source of back EMF, and that it was not merely a
passive retardation of the primary EMF.

In the 1830s and '40s a series of chemists continued Davy's
defense of the chemical model of polarization and further
demonstrated that polarization is caused by the creation of a
source of counter EMF—for example, Matteuci (1838);
Schönbein (1839); Daniell & Wheatstone (1842); de Heer (1840);
Poggendorff (1844); Lenz & Saweljew (1846).[25] Even de la Rive

---

[23] Davy himself does not use the word polarization, since it had not been coined at
the time.

[24] Davy, *Works*.

[25] *Comptes Rendus* 7 (1838): 741; Ostwald, *Elektrochemie*, p. 675; *Pogg. Ann.* 61 (1844):
586; 67 (1846): 497; *Comptes Rendus* 16 (1843): 772; Hoppe, *Akkumulatoren*, p. 90.

was eventually won to the chemical cause. One of the chief reasons for the victory of the chemical model after the early 1830s and the readiness with which chemists used it to create theories of electrochemical action was the growing influence of Michael Faraday's laws of electrolysis (1832) as the fundamental touchstones for interpreting electrochemistry.[26] This growing influence can be seen clearly in a series of papers published by Schönbein in 1838-1839.

Christian Schönbein, professor of chemistry at the University of Bassel, decided:[27]

> . . . . to put the question, whether a polarized condition (Zustand) of the electrodes is produced under the influence of the current (i.e. non-chemically), or, contrarily, whether their capability (Fahigkeit), to stimulate a secondary current is not based on the accumulation of the constituents of an electrolyte, e.g. of water or sodium chloride (Salzsäure), around the electrode wires. Let me put it in different words: is not the secondary current a normal (gewöhnliche) chemical action. . . .

Schönbein then describes a series of experiments designed to prove the last sentence. These experiments were not dissimilar to those of Gautherot forty years earlier. For example, Schönbein shows that a wire of platinum which had been positively polarized loses its polarization when dipped in a beaker of chlorine or oxygen gas. Similarly, a negatively polarized wire loses its polarization when plunged into hydrogen. On the other hand, no loss occurs if either a negative or positive wire is dipped into a chemically-inert gas such as carbon dioxide.[28]

Schönbein used this lengthy series of experiments to propose a number of general laws of polarization, among which are:[29]

> There does not exist actual Voltaic polarization (i.e. that caused by purely physical buildup of electrical atmospheres) either of the electrolyte or of the electrodes (der festen noch der flüssigen Leiter), and all secondary currents, which are excited (erregt) by means of so-called polarized bodies, originate in a normal chemical action, which consists either of the combination of ele-

---

[26] A.J. Berry, *From Classical to Modern Chemistry* (New York, Dover, 1968).
[27] *Pogg. Ann.* 47 (1839): 101.
[28] *Comptes Rendus* 7 (1838): 1065-1068.
[29] *Pogg. Ann.* 47 (1839): 116-117.

ments, or the decomposition of compounds. . . . With electrolytes, the conduction of currents and electrolyzation of them *is
the same thing* (my ital.)

This last sentence is essentially a statement of Faraday's first
law of electrolysis.

The importance of the work of Schönbein and that of the
other scientists mentioned above for the future development of
the storage battery is its creation of a mirror-image model of
the relationship between primary battery current and polarization. According to this model, a primary battery discharges by
the destruction of chemicals, but the flow of current thus produced causes the creation of new chemicals, which set up a
source of counter EMF. This counter EMF produces a counter
flow of current when the new chemicals decompose in turn.
The invention of storage batteries is based on this perception.

The theoretical studies of the chemical origin of polarization
began bearing practical fruit in the mid 1830s. The lack of good
measuring instruments was one hindrance to the advance of
electrochemistry before the 1820s, but a second major problem
was polarization itself. Before the mid 1830s no primary batteries were equipped with cathodic depolarizers. Therefore,
the current fell off precipitously as soon as discharge began,
making it very difficult to do any accurate or sustained work
with such cells. In effect, such early primary batteries discharged like a kind of slow-motion capacitor.

As soon as chemists became convinced, however, that
polarization is a chemical phenomenon, whose specific problem was the buildup of hydrogen on the cathode, the solution
was self-evident: hydrogen is electropositive; therefore, en-
case the cathode with something electronegative to react with
it.[30]

Ever since the late eighteenth century, chemists had known
of the varying electropotentials of different substances. As
soon as the primary battery was invented in 1800, they had
experimented with various combinations of electrode materials to find the most strongly electropositive or electronegative
ones. In combination, the most strongly opposed materials
would produce the greatest EMF. Alessandro Volta himself, for
example, discovered in 1802 that manganese dioxide (MnO_2),
used as cathode of a battery against a zinc anode in saline

---

[30] Ostwald, *Elektrochemie*, p. 740.

electrolyte, gives a much higher voltage than if a simple copper cathode were used.[31]

Yet although the strongly electronegative nature of materials such as MnO_2 were known, few attempts were made to use them as cathodic materials until the mid 1830s. This was because chemists did not appreciate the fundamental difference between the use of a copper or silver cathode, which plays a purely passive, electrical-conductive role, and materials like MnO_2, which are active oxidizers. Moreover, the use of such active materials added complications to battery design; MnO_2 is a poor electrical conductor, in contrast to the excellent conductivity of copper or silver.

Only after the chemical mechanism of polarization was "understood" were chemists able to appreciate the real advantage of placing strongly electronegative materials on the cathode, that is, they realized that such depolarizing substances made primary battery current strong and *constant;* Volta had only believed that they gave the cell higher voltage.

Although it is beyond the scope of this study to mention the variety of new primary battery designs proposed after the mid 1830s as chemists experimented with various cathodic oxidizers, one specific group of oxidizers is of central interest for the history of the storage battery—the peroxides. Volta put MnO_2 at the extreme end of the scale of electronegative substances. This fact stimulated chemists in the late 1830s to pay close attention to the entire class of metallic peroxides. As a result, they found that lead peroxide was both more electronegative and a better electrical conductor than manganese dioxide.

The German chemist P.S. Muncke was the first to observe this. In an 1835 paper, Muncke noted that little work had yet been done to determine the "electrical characteristics" (that is, electrochemical potentials) of many chemicals, leading him to explore this "weites Feld."[32] His attention was drawn to the peroxides because:

Manganese dioxide is known among all previously studied materials as the one which produces the strongest negative elec-

---

[31] *Pogg. Ann.* 111 (1835): 46.

[32] Ostwald, *Elektrochemie,* p. 741; Hoppe, *Akkumulatoren,* p. 120, 153; *Phil. Mag.* 12 (1838): 225-229. Schönbein, after noting that PbO_2 is more powerful than MnO_2, goes on to state that ". . . . peroxide of silver proves that most powerful means for exciting in iron its peculiar voltaic condition. It surpasses in this respect even the peroxide of lead." This statement is of interest for the researches discussed in chapter V. *Phil. Mag.* 10 (1837): 425-429.

tricity, and I thought therefore, that other peroxides might have singular (sonderbare) electrical characteristics also.

Muncke used manganese dioxide and lead peroxide as cathodes opposing zinc anodes with paper discs in between. The PbO_2 gave a deflection of the galvanometer needle one and a-half times that of the MnO_2.

A number of contemporary chemists were impressed with these results, including Faraday, de la Rive, Schönbein, and Gmelin. They all reproduced the experiment and confirmed Muncke's result.[33] Gmelin's comment reveals the connection with contemporary thinking on polarization:

> Manganese and lead peroxide behave therefore as very electronegative materials, because they yield oxygen to hydrogen, . . . and because this affinity of oxygen to hydrogen favors the transference of one to the other; the more or less reduced manganese or lead then takes up the electricity.

Neither Muncke nor Gmelin, however, had actually designed a battery to use PbO_2. De la Rive first did this in 1843. Describing his motivation for building the cell, de la Rive indicates his adherence to the chemical model of polarization:

> Convinced. . . . that the true obstacle to the transmission of the current of one pair (of plates) through a liquid conductor is the formation of the primary gaseous layers on the surface of the electrodes, I sought means to get rid of this obstacle by removing these layers.

De la Rive's 1843 PbO_2 cell was a modification of William Grove's famous nitric acid battery. In 1838, as part of the abovementioned movement to design depolarized batteries, William Grove had proposed a cell with a zinc anode, platinum cathode, and a porous diaphragm to separate the two electrodes. The electrolyte on the anode was the usual sulphuric acid, but on the cathode nitric acid was used. This acid acted as the depolarizing substance. The Grove nitric acid battery became popular both for scientific and telegraphic purposes for the remainder of the century, and many inventors used it as the starting point for the invention of new cells. In de la Rive's version, the usual nitric acid was replaced with either MnO_2 or PbO_2; this would do away with the need for two electrolytes

---

[33] *Electrical Mag.* 1 (1843-45): 28-41; *Comptes Rendus* 16 (1843): 772.

and make the diaphragm unnecessary. De la Rive found that the current was five times greater with PbO_2 than with MnO_2.

Charles Wheatstone conducted similar experiments in 1843 in England, but whereas de la Rive attempted to improve the Grove cell, Wheatstone was attempting to improve the Daniell cell (zinc anode, H_2SO_4 as anodic electrolyte, porous diaphragm, $CuSO_4$ solution as cathodic electrolyte, copper cathode). He compared Daniell batteries with a series of peroxide-depolarized cells. Again, Wheatstone tried MnO_2 and PbO_2, and the results demonstrated the superiority of PbO_2 as a depolarizer.

2. Phase Two: Polarization as Technology

The first half of the nineteenth century saw the creation of the earliest prototype storage battery by Ritter. A few chemists, such as Davy, studied these cells during this half century. Much work was also done in studying the phenomenon of electrochemical polarization and in appraising the value of lead peroxide as a depolarizing agent. During the 1850s the results of these two separate but related studies in electrochemistry were combined in the invention of the lead-acid storage battery.

The lead-acid battery is what Humphry Davy would have called a single-metal battery; one in which the same metal is used for both electrodes. The base or support metal for both electrodes is metallic lead. The positive electrode active material is PbO_2, while that of the negative is metallic lead in a porous or "sponge" form—to use the battery engineer's term. The reactions are:

which can be written in the following half reaction form:

Cathode:

Anode:

These reactions reveal the influence of the prior theoretical research; the concept of a rechargeable electrochemical system came from the polarization studies and the choice of lead as the basis for the specific system used came from the work of Muncke, de la Rive, Wheatstone, et al on the power of PbO_2 as a depolarizer. Theoretical science, therefore, was one of the parents of the lead-acid storage battery; the other parent was necessity—the need of scientists and engineers for a new tool, chiefly for the technology of telegraphy. As a result of this double parentage, the inventors of the storage battery were all men interested in technology, but who were also scientifically educated. They will be considered in the chronological order in which they became involved in storage battery development.

The first to take up the work was Sir William Siemens. In 1852, C.W. Siemens was trying to develop an electric power source which would provide much higher intensity currents than could contemporary primary batteries. His motivation for this work was most probably telegraphy; during the 1850s, Siemen's chief commercial interest was in telegraphic equipment. This was a period of rapid expansion of telegraphic lines, particularly on the continent. Telegraphic engineers steadily increased the lengths of the circuits and began to lay cables under large bodies of water. Consequently, the first problems arose with the ability of ordinary primary batteries to transmit signals through these more resistant conductors.

Siemens began in 1852 with the idea of developing a high voltage, rechargeable form of the Grove gas battery. The gas battery was an early fuel cell developed by William Grove between 1839 and 1842.[1] It was not designed for practical use, but rather was yet another offshoot of the electrochemical research into gaseous polarization. Grove reasoned that since it was possible *to produce* oxygen and hydrogen on the opposite platinum plates in a decomposition cell by means of current, it should follow that the *consumption* of these gases on the same plates to reform water should produce a flow of current, but in the opposite direction. This would be further proof of the chemical model of polarization and would also be supporting evidence for the concept of conservation of energy, of which Grove was a pioneer.

---

[1] George W. Heise & N. Corey Cahoon (eds.), *The Primary Battery*, p. 34.

He prepared two inverted glass tubes with platinum plates in acid solution and injected hydrogen and oxygen gas into the alternate tubes. Grove designed this cell to demonstrate "a beautiful instance of the correlation of natural forces," that is, conservation of energy.[2] To C.W. Siemens, however, it represented a potentially powerful source of useful current.

Grove's gas batteries gave only weak current, but Siemens reasoned that this was because the surface area of the electrodes was small. Therefore, Siemens used his knowledge of telegraphic technology to increase the electrode surface area enormously. The Grove nitric acid cell was the favorite battery in contemporary telegraphic practice, largely because of its high current output. These commercial cells differed in one important respect from the original Grove design; the platinum cathode was replaced by a porous carbon plate. This innovation not only reduced initial cost, but also increased the surface area. Robert Bunsen proposed this idea in 1841 and it soon became standard practice.[3] The cathodes were usually prepared by scraping the graphitic carbon deposits from the walls of illuminating gas retorts, mixing this material with a binder such as tar, and baking the mass in a mold.[4]

Siemens used this technique to produce large, hollow, cylindrical electrodes for his high-output gas battery. To increase their effectiveness still further, he gave the porous carbon an internal microscopic electroplate of platinum, since this helps catalyze the electrode reactions. Siemens's early experience as a commercial electroplater may have suggested this process. The cylinders were then immersed in dilute acid and surrounded by atmospheres of oxygen and hydrogen. Siemens reported that these cells gave "considerable power." He abandoned them, however, since the current was still "insufficient for my purpose," whatever the purpose may have been.[5]

Siemens's original idea was to develop a rechargeable gas battery; in practice, the gases were to be generated in the carbon cylinders by means of an external current source. When the gaseous system proved insufficient, Siemens turned his attention to solid polarization products. Therefore, he explained, "with a view of increasing the potential of the cur-

[2] Ibid., p. 35.

[3] *Pogg. Ann.* 54 (1841): 416-430.

[4] This technology was later to form the basis for the manufacture of carbon pencils for the arc lighting industry.

[5] *The Telegraphic Journal,* 1 October 1881: 376-377.

rents, I directed my attention to the *peroxides* (my ital.) of the metals, and soon found that peroxide of lead was the one giving the greatest promise of results." Siemens does not indicate the source of his interest in lead peroxide, but this doubtlessly was the scientific literature mentioned in the last section. Siemens had a doctorate in chemistry, and, given his electrochemical interests, he surely was aware of the work of Muncke, de la Rive, Faraday, Wheatstone, et al.

Siemens used the same cylindrical carbon electrodes for the lead battery, but instead of plating them with platinum, he dipped them in a lead acetate solution. The cylinders were then dried, heated to redness, cooled, and the process repeated until a considerable buildup of lead oxide was obtained. The prepared cylinders were immersed in a dilute sulphuric electrolyte and a strong current passed through them, converting one to lead peroxide and the other to metallic "spongy" lead. This cell was the first prototype of the lead-acid storage battery. Prior primary batteries had used lead peroxide and other batteries had been made rechargeable, but these two characteristics had not been combined in one cell.

Nevertheless, William Siemens does not deserve to be called father of the storage battery. He soon abandoned the lead peroxide work and published nothing on his results until 1881, when he did so for purely historical purposes. Siemens's lead batteries provided powerful currents, but he found the process of charging the cells from Grove nitric acid batteries "too expensive to render the secondary battery available for practical purposes." If Siemens had possessed sufficiently reliable and efficient electromagnetic generators in 1852, he might have carried on with this work. We do know that the second man to invent a lead-acid storage system, Dr. Josef Sinsteden, did so because of his invention of just such a generator.

Josef Sinsteden was a Berlin surgeon with interests in electromagnetic generators. During the early 1850s the Prussian state telegraphs were undergoing rapid expansion and Sinsteden conceived the idea of using generator current to power this increasing traffic. As he observed in the introduction to his 1854 paper, such generators can produce the high currents and voltages which only impractically large combinations of primary batteries can provide.[6] He contended that telegraphy can be operated practically by battery power only as long as the

---

[6] *Pogg. Ann.* 92 (1854): 1-21.

circuits are short. On long lines, however, the greatly enhanced voltage of the "magneto-elektrischen Rotations-Apparats" was needed to operate magnetic relays in the face of the "weakening of the magnetic effect" of lengthy circuits. This kind of "brute force" approach to long-distance telegraphy was popular during the 1850s; the concept being that short but very sharp voltage surges were necessary to force a signal through long wires, particularly where capacitive loading was a big problem. The idea was abandoned after it was discovered that such powerful surges destroyed cables.[7]

About 1851 Sinsteden "undertook to build an electromagnetic machine of great size to provide sufficient current strength to operate our longest telegraph lines." He demonstrated his machine on a number of lines, where it apparently worked successfully. To demonstrate the power of his machine, Sinsteden performed a series of laboratory tests. To measure the current strength he charged a series of secondary batteries and measured the polarization products obtained. Sinsteden began with the traditional gas measurement technique; current was passed between platinum plates in sulphuric solution and the volume of gases evolved measured. But then Sinsteden varied the technique by passing the current between pairs of lead, nickel, or silver plates in dilute sulphuric or zinc-saturated potassium hydroxide electrolytes.

Sinsteden remarked enthusiastically that he had obtained "ganz ungewöhnlich starke Polarisationserscheinungen" from all these combinations. The plates of the silver cell quickly became covered with a thick coat of oxide, which, on discharge, decomposed water and melted metal wires. Sinsteden was the first to notice that the surface of the negative plate of a storage battery soon becomes covered with a porous but coherent layer of the reduced metal, even though the negative electrode may have begun as a smooth, non-porous plate. This so-called metallic "sponge" greatly increases the capacity of energy storage.

Of all the systems Sinsteden tried, silver in acid solution was his favorite. It was easier to form a current-storing layer of silver salt than lead peroxide. Nevertheless, he also noted that a cell made of two lead plates in acid solution is effective in storing current and gives a good discharge. Of nickel Sinste-

---

[7] This was probably a major cause for the failure of the first successfully laid Atlantic cable in 1858.

den said little, although he noted that he got a strong current. Zinc in alkaline solution also gave a strong current, but Sinsteden observed the difficulty of getting the dissolved zinc to plate-out of the solution properly on recharging.

Josef Sinsteden's work was of greater importance for the evolution of the storage battery than that of Siemens, since his published results were used in directing the work of subsequent researchers. Nevertheless, his work was no more extensive than that of Siemens. Unlike the latter Sinsteden was not trying to design a new power source, but rather was proposing a new analytical technique for measuring large currents. The unity of Siemens's and Sinsteden's work lies in the mutual telegraphic motivation. After Sinsteden published his 1854 paper, he let the idea drop. Siemens and Sinsteden were men whose interests and knowledge of the scientific literature led them to invent the first lead-acid storage batteries, but because their interests were limited, they did not carry their inventions forward to a full elaboration.

The French chemist Gaston Planté was the first inventor to carry the lead-acid battery to a full elaboration. In fact, Planté spent thirty years (from age 25 until his early death) developing the design of the invention; Planté did not originate the lead-acid battery; he was anticipated by Siemens and Sinsteden and has read the latter work. Yet, in a fuller sense, Gaston Planté deserves the reputation he has received in the historical literature as the father of the storage battery, for he showed how to convert the physical phenomena described by Sinstenden into a set of techniques for the manufacture of a useful electrical storage device.

Planté, born in 1834 was educated as a chemist.[8] After graduating from the Faculty of Letters and Sciences of Paris in 1853, he was employed as a lecture assistant of Edmund Becquerel, whose commitment to the theoretical study of science as well as its commercial applications he shared. After 1858 he was connected with various commercial firms specializing in electrical technology, and it was this association which provided his first incentive to develop the lead-acid battery. From 1858 Planté was chemist for the Paris electro-metallurgical firm of Christofle et Cie.[9] He also had connections with the scientific, electrical equipment, and telegraphic apparatus manufacturing firm of Breguet. Planté began the

---

[8] *The Electrician* (London), 31 May 1889: 89.

[9] *Electrical Engineer* (London), 31 May 1889: 436.

storage battery research shortly after joining Christofle, when he was motivated by an idea of Professor Moritz Jacobi to replace primary batteries in telegraphy.

In 1850 Jacobi had proposed the use of large Grove gas batteries to send signals through long lines. Electromagnetic generators were to charge these batteries, which would consist of large platinum plates covered with finely divided platinum to increase their surface area. The similarity of this idea to that of Siemens and Sinsteden is clear and, indeed, Jacobi shared the same background as they. Scientifically trained, he was professor of civil engineering at Dorpat and St. Petersburg. Jacobi was an early inventor of electric motors and generators and was actively involved in promoting the two earliest forms of electrical technology—telegraphy and electroplating.

Planté was stimulated by Jacobi's suggestion, and, like Siemens, he undertook to develop a rechargeable cell. Unlike Siemens, however, he decided from the first to develop a system based on *solid* polarization products. Planté began with an exhaustive search of the literature on polarization and depolarizing agents.[10] This search revealed the promising characteristics of such metals as manganese, lead and silver, but, as he correctly pointed out, no one had yet systematically studied how the various metals behave if used in rechargeable cells.[11] Indeed, Sinsteden was the only one Planté knew of who had studied the situation at all, and his work could hardly be described as exhaustive. The latter wanted to make sure he chose the most suitable metal for his secondary cells.

Consequently, in 1859 he carried out a systematic search of all those metals which he felt were possible candidates for storage battery use. It would still not be accurate to call this work exhaustive, since a number of the common metals such as nickel were ignored. Moreover, Planté restricted his attention almost exclusively to sulphuric acid electrolytes. Nevertheless, for the metals which he did examine, Planté's work was the first serious attempt to evaluate the performance characteristics of each polarized system.

Planté used the same type of apparatus that had been used since the 1820s for polarization experiments. A pair of identical wires of the test metal were suspended in a beaker of 10 percent sulphuric acid. This system was then polarized by Grove

---

[10] Gaston Planté, *The Storage of Electrical Energy*, pp. 3-5.
[11] Ibid.

nitric acid cells and discharged through galvanometer of varying degrees of sensitivity.[12] Planté recorded the charge and
discharge currents for each metal, as well as the physical appearance and behavior of the wires. As a result, he was able to
make a number of fundamental discoveries about polarization.
For example, according to the chemical model of polarization
chemists had usually assumed that the *sole* reason for the
weakening of an electric current passing through a passive
electro-chemical system was the creation of a counter EMF due
to the buildup of chemical deposits on the electrodes. As a
corollary of this rule, it was concluded that those materials
which, when polarized, weaken the primary current *most*
should give the *strongest* secondary currents on discharge, just
as those which weaken *least* should give the weakest.

Planté found, however, that of all the metals he tried, a passive system with aluminum wires weakened the primary current most. Aluminum also gave the weakest secondary current.
He noted that this was caused by the formation of a layer of
non-conducting oxide on the cathode. On a scale of decreasing
secondary currents and decreasing power to resist the charging
current, Planté ranked the metals he tried:[13]

Secondary Currents (strongest at top)	Resistance to Charging Current (strongest at top)
Silver	Aluminum
Lead	Tin
Tin	Lead
Copper	Copper
Gold	Silver
Platinum	Gold
Aluminum	Platinum

Though silver produced the most powerful secondary current, it was unsuitable for a practical battery, Planté found that
as soon as a layer of silver oxide was formed on the cathode
during charging, it began to dissolve away into the electrolyte.
In the case of silver, tin, or copper wires, Planté watched the
colored metallic ions pouring uselessly off the electrodes as
soon as charging was complete.

---

[12] Ibid., pp. 5-6.
[13] Ibid., p. 7. It is curious that Planté did not try nickel, since he had studied
Sinsteden's work. *Comptes Rendus* 49 (1859): 403.

Platinum and gold were rejected: they were too expensive and they did not form very thick deposits of oxide. Lead, therefore, was the only metal left, and Planté found it acceptable. Neither it nor its oxide were soluble in dilute sulphuric acid. The peroxide was both a good conductor and strongly adherent to the lead plate from which it was formed.

Having found an acceptable substance for his planned storage cell, Planté set about the job of designing a practical battery. For the physical design, he copied a favorite pattern among early nineteenth century primary batteries; a sheet of felt was sandwiched between two thin lead sheets and rolled into a cylinder. This assembly was then inserted into a glass jar containing 10 percent sulphuric. Planté presented a set of nine such cells, having an aggregate surface area of 10 square meters, to the French Academy of Sciences on 26 March 1860:[14]

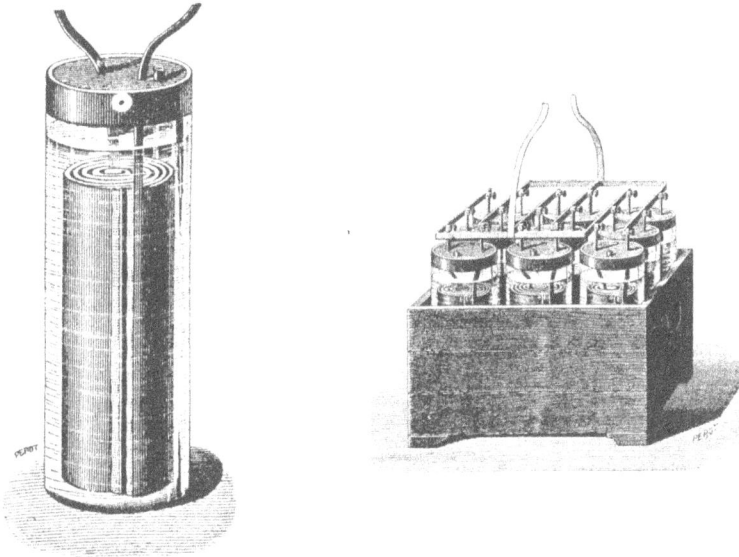

Planté's most important contribution to storage battery evolution was his invention of techniques for increasing the storage capacity of the lead plates. He spent many years developing these techniques which became the basis of one of the two commercial manufacturing processes adopted by the first storage battery firms in the 1880s. Although the original

[14] Ibid., pp. 34, 30, 31.

methods developed by Planté have been modified substantially over the past century, they still form the basis of
some commercial processes used today in battery
manufacture.

The problem which faced Planté in 1860 was that the
energy-storage capacity of a pair of flat lead plates is not great.
Lead is non-porous and does not allow the sulphuric acid electrolyte to penetrate below its surface. Consequently, only a
microscopically-thin layer of active lead peroxide forms on the
cathode during charging. Moreover, on discharge the sulphuric acid must attack the anode, converting its surface into
lead sulphate, but on both anode and cathode the active material layers are only atoms deep. Planté's problem was to make
the anode's metallic lead and the cathodic lead peroxide sufficiently porous to allow the electrolyte to penetrate.

He solved the problem by making a virtue out of one of the
worst defects of primary batteries; their tendency to discharge
themselves. Self-discharge occurs when chemical reactions
take place in a battery *without* a transfer of electrons to the
electrodes, as, for example, when the copper sulphate solution
surrounding the cathode in a Daniell cell diffuses to the zinc
anode and dissolves it:

$$Zn + CuSO_4 \longrightarrow ZnSO_4 + Cu$$

This same reaction also produces the normal electrical discharge of the cell, but here the electrons involved are transferred directly from ion to ion, without being made to move
through an external circuit. Planté, as an experienced electrical
experimenter, understood the phenomenon of self-discharge.
He turned it into a virtue in the following manner: The normal
lead-acid battery reaction on discharge is:

$$PbO_2 + Pb + 2H_2SO_4 \longrightarrow PbSO_4 + PbSO_4 + 2H_2O$$

with the PbO_2 and Pb representing the active materials on two
separate electrodes. Yet note that on the cathode *both* substances are present, since the lead peroxide is generated from
the lead plate. After a set of plates is polarized for the first
time, the new cathode will have a microscopically thin layer of
peroxide on it, and this substance, unlike the lead from which
it is formed, is porous. Consequently, there will exist on the
cathode an interface between PbO_2, Pb and H_2SO_4. If a set of
plates is polarized, therefore, and then *not* electrically discharged, the cathode will nonetheless perform the discharge
internally from its own lead, lead peroxide and sulphuric acid,

thus increasing the amount of metallic lead converted into porous lead salts. If the process is now reversed, it can also be used to develop a thick, porous layer of lead sponge on the anode. Planté's technique, therefore, was as follows: polarize a set of plates by charging in one direction; allow self-discharge; charge in the opposite direction; self-discharge; and so forth, until sufficiently thick layers of porous lead and lead sponge are produced. Planté called this process "formation" and the word has been used ever since as a technical term for the electrolytic preparation of storage battery electrodes.

"Formation," as Planté practiced it, was a tedious, time-consuming process. Self-discharge occurs rapidly on fresh plates, but as the active material beds deepen, the sulphuric electrolyte penetrates with increasing difficulty, slowing the rate of self-discharge. Gaston Planté was to spend the remainder of his life developing elaborate timetables for the months-long "formation" process. When it is remembered that Planté had no automatic apparatus to perform these cycles, and no other source of current except Grove nitric acid cells, it is apparent that the tedium of this work must have been enormous. In the next section, we will see what kept him at this work for so long.

3. Motivation for Innovation

The work of Gaston Planté raises a number of questions: What was the motivation for his lengthy development of the "formation" process? Why did he never attempt to shorten this incredibly tedious process by the seemingly obvious method of mechanically applying chemically-prepared active materials to the electrodes?

To the first of these questions there are two answers: Planté's motivation for the "formation" process was in part economic and part scientific. Dealing first with the economic motivation, Planté's battery was the first storage cell manufactured commercially. Beginning in the 1860s a number of telegraphic equipment and scientific apparatus supply firms offered Planté batteries in their catalogs. In France, the chief supplier was the Breguet firm and Planté had a close relationship with this company—working in its shops and collaborating with its engineers.

Ever since Volta's invention of the battery in 1800, the makers of scientific instruments had added batteries to their stock of goods. After mid-century, a number of these companies diversified into the growing field of telegraphic equipment since the components used in making scientific electrical apparatus and telegraphic equipment were very similar. The Breguets in Paris was one such firm which made this transition. By the 1860s and 70s, the larger of these firms in the United States, Britain and Europe, carried many models of batteries in stock.

Planté saw that some of the properties of his cells were unique, and that these properties might find them a market. It is crucial to note, however, that the cell's ability to store energy was *not* one of the unique properties that he valued. That is, that property which has given the lead-acid storage battery its chief technological value during the past hundred years was of only secondary importance to Planté. Indeed, given the very thin coating of active materials even on fully "formed" Planté plates, contemporary primary batteries had a much greater capacity for storing energy than the Planté cells.

The unique property of the lead-acid system that Planté valued most was its extremely low internal resistance. Contemporary primary batteries had internal resistances of about 1 ohm; Planté's cells had resistances around one thousand times less. This extraordinarily low resistance, combined with the cell's high voltage (2 volts), made it into a kind of hybrid primary battery and capacitor. Accordingly, the Planté cells were able to absorb and discharge their energy at a very much faster rate than primary cells.

Between 1860 and 1880 Planté took out a series of French and British patents on specific applications of his batteries. All of these inventions centered on the cell's low resistance and high voltage. Among other things, he suggested the firing of explosives, an electric lighting scheme, and two medical applications. Both medical ideas relied upon heating fine platinum wires to incandescence. The first scheme was to light up body cavities, but this proved impractical since Planté did not think of isolating the filament from the atmosphere. Nevertheless, this idea was to bear fruit in the 1880s when miniature incandescent bulbs powered by storage batteries were used in the first polyscopes. Planté's second medical idea was electrocauterization—fine platinum wires were to cauterize tissues more precisely than traditional methods. Practical applications of this technique began as early as 1872.[1]

---

[1] Planté, *Storage*, p. 90.

Planté also suggested a scheme for the electrification of lighthouses. As Roy MacLeod has shown, European interest in increasing the power of lighthouse illumination grew rapidly after 1850 as a result of the growing number of wrecks produced by increased shipping.[2] Indeed, the chief stimulus for the design of large electric generators in the 1850s and 60s was the idea of replacing lighthouse oil lamps with electric arcs or lime lights. Planté's 1868 French patent on electric lighting was part of this development.[3]

The Planté idea was to use a carbon arc light powered by his batteries. These were to be charged from Grove nitric acid cells. The heart of the plan was an electric motor-driven commutator, which rotated many times a minute, alternately connecting the lead-acid cells to the Groves for charging, and to the arcs for discharging. Thus, the lighthouse would send out a rapidly flashing but brilliant light.

The use of Grove cells made this uneconomical for large-scale illumination. In the early 1870s, however, Alfred Niaudet, the chief electrical engineer of the Breguet company, and Planté, collaborated on a plan to use the new Gramme generator to charge the lead-acid cells. Niaudet was chiefly responsible for introducing the Gramme machine into France.[4] In 1873 Niaudet and Planté published a joint paper on the charging of storage batteries by generators, and they also began using a Gramme machine at the Breguet factory for charging and "forming" the Planté batteries the company was making for sale.[5] In 1874 Niaudet proposed the use of Gramme machines and Planté batteries in lighthouses and aboard ships to avoid accidents.

The Planté-Niaudet scheme was designed for 40 lead-acid cells. These were wired to a motor-driven commutator, which shifted the cells from parallel to series connection with each revolution. When in series, the cells discharged through the arcs, and when in parallel, they were charged from the generator. The commutator revolved several times a second, thus producing an almost continuous light.

In addition to its use in lighthouses, Planté suggested this technique as a solution to the problem of "subdivision of the

[2] Roy M. MacLeod, "Science & Government in Victorian England: Lighthouse Illumination and the Board of Trade," *ISIS* 60 (1969): 5-38.

[3] French patent dated 29 February 1868.

[4] Malcolm MacLaren, *The Rise of the Electrical Industry during the Nineteenth Century* (Princeton, Princeton Univ. Press, 1943), p. 68; *Teleg. J. & Elec. Rev.* (hereinafter: *TJER*) 7 (1879): 10; 15 January, 225; *La Lumiere Électrique* 1 (1879): 13.

[5] Planté, *Storage*, 104; *La Nature*, 27 June 1874.

electric light" that is, reduction of the intensity of arc lamps so that they could be used for indoor illumination. He believed that by using sufficiently small plates in the battery, the current would be reduced, and thereby the arc's intensity. He admitted, however, that[6]

> This solution . . . appears rather complicated. Still, it is not impossible to carry out; for commutators like those described do not require a high EMF in order to keep them going; they could be arranged so as to turn like the bobbins of spinning looms. . . .

This method of electric lighting was never used commercially. Nevertheless, the basic methodology of the plan resembles that of most early electrical engineering invention—that is, the combination of preexisting components in a new way. All the components of the system were either manufactured directly in the Breguet factory or could be purchased elsewhere. The commutators, for example, were manufactured by the Breguets for telegraphic practice; they needed only the addition of motor-drive. It is probable, therefore, that the details of this plan were dictated more by the nature of standard electrical equipment then being produced by the Breguet factory than by the innovative ingenuity of Planté or Niaudet.[7]

In the plan outlined above, the storage battery's function was not that of a current *storing* device, but of a current *conversion* device. This function can also be seen in another early application of the Planté cells; for electric brakes on trains. In 1856 the French inventor Auguste Achard invented an electromagnetic brake.[8] To provide sufficient stopping power, the coils of the brakes required intense currents, albeit for only brief intervals. Large Daniell batteries were used at first, but these did not produce sufficiently strong currents. In the late '70s, Achard brakes were equipped with a combination of Planté batteries and Daniell cells; the Daniells being used to keep the Plantés in a charged state for the intermittent discharges. These brakes were tried on a number of French and Italian trains in the 1880s, but with indifferent results.[9]

Yet another example of the current-conversion use of the lead-acid battery was Saturn's Tinder Box; an invention Planté

---

[6] Planté, *Storage*, 105.

[7] Despite the fact that the scheme was never commercialized, a number of other inventors worked on it; for example J. Morin in 1875. Ibid., p. 236.

[8] *La Lumiere Électrique* 4 (1881): 121.

[9] *The Electrician* (New York), May 1882, p. 97; June 1882, pp. 112-113.

patented in 1873.[10] This was a combination door bell and lamp lighter. It consisted of a wooden box holding a small Planté cell and topped with a thin platinum wire which was raised to incandescence by touching a switch:[11]

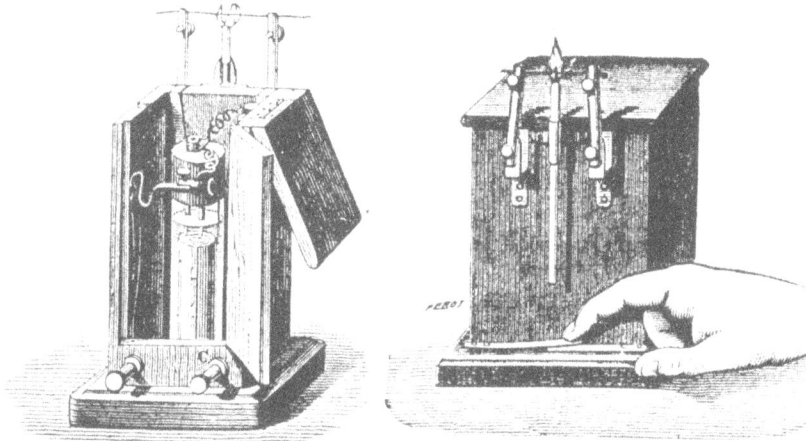

Saturn's Tinder Box was designed for use in those homes which had been wired for electric doorbells. The technology of the telegraph began to find its way into the consumer market in the 1860s, as telegraphic engineers converted their signalling devices into front-door bells for private homes, room-service buzzers for hotels, and fire and burglar alarms. The current was supplied from primary batteries. Plante's Tinder Box was connected in parallel across the terminals of these bell batteries, where it could absorb current from them. Its primary function was to provide an occasional high-intensity discharge to light a candle, lamp or cigar, but it would also discharge when the doorbell was rung, giving a much louder ring than would be possible with the primary cells alone. As in the case of the Achard brake and lighthouse illumination, the function of the lead-acids was not to store, but to convert electrcial energy.

Gaston Planté was the most famous of the men to work on applications of the lead-acid storage battery in the three decades following 1850, but a number of other inventors followed this path as well. Their primary goal was the solution of telegraphic problems. These inventors had two particular solutions in mind: first, the already-mentioned idea of using the

---

[10] British patent #1713—10 May 1873. The name comes from the alchemical linkage of lead with Saturn.

[11] Planté, *Storage*, p. 97.

battery's high voltage and low internal resistance to force signals through long circuits, often in connection with electromagnetic generators as charging devices; second, the use of
polarized systems to discharge capacitive loading in long lines
or cables. As lines grew longer, the capacitive effect of their
length increased and blurred the dots and dashes, thus slowing down the speed at which messages could be sent. But if the
signal could be made to produce a kind of "echo" of itself in
the wire, traveling in the reverse direction to the signal, then
the capacitive effect would be self-discharging and the speed
of signalling increased. It was reasoned that this electromagnetic "echo" could be induced by placing a pair of polarizable
plates in the circuit. In 1860 Jacobi used a pair of lead plates in
sulphuric acid for this purpose.[12]

In 1875 the English inventor Henry Highton took out a
British patent on an invention for the capacitive discharging
function of lead-acid cells. The following quote from this patent reflects the chief way in which inventors *prior to the 1880s*
looked upon the virtues of the secondary battery:[13]

> The object of this invention is to obviate the phenomenon called
> retardation in submarine, subterranean and other cables. For
> this purpose I place at either . . . end of a line, and also if
> possible at any convenient intermediate point in the line circuit
> pairs of plates of metal or carbon arranged in the manner of
> plates in a galvanic battery, excepting that only one conducting
> liquid is used, and the two plates in each pair are similar to one
> another. The electrical effect of this is similar to the mechanical
> effect of a spring in machinery, and immediately the electrical
> impulse ceases the original electrical condition of the line is
> restored by the reaction of the plates so as to be ready to receive
> a fresh impulse.

The last sentence particularly reflects the battery-capacitor
hybrid image of Planté and the other pre-1880 developers.

In 1869 Octavius and Frederick Varley took out a British
patent for the use of lead-acid storage batteries as high voltage
sources to replace primary batteries. The cells were to be
charged from a mechanical generator.[14] This was a different
function than Highton had proposed, and yet the Varleys, no
less than Highton, stressed the energy *conversion* function of
the system. Indeed, their concentration upon the high voltage,
low resistance properties of polarized plates is so strong that

---

[12] Ibid., p. 107.
[13] British patent #439—3 February 1874.
[14] British patent #2525—25 August 1869.

they refer to the plates as "induction plates" (i.e. a capacitor), although it is clear that they realize that the action is chemical rather than physical:

> We construct the induction plates of two or more pieces of carbon and palladium plates inserted in a cell filled with acid and water, or plates of carbon and metallic arsenic, or platinum substituted for the palladium or arsenic plates, or two lead or other metal plates, or two carbon plates. The current is made to generate hydrogen on the palladium, arsenic (or) platinum plates and oxygen on the carbon. The capacity which palladium and arsenic have for storing hydrogen render them very efficient induction plates, becoming when charged a very powerful gas battery. We increase the capacity of the carbon plates for oxygen by rendering the plates porous. . . . (a technique similar to that of Siemens is described)

In 1866 Georges Leclanché, the inventor of the prototype of the modern dry cell, obtained patents for a reversible gas battery for use in telegraphy. The two plates of carbon or "inoxidizable metal" were placed in K_2CO_3 solution and surrounded with granulated carbon.[15]

The earliest patent for a storage battery was by Charles Kirchhof of New York in 1861. The design was for a lead-acid cell:[16]

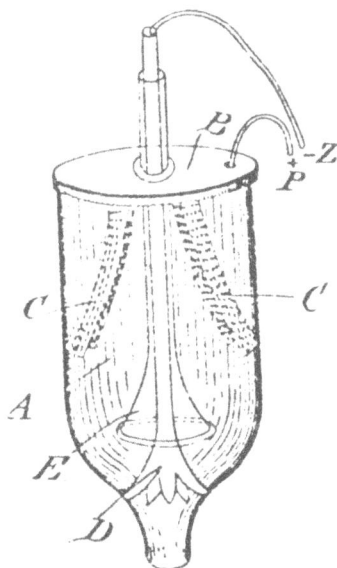

---

[15] British patent #2623—10 October 1866 (taken out by Alexander H. Brandon for Leclanché; U.S. patent #64,113—23 April 1867. The physical structure of the rechargeable cells was almost identical with that of the famous MnO_2 primary Leclanché; the lineal ancestor of the modern dry cell.

[16] U.S. patent #31,545—26 February 1861.

The glass vessel A contains platinum foil electrodes C and D separated by the glass tube E. An acidic solution of nitrate and acetate of lead is placed in the vessel and charging current from a generator causes PbO_2 to form on D and metallic sponge on C-C. This technique was derived from Wheatstone, who had used it to prepare the primary battery electrodes mentioned in section #1. Kirchhof refers to Wheatstone in his patent, along with Schönbein, Poggendorff, and other early nineteenth-century polarization researchers, but he was apparently ignorant of Planté's recent work.

Again, Kirchhof's plan was not to design an energy-storage device *per se*, but rather a device for converting currents. The patent is entitled: "Improved Method of Integrating Inconstant Currents"; the inconstant currents being the currents from electromagnetic generators. Kirchhof, like Sinsteden, wanted to provide stronger currents for telegraphic circuits by replacing primary batteries with generators. The insoluble problem which faced all early attempts to use generators in telegraphy, however, was the fluctuating, static-filled output of these crude machines.

Kirchhof designed his "regenerator" to convert the mechanical generator's unruly current into the smooth output of a battery. The process of "regeneration" was carried on continually in a manner similar to the Planté-Niaudet scheme of electric lighting:[17]

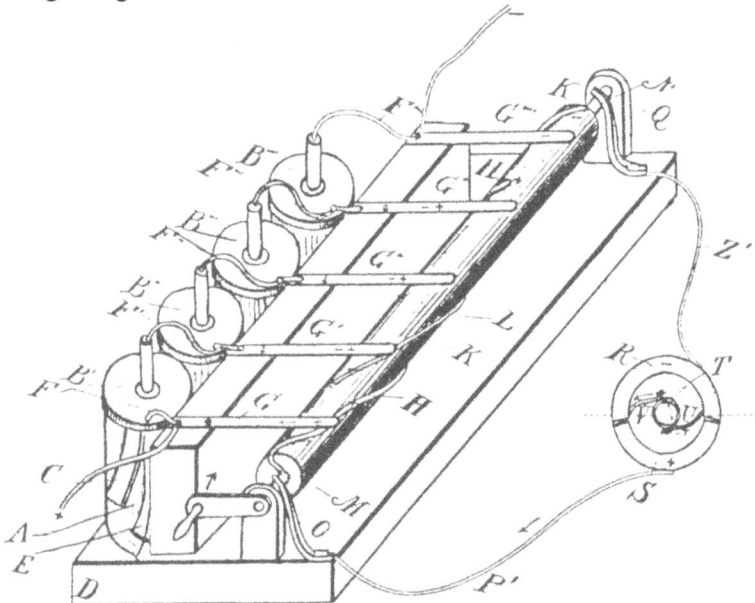

The bank of four "regenerator" lead-acid batteries is charged continually in series from the leads which go to the generator. The rotating commutator discharges each of the regenerators in sequence once each revolution, in an analogous manner to the staggered ignition of automobile cylinders by the distributor. Consequently, an almost perfectly smooth output is obtained.

The foregoing discussion of pre-1880 plans to use storage batteries, although not exhaustive, includes most of the proposed schemes, particularly regarding telegraphy. There is a unity to all these ideas which permits us to suggest an answer to the questions raised at the start of the section.

This unity lies in the inventor's image of the storage battery as a kind of slow-acting capacitor. Gaston Planté, like the other inventors, used this capacitor image a number of times: once, for example, in 1860: "This apparatus exactly performs, therefore, the role of the condenser; because it permits us to instantaneously collect the work performed by the (primary) battery over a period of time,"[18] (that is, that the discharging can occur much more rapidly than the charging); and again in 1879:

A secondary couple is, in fact, a kind of lever for dynamic electricity; because one can thereby obtain with a weak electrical power, an increase of this power, in such proportion as one wishes, on the condition of loss in speed, or a necessary sacrifice of time, in order to accumulate the effect.[19]

Alfred Niaudet observed much the same thing in 1875: "It is easily perceived that the secondary battery *can only produce effects of short duration* (my ital.), but in a very great number of cases such effects are all that is required" (*short* means a few seconds).[20]

In the next chapter we will see storage battery inventors from 1880 onward focusing their attention on the total energy storage capacity of the battery and paying increasingly less attention to the properties of high voltage and low internal resistance. For the 1860s and '70s, however, great storage capacity was of minimal importance, since the electromagnetic generators of the day were few and of low output. Little economic incentive existed before the 1880s to develop an electrochemical reservoir capable of storing large amounts of electrical energy. Indeed, as we saw in the case of Kirchhof and

---

[18] *Comptes Rendus,* 1860: 641-642.

[19] Planté, *Storage,* p. 69.

[20] *TJER,* 1 December 1875: 275; 15 January 1875: 225.

Niaudet-Planté, the problem facing early users of mechanical
energy generators was not that of storing their energy, but in
being able to use that energy *at all*. All existing electrical
technology had evolved to use primary battery current and this
current was much steadier than early generator current. Therefore, one of the functions of the first storage batteries was to
convert electro-magnetically generated current into ersatz primary battery current. The pattern of Gaston Planté's innovative activity is understandable against this background.
Planté's batteries, with their thin layers of active materials
intimately bound to the metallic lead plates from which they
were formed, were ideally suited to perform the function of
electrochemical "capacitors."

Planté's design of storage batteries had a scientific as well as
economic motivation. Planté thought that scientists could use
the lead-acid battery as a research tool in theoretical studies of
meteorology. Storage batteries produce high intensity discharges of electricity as does the atmosphere. Accordingly,
reasoned Planté, it should be possible to use the one to produce analog models of the other.

Interest in meteorology was common among eighteenth and
nineteenth-century chemists and physicists, the most famous
example being John Dalton's life-long attempt to create a
mechanical model of the atmosphere. Many nineteenth-
century treatises on chemistry or physics contain chapters on
meteorology, in which atmospheric phenomena are treated as
a special case of the book's major theme. Indeed, this tradition
is as old as Aristotle. The linkage between meteorology and
electricity, however, was a recent one.

Until the mid eighteenth century, such electrical phenomena
as lightning were assumed to be some sort of gaseous combustion. But during the eighteenth century electricity emerged as
a major field of scientific study. Scientists learned how to
magnify its effects, so that the miniscule cracklings of hair
combs were amplified into powerful and brilliant sparks. The
more powerful these sparks grew, the more obvious became
the analogy with lightning. Benjamin Franklin proved the
analogy in the 1750s with his famous kite experiments. These
experiments, along with many others carried out during the
succeeding century, showed that continual electrical potential
changes occur between the clouds and the earth, and between
the clouds themselves. In other words, just as the weather
changes continually, so does the electrical state of the medium
in which it occurs. This led to the theory that there exists a

cause and effect relationship between the two; we will call it "electric meteorology."

Franklin was the first to suggest an electric meteorological model,[21] one that was fairly simple: water is made of hard, Newtonian atoms. These atoms are close together when water is in the liquid state, since water atoms neither repel nor attract each other; therefore, they lie together like a mass of ball bearings. Heat and electricity are both subtle fluids which have an attraction for water atoms, but which are self-repellant. Consequently, water atoms can be given a coating of heat fluid, or electric fluid, or both, in which case the water atoms repel each other in proportion to the amount of coating they receive. This process of repulsion we know sensually as evaporation.

If we heat two pots of water and charge one with electricity as we heat it, the vapor which rises from the electrified pot will be less dense than that from the non-electrified pot, since the electrified vapor will have twice the internal repellant power. On the surface of the earth and the oceans the same processes occur, with some evaporating water receiving only heat, and some both heat and electricity. These different kinds of vapor will then form different kinds of clouds, and thus different kinds of rain, wind, and so forth.[22]

Franklin himself came to question this model. Among other things, the necessity of using *two* separate and distinct causes (electricity and heat) to explain *one* effect (evaporation) made the model suspect, and led to inevitable challenge from those who saw no reason to use more than heat to explain the phenomenon. Moreover, the model of eighteenth-century electric meteorology could be used to explain relatively little about atmospheric phenomena.

During the nineteenth century, however, increased knowledge of electrical phenomena provided electric meteorologists with greater scope for explaining atmospheric effects. For example, Faraday's discovery of the principles of electromagnetic induction was combined with growing empirical observations about the rotation of storms; the rotation was assumed to be due to the inductive effect of the earth's magnetic field upon the flow of current which was assumed to take place when rain fell.

---

[21] I. Bernard Cohen, *Benjamin Franklin's Experiments* (Cambridge, Harvard Univ. Press, 1941).

[22] This explanation of Franklin's model is, of course, greatly simplified. For details, see: Cohen, above.

Of greater importance for stimulating electric meteorology, however, was the developing concept of conservation of energy. The idea that heat and electricity are not separate fluids, but mutually interconvertible forms of the same universal principle, reduced one of the problems which made the original Franklin model suspect. Thus, for example, in 1832 an English amateur scientist named Thomas Pollock published an electric meteorological treatise based on the concept of the interconvertibility of forms of energy.[23] Pollock states that there is in nature a "universal principle" which can manifest itself as electricity, light, heat, gravity, magnetism, and so forth. The specific manifestation which the principle takes is determined by the nature of the material bodies in which it is contained. As the state of the material body changes, so will the manifestation of the principle change, and this will produce changes in the weather.

To take one example: when water evaporates, the principle uniformly fills the empty spaces between the water particles. This uniformly homogenous internal distribution of the principle is sensually perceived as heat. When the water condenses to form clouds, however, the principle is squeezed out and forms a shell on the exterior of the cloud. This new manifestation is perceived as electric charge. If the clouds condense still further, rain falls, and the shell of electricity, squeezed in upon itself, discharges to the earth, producing lightning.

Pollock's model stands at the extreme of nineteenth-century electric meteorological theorizing. It was an attempt to explain all weather as a continual interconversion between heat and electricity. The reasoning is broad but weak and amateurish. Nevertheless, a number of such theories linking electric meteorology with the new concept of energy were suggested during the nineteenth century.[24] Most of these theories were the work of amateurs and make broad use of analog reasoning with little critical analysis of the analogs.

At a more moderate level were the occasional speculations of more respected nineteenth-century scientists about the electrical origin of specific atmospheric phenomena. The auroras, for example, were universally assumed to be electrical; these ideas

---

[23] Thomas Pollock, *Attempt to Explain the Phenomena of Heat, Electricity, Galvanism Magnetism. . . ."* (London, 1832); see: *Wheeler Gift Catalogue,* Item #870.

[24] See, for example: G.A. Rowell, *An Essay on the Cause of Rain & Its Allied Phenomena* (Oxford, 1859).

came from contemporary experiments with the discharge of electricity in rarified gases. Another favorite topic for electric meteorological speculations was the tornado or water spout. In 1812, for example, Humphry Davy speculated that:[25]

> The water spout is probably the result of the operation of a weakly electrical cloud, at an inconsiderable elevation above the sea, brought into an opposite state (of charge); and the attraction of the lower part of the cloud for the surface of the water, may be the immediate cause of this extraordinary behavior.

Davy's reference to the "extraordinary behavior" of tornados was the chief reason for the widespread speculation about their electrical origin; the powerful lightning which accompanies tornados and the intense concentrated power was hard to associate with the power of evaporation alone. Some theorists saw the tornado as analogous to the rise of pith balls between charged plates, with trees and houses taking the place of pith balls. Others saw its whirling vortex as analogous to the rotation of current-carrying conductors around a magnet.[26] Either way, the whirlwind was a powerful symbol for electric meteorologists.

Without going into any great detail on the subject of electric meteorology, we may conclude that it was a theory toward which many leading nineteenth-century scientists were at least partially favorable and about which many amateurs and less than top notch professionals were highly enthusiastic. Gaston Planté was one of this latter group. With the exception of a few miscellaneous writings, Planté's published works fall into two categories: a set of empirical studies of storage battery design and a group of highly theoretical and rather fantastic speculations about the electrical operation of the universe. These speculations are an expansion upon the electric meteorological tradition.

According to Planté, the universe is a great electrical machine. All those phenomena which we today would call manifestations of energy are derived ultimately from electricity. Thus the sun is a great charged ball, whose light and heat arise from the internal movement of electricity, like hysteresis

---

[25] Davy, *Collected Works*, (Vol. 4—The Elements of Chemical Philosophy, New York, Johnson Reprint Co., 1972) p. 102.

[26] See, for example: *Sturgeon's Annals of Electricity* 2 (January—June 1838): 195-203.

heating. However, the sun itself does not generate its own electricity:[27]

> Thus we believe that the sun is electrified, but that it does not create the electricity which it possesses, any more than the light and heat which arise from it; it is a store received from the nebulous ring, of which it is but a brilliant particle, condemned to become extinct some day; this nebulous ring would be derived from another electrified wave and so on up to the primary cause, the creator of all power and motion. Taken from this point of view, the incandescence of the solar globe prolonged through a series of ages, would in itself be but a spark of short duration in the infinitude of time and space.

Planté was carried away here by an ecstasy of analogy. The poetic terms are undefined and Planté gives no indication of how the "nebulous rings," "electrified waves," and "primary cause" are to interact, nor even what they are. In fact, the sole evidence for the electrical theory of the sun lies in a rather prosaic laboratory analog demonstration.

In this demonstration, Planté fixed a length of thick iron wire between two movable binding screws. A high intensity current was then passed through the wire. As the metal melted, the screws were advanced towards each other, allowing the molten, glowing iron to form into a ball. As these balls formed, Planté watched them through a smoked magnifying gas. What he observed was the operation of the sun:[28]

> We have seen that these melted globules show brilliant eruptions in consequence of the disturbance in their interior substance under both the calorific and chemical action of electricity, that the jets of gas and incandescent particles. . . . made their way out by the cavities or perforations produced in the interior of the globule itself; and that these perforations, allowing the comparatively colder and less luminous interior of the globule to be seen, formed dark spots upon its brilliant and undulating surface. . . .

Not only the sun, but the galaxies themselves may be explained electrically. By Planté's day, astronomers had recognized the spiral form of several galaxies. Planté used one of the pieces of apparatus he had used in the original storage battery

---

[27] Planté, *Storage*, pp. 204-205.
[28] Ibid., p. 203.

experiments and allowed time for the stream of colored ions to separate itself from the wire. If a magnet was then placed outside the cell, the following effect would be noted: [29]

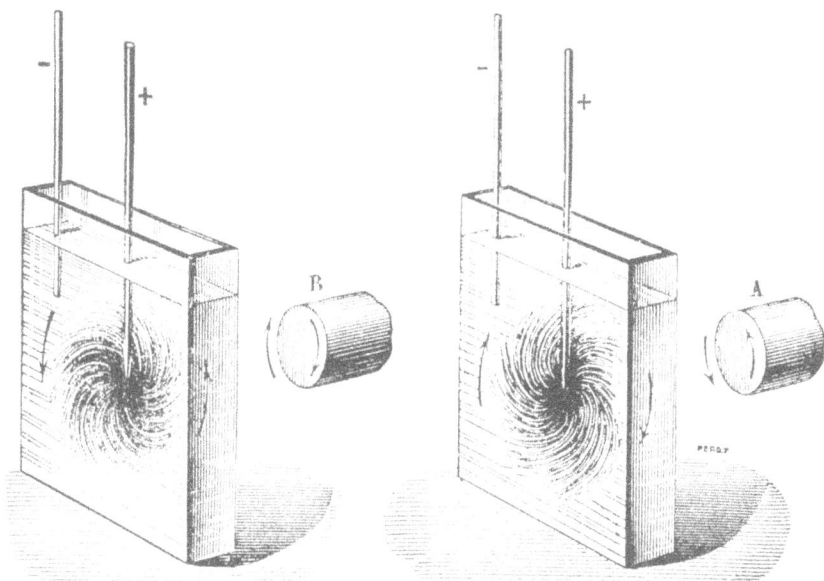

From this observation, Planté concluded that the galaxies are, in effect, giant electric motors:

> In view of so striking a similarity, may it not be reasonably supposed that the nucleus of these nebulae may be formed by a veritable electric furnace; that their spiral form is probably caused by the presence of celestial bodies powerfully magnetized, and that the direction of the curve of the turns of the spiral must depend upon the nature of the magnetic pole turned towards the nebula?

> It would be interesting to search, among the stars already known round these nebulae, those which, by their position, could exercise this magnetic influence, or to explore the celestial firmament, on the axis of which the spirals appear to turn. . . .

> In the case where a star might be known to fulfill these conditions, we might still examine, along the line passing through the center of the nebula, and the star itself, if there were not a second spiral nebula connected with the other magnetic pole of this star, the curves of which, turning in the opposite direction to the magnetic currents of this pole, might, nevertheless, ap-

[29] Ibid., pp. 142, 147.

pear to the observer to be directed in the same direction as those
of the first and the combination of the three bodies would thus
form a symmetrical stellar system.

The preceding theories are, of course, astronomical rather
than meteorological. Yet they originated in Planté's earlier
meteorological speculations. The galaxy model, for example,
was also used to explain the circular motion of storms, and,
particularly, that favorite topic of the electric meteorologists,
the water spout:[30]

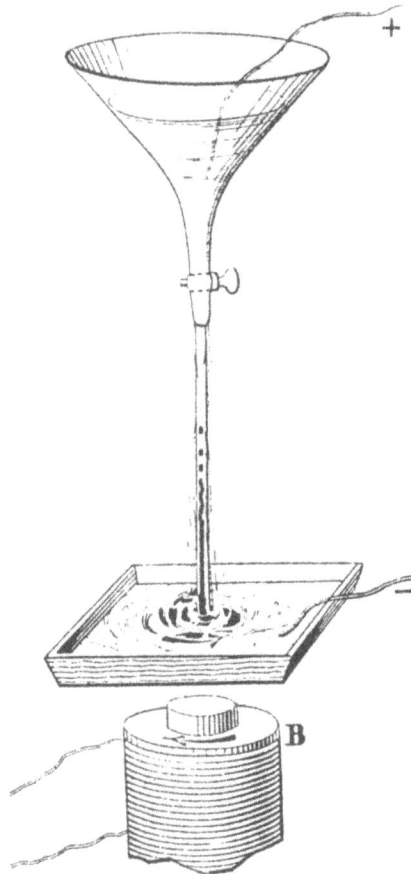

Saline solution was poured through the funnel while 800 volts
was maintained across the "cloud"-"ocean" gap. Planté observed all the phenomena about real water spouts which
sailors had reported, including the lightning.

---

[30] Ibid., p. 129.

The formation of hail was another topic with proponents of electric meteorology, since the suspension of large hailstones in the atmosphere seemed inexplicable except by electrical means. Thus both Volta and Jean Peltier created electrical models of hail formation.[31] Jean Peltier, who wrote extensively on electric meteorology,[32] described hail formation as being due to the discharge of two clouds into each other if the clouds have different charges and elevations. Particles of water or ice bounce back and forth between the two clouds like pith balls between capacitor plates. As they do, these particles combine until they become too heavy to remain suspended.

Years after Peltier proposed it, Planté used a variant on this model to create a physical analog of hail formation. Two electrical leads carrying a very high potential were thrust into a beaker of saline solution. A powerful spray of droplets was produced.[33] The beaker of solution represented one cloud and the electrical leads the second cloud at a different potential. In the atmosphere, the spray of droplets is thrown high into the cold upper regions, where they freeze and form hail.

The puzzling phenomena of ball lightning was similarly explained. This time only one high potential wire lead was immersed, while the other was suspended close to the liquid surface. Soon a brilliantly glowing ball of incandescent vapor formed between the two "clouds," that is, under the hovering wire lead.[34]

Planté continued in this manner, leaping from one far-fetched laboratory analog to the next. As a theorist, he was undisciplined and uncritical, but as an experimentalist, Planté had an extraordinary capacity for taking pains, and it is this faculty which represented his second motivation for developing the storage battery.

For many of his experiments, Planté needed a more powerful source of current and voltage than was available in the mid nineteenth century. Electrostatic machines and Leyden jars could produce high voltages, but without much current or duration. Primary batteries could yield current, but because of their internal resistance they were inadequate for producing the analog of lightning. Consequently, Planté developed a piece of apparatus which was a variant on the invention which

---

[31] Dionysius Lardner, *A Manual of Electricity, Magnetism and Meteorology* (Vol. 2—1844) pp. 241-267.

[32] J.C.A. Peltier, *Notice Sur la Vie et les Travaux Scientifiques de J.C.A. Peltier.*

[33] Planté, *Storage*, pp. 126-127.

[34] Ibid., p. 117.

he and Alfred Niaudet planned for lighthouses. A bank of 800
individual, test-tube sized lead-acid batteries was prepared.
Rotating commutators atop these banks of cells shifted them
from parallel (for charging by two Grove nitric acid batteries)
to series (for discharging). These 1600 volts were then further
increased by being used to charge up a glass and metal foil
capacitor bank of 80 plates. Again, this device was equipped
with a commutator to receive the charge in parallel and discharge in series. The result was a voltage of 128,000 volts, made
almost continuous by the commutator system.[35]

Planté's scientific and economic motivations are a reflection
of general developments in technology and science in the late
nineteenth century. At that time, in both the technology of
telegraphy and the science of physics, the generation of high
voltages became a problem of increasing importance. It is not
surprising therefore, that Planté's interest in both electrical
technology and electrical science, should have stimulated him
to spend much of his career developing an invention which he
believed would solve this mutual problem. The lead-acid storage battery was the result.

In 1879 the famous electrical inventors Elihu Thomson and
Edwin Houston patented a storage battery. At first sight, it
appears a rather unremarkable throwback to earlier ideas:[36]

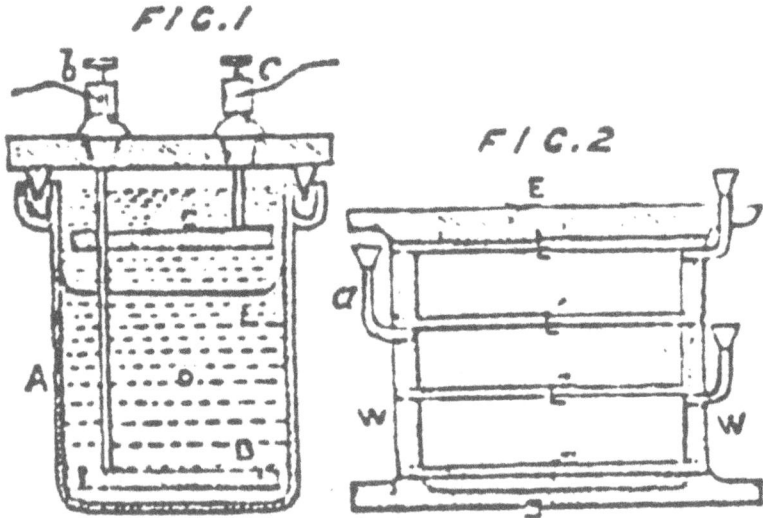

FIG.1

FIG.2

---

[35] Ibid., pp. 115, 217.

[36] U.S. patent #220,507, Thomson and Houston, 14 October 1879; La Lumiere Électrique 3 (1880): 68; TJER 8 (1880): 52.

Here B is a plate of copper: C a plate of zinc; and E a porous diaphragm. All are immersed in an acidic zinc sulphate solution in a glass vessel. This is nothing more than a Daniell cell which the patentees say can be run backwards for recharging.

Thomson and Houston proposed this invention either for telegraphy or for lighthouses; ideas which, like the battery itself, were very derivative of past inventors. And yet, this patent contains a fundamentally new concept not found in prior schemes. Like Kirchhof and Niaudet, Thomas and Houston proposed to charge up their batteries with dynamos, but their idea was to build batteries of large storage capacity to absorb a great amount of generator output at a time. Thus, the lighthouse battery would store current from a generator run by day, and discharge the current at night. Here we have the modern concept of the secondary cell as an energy bank.

Contemporary with the Thomson-Houston copper sulphate cell was the 1879 patent of d'Arsonval. D'Arsonval, like Siemens and Planté, was influenced by de la Rive; his storage battery was a rechargeable form of de la Rive's $Zn-H_2SO_4-PbO_2$ cell of 1843. The anode was a pool of mercury with zinc dissolved in it, and the cathode consisted of a lead plate covered with fine lead shot. The mechanical construction of this cell was obviously not for the purpose of producing capacitor-like effects, but, as d'Arsonval stated: "d'augmenter la durée de la charge accumulée."[37]

This changed perception of the function of storage batteries arose from a more fundamental change in the nature of electrical engineering itself. At the end of the 1870s, the dynamo ceased being an experimental device and entered the stage of commercial exploitation. Experimental arc lighting installations, which had begun around 1875, became a commercial element in the electrical industry during 1878-1879. The incandescent electric light, although not yet reduced to practice in 1879, was very much in the news at this time, and no perceptive electrical inventor would have failed to appreciate the value of storage batteries for this potential technology. Before 1878-1879, electrical inventors who were alert to the main chance in their business designed inventions chiefly for telegraphy. Then, quite rapidly, the emphasis shifted to power and lighting. As we will see in the next chapter, the design of storage batteries participated in this fundamental change.

---

[37] French patent #133,884—28 November 1879; *La Lumiere Électrique*, 15 February 1880: 78.

II. Commercialization Begins: 1880-1889

1. The Early Commercial Innovators

The commercial technology of electric lighting began at the end of the 1870s. Inventors had experimented with electric lighting since the beginning of the century, but only with the introduction of commercial, self-excited dynamos in the 1870s did electric lighting evoke widespread interest. Street lighting with carbon arcs was done for demonstration purposes beginning in the mid 1870s and the first commercial installation of street lights began in 1878.[1] This year also marked the beginning of serious attempts to develop a commercial incandescent electric bulb.[2]

These developments stimulated the creation of the storage battery industry. At the end of the 1870s, as rumors of the impending introduction of incandescent lighting spread, a number of European and American inventors perceived the important role that the lead-acid battery would play in the new technology. Consequently, these men began to develop storage batteries which they perceived to be adapted to the needs of a yet non-existent incandescent technology.

As we have seen, the storage battery had been suggested as an adjunct to electric lighting many years prior to 1878-1879. These suggestions, however, always pertained to *arc* lights, and, in practice, storage batteries were never used for this purpose. There were a number of technical reasons for this, the chief one being that the current consumption of arcs is so great that they would have exhausted the batteries too rapidly to be practical. Such systems of the Niaudet-Planté "slow-capacitor" scheme, while not subject to this difficulty, were not used because they were an unnecessary complication of arc light operation. Engineers found it much easier to raise voltage by redesigning the dynamo windings than by passing low voltage dynamo current through a complicated series-parallel storage battery bank.

For incandescent bulb lighting, the advantages of storage battery adjuncts were considerable. Incandescent lighting was

---

[1] Charles Singer (ed.), *A History of Technology* (London, Oxford Univ. Press, 1956), 5: p. 196.

[2] Arthur A. Bright, *The Electric Lamp Industry, passim*.

more compatible with batteries since bulbs consume less current than arcs and therefore can be run for significant periods from batteries. In contrast with the thick carbon pencils of the arc, incandescents used a delicate carbon thread filament; therefore, the life of the filaments was significantly shortened by passing fluctuating current through them. The advantages of using storage batteries to smooth the erratic output of early generators were seen as early as the 1861 Kirchhof patent, but with arc lights neither the life nor the performance of the lights were greatly diminished by fluctuating output. The popping of filaments and the change in intensity of the light were more serious problems with incandescents.[3]

Moreover, the economic importance of prolonging bulb life was greater in the early 1880s than today, since early bulbs cost over a dollar apiece.[4] While dynamos of the late '70s were greatly improved since Kirchhof's day, they were still far from reliable. The graph below shows the output of a power station of the 1880s, with and without use of a storage battery buffer:[5]

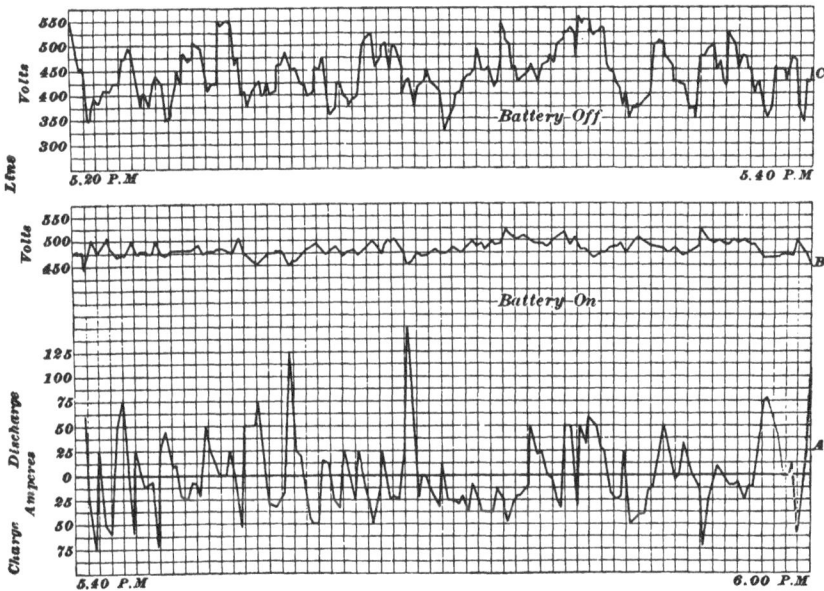

---

[3] *Electrical Engineer* (London) 4 (1889): 35-38.

[4] In May 1883 the London *Electrical Engineer* reported the prices of twenty-six common commercial bulbs made by Edison, Gatehouse, Lane-Fox, Maxim, SWAN & Woodhouse and Rawson. The cheapest of these cost a dollar, the most expensive $1.87, while most of the bulbs cost about $1.25; these were in the commonest 16-20 c.p. range. 5/1/1883, p. 15

[5] *International Correspondence School Reference Library*, (Volume on Storage Batteries, Incandescent Lighting, etc (Scranton, International Textbook Co., 1905), p. 63.

A second factor which made the battery assume great importance for incandescent lighting relates to the perception which
many electrical engineers had of the potential of this new lighting technology. During the early 1880s incandescent lighting
was generally perceived as a small-scale operation in which the
size of each individual unit of dynamos and lights would be
very limited. Incandescent lighting was often conceived as
something which each consumer would provide for himself,
with the dynamos and engines located on each consumer's
premises. Many of the earliest "systems" began in this way,
with a small group of neighbors or businesses pooling their
resources to install a small dynamo in a shed within a few
hundred feet of all the participants.[6]

Not all early promoters of incandescent lighting put their
emphasis on the dynamo-in-the-basement schemes, but even
the largest of early plans was limited to a few blocks.
Moreover, with the exception of the few early a.c. promoters,
none of the early lighting pioneers believed that systems of
central station distribution could be installed beyond the
limits of central urban areas. It was generally believed that
incandescent lighting in suburban and rural areas would be
restricted to private plant operation, with each home or factory
providing its own power.[7] This was because of the expense of
transmitting direct current over even limited distances.

Because of these considerations, the design and operation of
power stations was conceived in a far more restricted way
around 1880 than would be common only a decade later. For
example, it was uneconomical to provide full-time attendants
for a system which would serve only a handful of consumers.
In the larger systems, such as Edison's famous Pearl Street
station, a number of full-time employees were practical, but
Pearl Street was one of the few large systems. In the early days
of electric supply, the current was needed for only 4 to 6 hours
per day, but unfortunately those hours did not coincide with
those when workmen or servants would normally be up and
attending their duties. Therefore, the maintenance of dynamos
and engines as a part-time duty of otherwise-employed workers presented scheduling problems. This is where the storage
battery could provide flexibility.

The battery could store current produced by a generator
which ran during normal working hours. Therefore, operation

---

[6] Percy Dunsheath, *A History of Electrical Power Engineering* pp. 142-143, 148-149.
[7] R.H. Parsons, *The Early Days of the Power Station Industry* p. 136.

of the dynamos could be assigned as the duty of already-employed staff or workers. Moreover, the storage battery could add to the economy of lighting, since, if the generator was required to operate the lights directly, it had to be run whether one or the entire installation of bulbs were on. This is inefficient, and the smaller the size of the installation the greater is likely to be the inefficiency. The storage battery, however, will only discharge the amount of energy needed by the system at any time. In addition, there is the filament-saving factor of the use of non-fluctuating battery output.

The man who initiated this idea of using storage batteries was the French chemical engineer Camille Faure (1840-1898).[8] In October 1880 Faure took out a French patent proposing a new way of manufacturing lead-acid storage batteries.[9] In contrast to Planté's painstaking method of slowly building up the active material layers from the lead plates by electrolytic means, Faure proposed coating the fresh lead plates with a paste of powdered lead compounds and sulphuric acid. These pasted plates could then be "formed" into PbO_2 and sponge lead by a single charge. Not only did Faure's pasted process reduce the length of "formation" from months to hours, but it also allowed a much thicker layer of active materials to be prepared than with Planté "formation."

Faure's modification of the Planté process is understandable in the light of his motivation for the invention. The evolution of the dynamo and incandescent bulb made the development of a large scale electrochemical reservoir of electrical energy desirable. Planté and his contemporaries had never had such a motivation. For Faure, on the other hand, the low internal resistance inherent to Planté-type cells was of secondary importance to their storage capacity. Thus, he was willing to tolerate the increased resistance of his batteries caused by the thickness of the active layers and the less intimate connection between the active materials and the lead plates.

Camille Faure spent most of his life as an itinerant engineer in England and France. He received some formal technical education at the Ecole des Arts et Metiers, but most of his education came either from on-the-job experience or private reading. His early professional years, 1859-1872, from age 19 to

---

[8] Faure, like Planté, and a number of early battery designers, died very young. There is no proof, but the effects of lead poisoning cannot be excluded.

[9] U.S. patent #139,258—20 November 1880.

32, were spent mostly in France as a draftsman and superintendent of several iron works.[10]

From 1872 to 1880, Faure was employed by the Cotton Powder Company, for whom he designed and built a gun cotton plant in Kent. He remained at the plant as superintending engineer; a job which gave him leisure for his own experiments as well as providing him with a laboratory in which to carry them out. Faure began his storage battery work at the Faversham plant. Although the manufacture of nitrocellulose has little to do with electricity or storage batteries, the Faversham plant had plentiful stocks of the two basic materials for battery manufacture—sulphuric acid and lead. Lead was used to line the nitrating tanks.

Camille Faure had been interested in electricity since his youth, but the immediate stimulus for his decision to actively participate in the new field of electric power engineering and lighting came from his visit to the 1878 Paris Exhibition.[11] Here he was inspired by the displays of dynamos and electric lights. Faure also saw some Planté batteries on display; apparently these were of Breguet manufacture. He was immediately struck by the idea of using storage batteries as aids in the commercialization of incandescent lighting. The only problem was to increase the storage capacity and to reduce the "formation" time. Consequently, Faure returned to his Kent laboratory to design the prototype of his pasted plate cells.

Although Faure's immediate inspiration came from his experience at the 1878 Exhibition, this inspiration is an example of Louis Pasteur's observation that chance favors the prepared mind. From the beginning of his career, Faure had been interested in telegraphy and battery design. In 1859 at age 19 he designed a modified form of the Grove-Bunsen nitric acid battery which was sold by the Elliott Brothers firm of London; a large scientific instrument-telegraphic supply house.[12] Since at least 1871, Faure had investigated the use of lead compounds as depolarizers in telegraphic batteries and had studied the use of rechargeable lead cells in telegraphy.[13] The use of lead de-

---

[10] *Electrician Electrical Trades Directory* (London) 1894 ed., p. XXX; 1898 ed. p. XXXI; 1899 ed., obituary section.

[11] Ibid., 1898 ed., p. XXXI.

[12] This was sold under the trade name: Black Bottle Cell.

[13] Testimony of C. Faure, U.S. Patent interference case #8306 (Box #1162) (Keith vs. Shaw vs. Maloney & Burger vs. Brush vs. Faure), Federal Records Center, Suitland Md., p. 7 of Faure testimony.

polarizers in primary telegraphic batteries was a common innovation theme among inventors during the 1850s, '60s and '70s, having been stimulated by the work of Muncke, de la Rive, et al.[14]

Faure had other long-term interests in electrochemistry. He was one of the first, for example, to foresee the cheap production of chemicals by electrolytic means in areas with great hydro-electric potential.[15] One of his ideas, planned during the 1880s, was a scheme to use the water power of Dauphine for the manufacture of calcium cyanate, an artificial fertilizer. Faure also held patents on the electrolytic manufacture of aluminum.

Faure's background was typical of many men who entered the field of electric lighting and storage battery design in the 1880s. Although his chief employment prior to 1880 was outside of the area of electrical technology, Faure nonetheless was familiar with the two commercial electrical technologies of the pre-1880 period—telegraphy and electroplating. When incandescent lighting became a reality, Faure was able to apply to the new technology solutions which had been proposed for these earlier technologies.

In Europe and in Britain, all the promoters of the use of storage batteries for incandescent lighting purposes were stimulated by Faure's ideas. In the United States, however, at least two men claimed to have developed the idea independently and to have invented forms of the storage battery similar to that which Faure invented. The backgrounds of these two inventors—Charles Brush and Nathaniel Keith—are similar to that of Faure and suggest that their claims to independent invention may be valid.

Charles Brush (1849-1929) was one of the giants of early electric power and lighting technology. He was educated as a chemist, obtaining a bachelor's degree in chemistry in 1869.[16] Following graduation, Brush opened a consulting office in Cleveland and studied electricity in his spare time. In 1875 he entered into an agreement with a Cleveland telegraphic supply

---

[14] See, for example, Donato Tommasi, *Traite des Piles Electriques* (Paris, George Carre, 1889), for the Marie-Davy $PbSO_4$ cell (1859), the Becquerel $PbSO_4$ cell (1860), the Torregiani Pb cell (1866), and the Gaiffe Pb_3O_4 cell (1872).

[15] *Electrician Electrical Trades Directory* (London) 1898 ed.

[16] Harold Passer, *The Electrical Manufacturers, 1875-1900* (Cambridge, Harvard Univ. Press, 1953) p. 14.

house whereby they manufactured inventions he developed in their facilities. It was here that Brush did the early work on his famous dynamo and arc light designs.

Brush undertook some electrochemical researches at the same time, and these were to have a bearing on his later storage battery work. In 1876, for example, he patented a new type of primary telegraphic battery designed to replace gravity Daniell batteries. The gravity Daniell was a modified form of the original Daniell cell in which the porous diaphragm was eliminated, thus reducing the cell's internal resistance as well as its cost. The separation of the electrolytes was accomplished by taking advantage of their different densities. It was the most popular American telegraphic battery.

The copper and zinc electrodes were placed at the bottom and top respectively of the battery jar. The copper sulphate solution was then placed in the bottom of the jar and a dilute sulphuric solution floated on top of that. Although popular, the gravity Daniells were non-durable, since the cathodic depolarizer ($CuSO_4$ solution) is liquid, and the copper ions diffused readily to the zinc anode and caused self-discharge. This could be avoided by using a solid depolarizing chemical, and many inventors since the early 1850s proposed modifications of the Daniell cell with solid depolarizers. Brush replaced the $CuSO_4$ solution with a paste of lead oxychloride: $PbCl_2$ • $Pb(OH)_2$. In the construction, he copied the gravity Daniell pattern; replacing the copper cathode at the bottom of the jar by a lead strip surrounded by lead oxychloride paste.

During the 1880s, when Brush was fighting in the courts for the priority of his storage battery patents, he testified that he had experimented with the reversibility of his primary battery in 1876.[17] Unfortunately, there is nothing in the patent to support this contention, but given the extensive contemporary interest in telegraphic use of rechargeable batteries, Brush certainly must have been aware of these ideas and may well have tried them out. Moreover, an electric lighting patent which Brush received in 1879 indicates his familiarity with the inventive tradition of Kirchhof, Planté, and Niaudet.

This patent was one of many Brush took out after 1875, when he turned his attention away from telegraphy and towards arc lights and dynamos. The 1879 patent was part of a design for a self-adjusting arc lamp; an electrochemically polarized system

---

[17] U.S. Patent interference case #8306, see note 13 above, p. 6.

would be used to control the lamp's adjustment. Although never used in practice, the invention demonstrates Brush's debt to the work discussed in the last chapter.

Since arc lamps must be constantly readjusted, (the oxidation of the tips of the carbon pencils requires that they be advanced towards each other as they burn) some sort of self-regulating feed mechanism must be incorporated in the lamp. At the time Brush applied for his patent, the standard solution to this problem was to use an electromagnetic device. Brush chose to substitute an electrochemical mechanism for the electromagnetic one.

The arc circuit was placed in parallel with a "gas battery or secondary pile." Thus the electric current would move both through the battery and across the carbon arc. As the current passes through the battery, it becomes polarized, resulting in a counter EMF in that branch of the circuit. By selecting a suitable number of cells in series, claims Brush, we can produce just that amount of counter EMF which will oppose the dynamo current to the same degree that the arc branch will retard the current when the spacing of the carbons is just right. That is, a balance will be struck between the two branches of the circuit. Any variation in the spacing of the carbons will alter this balance, causing current to flow around the parallel circuit. This current can be used to activate the regulator which moves the carbons. What Brush describes, therefore, is a differential control device, analogous electrically to the mechanical differential gear.[18]

In this patent, Brush indicates his preference for the polarized gas battery (Grove gas cell), although he indicates that a lead-acid cell may also be used. The gas battery was to be made with large, porous carbon plates, similar to the Siemens 1852 prototype. Here Brush was thinking as Planté did of the storage battery as a slow capacitor, rather than as a reservoir of energy. It was not until June 1881 as the incandescent revolution got underway, that Charles Brush filed the first of his patent applications for a "reservoir" type storage battery—one with pasted, Faure-type electrodes.[19]

This date was about nine months after Faure was granted his basic French patent. Brush, however, always claimed that he conceived the idea of the pasted plate battery independently of

---

[18] U.S. Patent #219,212—2 September 1879—filed 14 January 1879.
[19] U.S. Patent #260,653—4 July 1882—filed 15 June 1882.

Faure. This is possible, since Faure did not file for an American patent until the same time as Brush.[20] More important than the question of priority, however, is the question of innovative inspiration. It is clear that it was the background of these two men, so similar in many crucial ways, that led them both to conceive of the new role for the storage battery and to redesign it accordingly. We shall follow this line of reasoning further with a third inventor.

Nathaniel Keith (1838-1925) was educated for medicine.[21] He never practiced, however, but put his scientific training to use in a career devoted to various types of engineering. For ten years, 1860-1869, Keith worked as a mining and metallurgical engineer in Colorado. For a year in the mid '80s he was scientific editor of the engineering journal *Electrical World*, at a time when he also served as one of the founders of the American Institute of Electrical Engineers. During the last fifteen years of the nineteenth century, Keith operated an electrical goods manufacturing company.

Keith's interest in electricity began in the early 1870s when his mutual interest in metallurgy and chemistry led him to take up the popular business of electroplating. In the mid 1870s, he broadened these interests by experimenting with the electropurification of metals, and particularly the recovery of trace quantities of rare metals from the impure anodes used in electro-refining. The motivation for this work was no doubt the appearance at this time of the first commercial dynamos designed specifically for electrochemical purposes; for example those of Edward Weston.[22]

In 1876 Keith began research on the electrochemical potentials of various metals and electrolytes, realizing that these constants determine the preferential transfer of ions from anode to cathode.[23] By the winter of 1877 he was concentrating his efforts on an electrochemical technique for the recovery of trace silver from crude lead. While conducting these tests, Keith noticed the tendency of the lead plates to return the current put through them; the plates would run his dynamo backwards as soon as he disconnected it from the steam engine.

---

[20] U.S. Patent #252,002—3 January 1882—filed 20 April 1881.

[21] *Who's Who in America,* 1879-1942 ed.

[22] David O. Woodbury, *A Measure for Greatness* pp. 46-47.

[23] Testimony of N. Keith, U.S. Patent interference case #8711 (Box #1266) (Keith vs. Brush) Federal Records Center, Suitland, Md.

For an experienced electroplater like Keith, there was nothing new in this. All electroplating baths become polarized and can reverse the plating dynamo's polarity when the machine is disconnected from its power source. In fact, until suitable protective devices were invented, electroplaters sometimes found that their dynamos were *removing* metal from wares that they were supposed to be plating, after the machine's polarity had been accidentally and unknowingly reversed by a polarized bath. Moreover, Keith knew of Planté's work and immediately associated it with his own accidental observations.

What did surprise Keith, however, was the extremely strong and constant secondary current he obtained from his experimental lead plates. In order to investigate the matter further, he ordered a specially-designed dynamo from Edward Weston.[24] Weston was an experienced designer of nickel-plating machines, but none of his standard dynamos were powerful enough for Keith.

Keith's idea for preparing storage battery plates differed from both that of Planté and Faure-Brush. The active materials were to be deposited in the lead plates from a plating solution of lead acetate. This was a very old technique, having been used by both Nobili and Becquerel in producing their famous colored effects, and by Wheatstone in preparing his PbO_2 primary battery plates.[25] With his scientific background and interests, it is doubtful that Keith was unaware of these sources. Moreover, this particular approach to storage battery manufacture is the kind of approach one should expect an electroplater to invent. The crucial point about this technique, however, is that, like the method of Faure and Brush, it was designed to produce a reservoir-type storage cell.

In early 1878 Keith abandoned this work. Although he recognized the applicability of his invention for electric lighting, the technology of the incandescent bulb was only then beginning to excite the interest of some inventors. In 1882, Keith again picked up the thread of this research and patented a storage battery based upon the manufacturing technique described above.[26] A company was formed which manufactured the cells both for electric lighting and streetcar propulsion, but it was not a commercial success.

---

[24] This testimony of Keith was borne out in court by Weston. See testimony of Edward Weston in case #8711, above.

[25] William Sturgeon, *Elementary Lectures on Galvanism* p. 197; Pogg. Ann. 9 (1827), 183-184; 97 (1856): 22-29.

[26] U.S. patent #273,855—3/13/1883.

2. The First Manufacturing Companies

The earliest pasted plate lead-acid batteries were not very prepossessing. The first Faure cells consisted of two lead plates, each of which had a paste of lead sulphate and sulphuric acid smeared on one side. The individual plates were then wrapped in an inner wrapper of parchment and an outer layer of felt. The plates were placed one atop the other, pasted sides facing, and the pair rolled up and inserted into glass jars a foot high and five inches in diameter:[1]

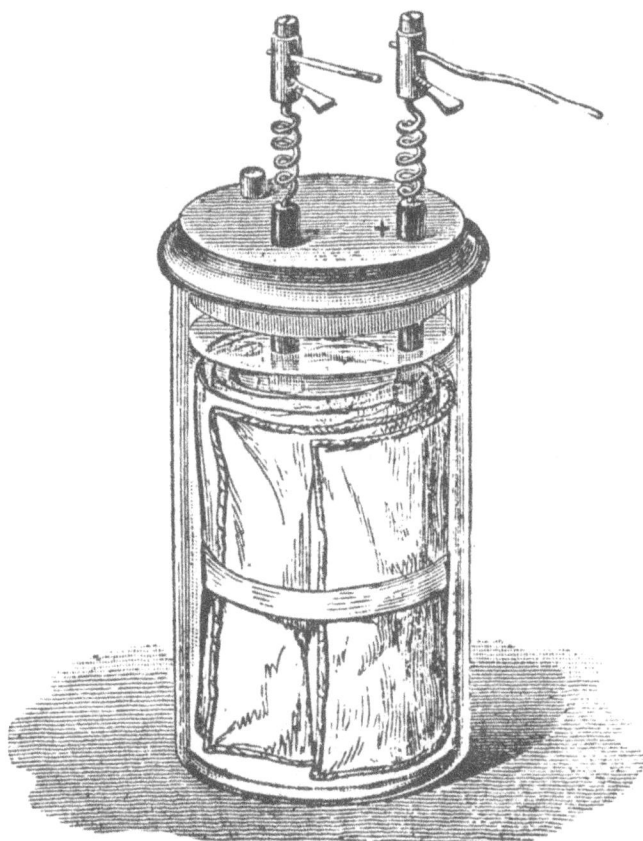

Although this jellyroll pattern had worked well for Planté, it presented problems for the pasted-plate version; the paste was

---

[1] *Journal of Institution of Electrical Engineers* (hereinafter: *JIEE*) 60 (28 February 1922): 463; E.J. Wade, *Secondary Batteries*, p. 105.

pushed out of place by the rolling. Faure used the rolled pattern for his first cells, but it was soon replaced in commercial manufacture by a flat-plate design—rectangular plates arranged in parallel inside rectangular wooden boxes. Aside from this structural change, no other alterations in the original Faure design were made in the earliest commercial pasted-plate batteries.

Initially, the commercial potential of the Faure cells seemed impressive. As adjuncts to incandescent lighting they were intended to provide both economy and convenience. The full storage capacity of the cell was developed almost from the first charging. In addition to this speed of "formation," the pasted batteries were easy to construct, requiring a minimum of machinery and no skilled labor. The raw materials were cheap and there were no moving parts to cause trouble; no great precision in manufacture was required either. In short, the Faure battery appeared to be a sure thing for making big, quick profits.

The first attempt to obtain these profits began in October 1880; the same month Faure obtained his basic French patent. S.A. La Force et la Lumiere (hereinafter: Force et Lumiere) was organized by the Belgian financier Gustave Philippart to work the Faure patent. Force et Lumiere, making neither large nor quick profits, was instead a commercial disaster. Nevertheless, this first company played an important role in the evolution of the storage battery. Gustav Philippart was a master promoter and it was largely due to his efforts that the storage battery, an esoteric device known only to telegraphic engineers, became a household word during the 1880s, attracting almost as much attention and investment as the incandescent lamp itself.

Philippart's greatest publicity coup was his recruitment of William Thomson (later Lord Kelvin) as an advocate of the batteries. In May 1881 a shipment of Faure cells of the jellyroll pattern was sent from Paris to Thomson's laboratory at the University of Glasgow. Thomson, ill at the time, had the laboratory staff drop whatever they were doing to test the storage batteries. The cells performed well, and, according to one of the assistants:[2]

> (Thomson) was intensely delighted, (and) said that at last a dream of his life had been realized in the most complete manner, and that now with the aid of secondary batteries great de-

---

velopments in practical applications of electricity were possible.
The laboratory was kept running both day and night. He lay in
bed and called for experimental results. . . . He sent a long letter
to *The Times* on June 9th, 1881, which attracted much attention,
and was the subject of a great deal of discussion.

Thomson was one of the most prestigious scientists of his
day, and his well-publicized recommendation of the "marvel-
ous box of electricity," as he called it, touched off a short period of enthusiasm for the storage battery. The June 9 letter
described the great amounts of energy which the battery could
store. Subsequently, Thomson proposed a number of impressive electrical engineering projects which the storage battery
would make possible. For example, in the fall of 1881, at a
meeting of the British Association for the Advancement of Science, he suggested a plan for the long-distance transmission of
power. Thomson put forward one of the first schemes for the
transmission of electric power from Niagara Falls to Buffalo.
Power was to be generated at the Falls at 80,000 volts, then
transmitted to Buffalo, where it would be received at a conversion station equipped with 40,000 Faure batteries in series.
These batteries would then discharge into the consumer's circuits in groups of 50, giving 100 volts.[3]

During the summer and fall of 1881 the Force et Lumiere
Company supplemented the publicity it received from Thomson by demonstrating the batteries before various public
gatherings, lighting incandescent bulbs from them. A number
of small, private lighting installations were set up with the
batteries. But then, in the winter of 1881, bad news began
appearing in the technical press. It was discovered that the
Faure batteries disintegrated after a few months of operation.
The felt and parchment decayed and fell apart, and the active
material masses crumbled, since they possessed neither coherence nor attachment to the lead plates. The Faure battery was
an obvious failure.

The first promoters of the storage battery were confronted
with an unexpected reality about their new technology: storage battery manufacturing, far from being an easy way to make
a profit, was a most complex business, requiring the development of elaborate techniques for the manufacture of practical,
durable batteries. The simplicity of the storage battery concept
masked the difficulties of turning the concept into a commer-

---

[3] *Trans. BAAS*, Section A, 1 September 1881, p. 518.

cial reality. It was not to be until the late 1880s that a few companies, after considerable research and expenditure of capital, were able to market a commercially successful battery.

One reason for the overly optimistic outlook of early storage battery designers towards the difficulties of producing a good battery was the influence of past battery manufacturing methods. In 1880 the commercial production of electric batteries was an old, established industry. Methods were simple and designs uncomplicated. For example, the manufacture of a gravity Daniell required nothing more complicated than a pot to melt the zinc, an open mold to cast it, and a tin snips to cut the copper cathode. Of all aspects of pre-1880 electrical equipment manufacture, the manufacture of batteries was by far the easiest.

As we have seen, a number of the men who pioneered in the manufacture of the first pasted-plate storage batteries had either been designers of telegraphic primary batteries themselves (Faure and Brush) or were at least familiar with the methods of such manufacture. A great many of the first lighting and power engineers began as telegraphic engineers. It is not surprising, therefore, that they initially underestimated the difficulties of producing a practical storage cell. In fact, a number of the early storage battery manufacturers, such as Charles Brush, attempted to carry out this manufacture as a sideline to their main business, which in Brush's case was the manufacture of dynamos and arc lights. This, again, was a pattern of manufacture carried over from the pre-1880 period, when telegraphic supply houses and scientific instrument makers produced primary batteries as a sideline to their more complicated and expensive products.

No manufacturer who tried to make storage batteries as a sideline to his other manufacturing ever developed a successful one. Those firms which did succeed gave their full attention to the project. The first company to do this, one which pointed the way for subsequent storage battery manufacturers, was the Electric Storage Battery Company (hereinafter: EPS).

EPS was created in London in March 1882.[4] The moving force behind this company was a metallurgist named John Scudmore Sellon (1836-1918).[5] In the summer of 1881, about

---

[4] *The Electrician* (London), 8 April 1882, p. 330.

[5] Not to be confused with Robert Percy Sellon, another early and more famous British electrical engineer. *JIEE* 56 (1918): 543.

the time of the William Thomson letter to the *Times*, Sellon
began working on storage batteries, and in September 1881, he
received the first of a series of storage battery patents.[6]

Sellon's designs represented a distinct advance over the
original Faure battery. In the Faure cells, the active materials
had simply been smeared on the surface of the flat lead plates.
Sellon proposed to perforate the plates and insert the active
paste into the perforations. These holes would have a convex
cross-section to retain the materials more firmly:[7]

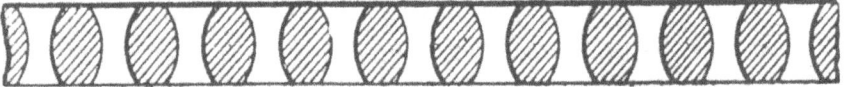

This improvement was rather obvious; the patent offices of
the United States and Europe received numerous perforated
plate designs beginning about this time. Faure himself soon
began using perforated plates.

Equal in importance to his perforated-plate patent, but considerably less obvious, was a patent Sellon obtained five days
after the first.[8] This invention was for a lead-antimony alloy to
replace the pure lead used in the Faure plates. It was possibly
Sellon's background as a metallurgist which caused him to
realize that if lead grids were to be used as electrode supports
instead of solid lead plates, they would be too weak mechanically. The use of antimony as a hardening agent for lead is an
age-old technique. As an unexpected benefit, it was later
found that the antimony not only hardened the lead, but also
made it resistant to electrochemical attack. Consequently, storage battery designers were able to produce increasingly
thinner-walled grids for their plates without worrying that
unintentional Planté-type "formation" would unduly weaken
them.

Sellon, however, was not the only inventor working on storage batteries in England. His leading potential competitor was
Joseph Swan, the man who developed the first practical incan-

---

[6] British patent #3926—10 September 1881. Also: #3987—15 September 1881;
#4005—16 September 1881; #4632—22 October 1881; #5631—23 December 1881;
#319—21 January 1882; #2818—15 June 1882; #217—3 January 1883; #1644—2 April
1883; #5069—24 October 1883; #5741—14 December 1883; #6228—10 April 1884;
#1764—9 February 1885.

[7] Wade, *Secondary Batteries*, p. 20.

[8] British patent #3987—15 September 1881.

descent bulb concurrently with Edison. In 1881 Swan set up the Swan United Electric Light Company to commercialize his bulb. At the same time, he formed an alliance with R.E. Crompton, whose company had been manufacturing dynamos and arc lights since 1878.[9] The function of the alliance between Swan and Crompton was for the latter company to install a combination of their own dynamos and Swan's bulbs in private lighting installations.

Swan and Crompton were both men who kept a careful watch on innovations in the electrical field. They recognized the value of the new storage batteries for incandescent lighting, and Swan undertook to develop one. He received his first patent on 24 May 1881; this was for a perforated-plate battery similar to that of Sellon.[10] After being cast, the holes in Swan's grid plates were packed with finely-divided metallic lead and "formed" rapidly.

Meanwhile, yet another potential English competitor was working on the idea of grid-type plates to hold the active materials. I. Volckmar was a French inventor who came to England in 1881. In October of that year he patented a grid similar to that of Swan, although with a slightly more open design.[11] Volckmar is one of the more shadowy figures in the early history of electrical technology. In early 1881 he was employed as a foreman at the Paris workshop of Force et Lumiere. At this time, the founder of the firm, Gustav Philippart, and one of his sons, Simon, were in England trying to establish a British branch of the company. Camille Faure may also have been with them. Volckmar, along with another Philippart son, were left behind to mind the Paris operation.[12] These two used the opportunity to dispose of some of the company's patents for their own advantage. They were detected and fled to London, although the son apparently soon returned and reconciled with his family.[13]

In December 1881 John Sellon met Volckmar and observed that: "In him I recognized a gentleman of great power of work and perfection of detail."[14] Be that as it may, Sellon certainly

---

[9] Dunsheath, *History*, p. 131.

[10] British patent #2272—24 May 1881.

[11] Hoppe, *Akkumulatoren*, p. 133.

[12] M. Barak, "A History of the Chloride Group" (typescript dated 9/1973) p. B2.

[13] Copy of charges filed in the Paris court, reported in the London *Electrician* 15 April 1882, p. 358.

[14] Ibid., 8 April 1882, p. 343.

recognized that Volckmar's October patent was similar to his own. Consequently, the two men decided to pool their interests for a joint venture. At the same time, Sellon approached Swan about his battery patents. Swan admitted that he hadn't done anything to reduce the patents to practice, and agreed to license them to Sellon.[15] Here we see the beginning of a pattern of industrial practice which was to typify the British storage battery industry, and which was in marked contrast to the early American practice; that is, a pattern of consolidation of competing interests without preliminary and destructive patent litigation.

The British storage battery industry was created in March 1882, with the start of the first two companies. On March 4, the Faure Electric Accumulator Company (hereinafter: FEA) was incorporated.[16] This company was the outcome of the previously-mentioned British trip of Simon and Gustav Philippart. The Philipparts were careful to make the company appear to be a British operation; the prospectus, for example, listed only Englishmen on the board of directors, and the only Frenchman whose name appears was Faure, who shared the position of consulting engineer with the distinguished English physicist William Ayrton.[17] The name Philippart did not even appear. The fine print, however, indicated that Force et Lumiere was to be paid with half the total issued shares of FEA in return for its patents and processes. In short, FEA was to be controlled by Simon Philippart.

The second company formed in March 1882 was EPS— almost three weeks after FEA. Although EPS actually evolved as an independent company, it began as a branch operation of the Anglo-American Brush Company. Anglo-American Brush was a firm created in Britain to work the patents of Charles Brush, that is, it was involved in the manufacture and installation of electric lighting plants. Brush was an exponent of the use of storage batteries as incandescent lighting adjuncts, as were a number of the members of the board of Anglo-American Brush. One of these members was John S. Sellon. Consequently, the initial capitalization of EPS came from Anglo-American Brush, which purchased all the shares not given to the patentees for their inventions.[18]

---

[15] Ibid., and 23 September 1882, p. 453.

[16] Ibid., 4 March 1882, p. 243.

[17] Ibid., p. v.

[18] Barak, "History," p. B2.

Unfortunately for EPS, however, Camille Faure had received his basic pasted-plate patent in Britain in January 1881—at least four months before any of the EPS-owned patents.[19] The Philipparts, therefore, threatened legal action to stop EPS from manufacturing. The atmosphere was further strained by Faure's belief that not only was the EPS infringing his patent, but that they were actually using techniques stolen from him by Volckmar. During early 1882 Faure had Volckmar charged before the Paris courts, claiming that after Volckmar had been hired by Force et Lumiere:[20]

> . . . it was not long before he grievously abused the confidence which had been placed in him, that, in fact, he used his influence to lead astray from his duties Mr. Gustav Philippart, another son, who was at that time a minor, and succeeded in inducing this inexperienced young man to cheat his father and the company in whose service he was . . . and that since then the two have used all their endeavors to corrupt the workmen of La Force et la Lumiere Co., and to compete treacherously against their former masters.

A heated battle of words took place during early 1882 in the letters columns of the technical press, with Sellon, Faure, Volckmar, and Simon Philippart exchanging insults and threats.[21] The matter came to a head on 10 May, when FEA sued EPS for infringement of the Faure patents, but almost immediately, the principals of both firms met and settled the matter out of court.[22]

According to the settlement, EPS and Force et Lumiere were to divide up the international storage battery market. The agreement was complex, but fundamentally it gave EPS control of most of the battery business in Britain, with Force et Lumiere taking the Continent, and both a share of the United States market.

With this split of interests between EPS and Force et Lumiere, the shareholders of FEA soon began to worry what field was left for them.[23] They were soon confirmed in their fears. Although FEA retained some of its manufacturing rights to the original agreement with EPS, by November 1882 Simon

---

[19] British patent #129—11 January 1881.

[20] Faure vs. Volckmar, Civil Tribunal, First Instance of the Seine, 28 March 1882.

[21] For example, *The Electrician* (London), 22 April 1882, p. 373; 8 April 1882, p. 39; *New York Review of the Telephone & Telegraph* 15 May 1882, p. 98-99; 5/1/1882, p. 1.

[22] *The Electrician* (London), 20 May 1882, p. 1.

[23] Ibid., 1 July 1882, p. 165.

Philippart revealed his plan for FEA to end all manufacturing.[24] Henceforth, FEA was to be the holding company for the European rights to patents worked by two new companies: the French Electric Storage Company, which would manufacture batteries, and the Metropolitan General Electric Company, which would build self-propelled electric streetcars to use the cells. These patent rights were paid for out of the assets of FEA.

When the British shareholders in FEA bought their stock in early 1882, they believed that they were financing the creation of a British manufacturing business over which they would have control. By November, however, it was clear that they were in the middle of a pyramiding operation designed chiefly to finance Force et Lumiere. Considerable protest arose from the shareholders and the board of directors all resigned, but Simon Philippart made it clear that he controlled the votes and would use FEA's assets as he saw fit.[25]

At the time, the Philipparts were regarded as little better than swindlers trying to make a fast buck from worthless patents. The poor performance of the first Faure batteries helped support this view. Although it probably would be an exaggeration to accuse the Philipparts of outright dishonesty, it is nonetheless true that they spent more time creating financial card houses than on battery design. Nevertheless, the Philipparts do appear to have been serious about the actual manufacture of storage batteries. One of the chief problems facing their efforts, however, was Simon Philippart's faulty perception of the commercial potential of the storage battery. Force et Lumiere concentrated its efforts on streetcar propulsion rather than incandescent lighting. This application never proved commercially feasible and was particularly disastrous for the early batteries of the 1880s. Force et Lumiere conducted trials of self-propelled electric streetcars until 1884.

The Faure companies declined during 1883 and collapsed in 1884. On 7 April 1884, Force et Lumiere, French Electrical Power Storage, and Metropolitan General Electric all announced bankruptcy.[26] FEA struggled on until July, trying vainly to salvage something.[27] The original Faure companies

---

[24] Ibid., 9 December 1882, pp. 95-96.

[25] Ibid., 1 July 1882, p. 165.

[26] Ibid., 19 April 1884, p. 551.

[27] *The Times*, 10 July 1884, p. 11.

were an unfortunate attempt to make too much money too quickly with too little technology.

By contrast, EPS developed into a successful company. Although it lost money steadily until 1886, EPS continually improved its economic position by improving the quality of its cells and by increasing the public's acceptance of them. Indeed, during the first few years of the evolution of the commercial storage battery this latter activity was all that was possible for the manufacturers—the idea of making profits was a dream for the future. The complete failure of the first Faure batteries after the enthusiastic puffery of 1881 had created a bad impression in the public mind, and it required a good deal of effort and time to overcome this. It took a number of years of hard research for the EPS engineers to develop a successful battery. Therefore, even though EPS made no profit until 1886, the company's success in slowly breaking down the anti-battery prejudice of electrical engineers was a considerable achievement.

A few electrical engineers defended the storage battery even during its darkest days in the early 1880s. These men claimed that a good, durable storage cell would soon be available to solve many of the problems of incandescent lighting. Nevertheless, a majority of engineers were dubiouʳ if not hostile towards batteries; such attitudes were common until the late '80s. Stories abounded of batteries being installed, performing well for a month or two, and then losing capacity, collapsing, or going mysteriously inert. This occurred not only with the Force et Lumiere or EPS cells, but also with the batteries of the other manufacturers who appeared after 1881.

A few quotes from the technical press illustrate the general opinion of electrical engineers towards storage batteries in the 1880s:

> May 1883—. . . the storage battery is still considered as an expensive toy, containing perhaps the germ of a great discovery, but as yet unfit for practical work.[28]

> Feb. 1883—Various persons have from time to time announced the invention of storage batteries which could be used in electric lighting, but none of these have had other than scientific value.[29]

---

[28] Edison Bulletin #18, 31 May 1883, p. 9 (330) *New York Evening Post* 1 May 1883.
[29] Edison Bulletin #16, 2 February 1883, p. 31 (275). *Boston Herald*, 28 January 1883.

> Dec. 1886—(in delivering a paper) I approach the (system of using storage batteries) with some degree of nervousness. In more than one quarter the man who comes out with a kind word for the secondary battery is set down either as a knave or a fool.[30]

In 1922, Gisbert Kapp[31] gave some recollections of the early 1880s:[32]

> Especially the distrust against batteries was great in those days. I remember Mr. Mordey,[33] after eulogizing the various advantages of alternating current, telling me that in his opinion the greatest of them was that batteries could not be used with A.C. supply.

Thomas Edison provided the best quote in a January 1883 interview:[34]

> The storage battery is, in my opinion, a catch-penny, a sensation, a mechanism for swindling the public by stock companies. The storage battery is one of those peculiar things which appeal to the imagination, and no more perfect thing could be desired by stock swindlers than that very self-same thing. . . . Just as soon as a man gets working on the secondary battery it brings out his latent capacity for lying. . . . Scientifically, storage is all right, but, commercially, as absolute a failure as one can imagine.

Actually, this outburst was in part a calculated attack by Edison against the incandescent system of Charles Brush, which was based on the use of batteries. Nevertheless, Edison's suspicion of the inherent unreliability of storage batteries was genuine, and the statement reflects an attitude with which many contemporary electrical engineers would have agreed. As late as 1903, an American summing up the history of storage batteries said:[35]

> The amount of money which has been lost in the endeavor to force upon the market secondary batteries having little or no

---

[30] *The Electrician* (London), 24 December 1886, p. 151.

[31] Gisbert Kapp. British electrical engineer, 1852-1922. Began work in electrical distribution in the employ of Crompton, but shifted his interests to A.C. in the 1880s.

[32] *JIEE* 60 (28 February 1922): 473.

[33] W.M. Mordey. British electrical engineer, 1856-1938. Early exponent of A.C. and designer of alternators.

[34] *The Electrician* (London), 17 February 1883, pp. 329-331.

[35] *Trans. Am. Electrochem. Soc.* 3 (1903): 169-170.

intrinsic value, to say nothing of the heartaches to honest investors from unfulfilled dreams, makes the history of the storage battery's development almost tragic. . . .

The preceding comments give some impression of the difficulties faced by the EPS during the early 1880s when public opposition to batteries was greatest. Still, the company was successful in convincing some people to install battery lighting systems. That the nature of these installations was extremely limited can be attributed not only to public prejudice against batteries, but also to the technical and economic problems facing early incandescent lighting technology. Some of these problems were international, while others applied only in the British context. We will examine these problems next.

The earliest schemes for battery use in electric lighting predated the formation of EPS. The first proposal, suggested in 1881 soon after Faure announced his invention, was to deliver charged storage batteries to the homes of consumers each morning like bottles of milk. The subscriber to the service was to place his discharged cells on the doorstep each morning, where they would be picked up and returned to the power house for recharging.

Competent engineers ridiculed this plan, and apparently it was never tried in practice. Nevertheless, the "milk bottle" battery scheme was not as absurd as it appears from our technically advanced perspective. During the first years of its existence, incandescent electric lighting was a technology for the well-to-do only. The earliest distribution systems, such as Edison's Pearl Street, were established in small areas with a high density of wealthy residents. This was for a number of reasons.

Most importantly, the cost of early electric lighting made it a luxury. This was particularly true for Britain, where gas lighting had been installed on a more extensive scale than in the United States and was therefore cheaper.[36] One of the reasons for the more rapid progress of incandescent lighting in the United States, was the lack of preexisting gas lighting systems in many localities. While British statistics on the cost of early electric lighting vary widely, they confirm that gas was considerably less expensive than electricity and that electricity stood no chance against gas except on such non-economic grounds as convenience, brilliance, and novelty.

---

[36] *The Electrician and Electrical Engineer* (New York), 3/1885, pp. 105-106.

In 1885, for example, Sir William Preece, head of the British telegraphs and an authority on many aspects of electrical engineering, toured the United States and reported his findings to the London Society of Arts. He compared the economic problems of incandescent lighting in the two countries:[37]

> The great mistake that has been made there, as it has been made here, has been to enter into competition with gas. The Edison company started on the basis of charges made for gas and the result of two years working has been to show that this was a financial mistake. Electricity must earn its future success *entirely upon its own merits, and not on its economy.* (my ital.)

> Even now in England it is possible to make a (incandescent) system pay at the rate of ½ penny per glow lamp per hour; and no one who has experienced the charm and comfort of electric light, and who has electricity brought to his door, and supplied to him at this rate, would hesitate to pay that figure for the use of electric lamps, although he could obtain the same light by gas at *less than half the price.* (my ital.)

Such cost variables would not have made electric lighting appealing to the poor or lower middle class, particularly in Europe, where the extra heating benefit of the gas light was welcome. Moreover, these comments by Preece were made before the introduction of the Welsbach mantle, which greatly cheapened the cost of gas lighting, and before the introduction of chain-pull automatic gas lighters, which eliminated one of the convenience factors of electricity.[38]

At the opening of the Notting Hill (London) Electric Light Station in 1891, Sir William Crookes, the chairman of the company, was asked if the light was cheaper than gas.[39] He replied that it was two and a half times as expensive as gas, and, like Preece six years earlier, emphasized the non-economic benefits of the light. For 1891, this position was a rather extreme one against the economics of the new light, although Crookes cannot be considered as anti-electricity by any means; in the same year, for example, William Preece prepared an elaborate analysis of the economics of gas versus electricity, in which he softened his criticism of 1885.[40] Here he predicted that in the

---

[37] Ibid.

[38] The increase in lighting efficiency as a result of the use of Welsbach mantles, as well as from improvements in the composition of gas itself, was from 4-6 c.p./ft³ to 60-70 c.p./ft³. Bright, *Lamp Industry*, p. 212.

[39] *Electrical Engineer* (London), 5 June 1891, pp. 548-549.

[40] Ibid., 3 July 1891, pp. 14-16.

future electric light would probably be made at least as cheap as gas, but for actual 1891 conditions, Preece still found that gas had a 9 to 50 percent cost advantage over electricity. Moreover, Preece's analysis ignored one important variable which added considerably to electric lighting costs—the replacement of bulbs. Early carbon filament bulbs were short of life and high of cost.[41] Moreover, because of their carbon filaments, the energy efficiency of these bulbs was low—only about 1/3 that of a tungsten bulb.[42]

In 1892, Arthur Guy, after going through a lengthy economic balance, came to the conclusion that " . . .it is seen that electricity in the form of glow lamps costs more than double gas."[43] Although costs of electricity versus gas varied in England in the pre-twentieth-century period, it is definite that gas was considerably cheaper, probably in the 75 percent to 100 percent range. Unfortunately, such cost breakdowns were not made in the United States, but it is probable that the economic situation was not more favorable to electricity there, particularly in big cities where gas systems did exist.

The small size of early electrical distribution systems kept costs per kilowatt high, since there could be no economies of scale. There were a number of reasons for the size limitations: the initial difficulties of designing large dynamos; the use of low voltage d.c., which made distribution of current more than a mile from the power house uneconomical;[44] and the prudence of early promoters, who deliberately kept their first systems small until they understood the problems and parameters of design.

---

[41] The first Swan lamps of 1879 cost 25s ($6.25). By 1891, this price had been reduced to $1.25, which was about the same as that for the contemporary Edison bulb in the United States. By 1892, a typical 16 c.p., 54 watt English bulb (the most commonly used size) cost about 95¢, and had an average life of 1000 hours. *JIEE* 60 (28 February 1922): 466; *Electrical Engineer* (London); 16 September 1892, pp. 279-281.

[42] Carbon cannot be made to incandesce at more than 1600°C without a too rapid evaporation of the filament. At such a relatively cool temperature, almost all of the energy radiated by the filament is in wavelengths below the visible spectrum. If the temperature can be raised, however, by using osmium, tantalum or tungsten, the ratio: energy radiated in visible spectrum/total energy radiated will increase.

[43] *Electrical Engineer* (London), 16 September 1892, pp. 279-281.

[44] Here the problem of economic feedback entered; since the early carbon lamps were energy inefficient, they consumed much more current than modern lamps, and accordingly, more copper was needed in the feeders. And since the early distribution systems were D.C., they required a large amount of copper anyway. These two factors, therefore, increased each other in determining the total quantity of copper required.

Furthermore, there was a uniquely British reason for the small size of early electrical distribution systems. On 2 August 1882 Parliament passed the ill-fated Electric Lighting Act. This piece of legislation was an attempt to encourage the development of electric lighting and to insure at the same time that this development served the entire population and not just a handful of investors. The Act failed on both counts. As far as encouragement of lighting was concerned, the Act gave any company which wanted to set up a central station distribution system permission to dig up the streets to install the distribution feeders and mains, granted that it submitted a plan to the local borough authorities and got their approval. But the Act also specified that the borough authorities could purchase the entire system from the private owners at the end of twenty-one years.

Two years after the Act was passed, 120 applications for distribution systems had been submitted to local authorities, of which 73 were approved. But of these 73 plans, not one had been begun by 1888, when the Act was amended to extend the twenty-one year period to a more reasonable forty-two years.[45] Before 1888, investors were not willing to risk their money on a system which was likely to be nationalized in less than a generation.

Consequently, those British central stations built between 1882 and 1888 were of two kinds; a few huge a.c. systems, which covered so large an area that they extended across several municipal boundaries and therefore could not be bought up, and a number of minuscule d.c. systems which were so small that all cables lay in private property and therefore were equally immune to takeover.[46] Most of these systems were little bigger than large private lighting installations.

Even after the amendment of the Lighting Act in 1888, however, British lighting promoters still had good reason to restrict installation to well-to-do neighborhoods. This was in order to entice investors by assuring them that there would be plenty of subscribers to the service in the area to be supplied. Sir Thomas Callendar, who was responsible for installing one of the first British central station distribution systems in 1889, remarked on the necessity of using this tactic: "It was impossible

---

[45] Dunsheath, "History," p. 146-147; Thomas P. Hughes, "British Electrical Industry Lag; 1882-1888," *Technology & Culture* (1962): 27-44.

[46] Of the former case there was only the Ferranti Deptford installation.

to get people to back up the electric light with hard money. The company endeavored to supply a rich neighborhood, where there was plenty of money and a great desire for the light and yet we could not get enough financial supply."[47]

The foregoing discussion shows why the "milk bottle" battery lighting plan was not as impractical as it first appears. Due to the weight and bulk of the batteries, the system would have been restricted to a *small* area. Because of the labor involved, it would have been confined to a *rich* area, in which people could be expected to pay a premium for the convenience, brilliance, and social prestige which comes from possessing the very latest technologies. The inconvenience of shifting the batteries each morning also would have presupposed a well-to-do district, where servants could accomplish the task.

Even if we grant these factors, however, the "milk bottle" scheme was still too cumbersome to have been a commercial success. Nevertheless, a close variant on this system was successfully commercialized and provided an important early stimulus to a few battery companies, most particularly the EPS. This was the business of "electric catering."

A number of American and European firms engaged in the catering of electricity in the 1880s and early '90s, but the EPS was the pioneer. The practice began sometime during 1883. The company offered to provide electric lighting for balls, weddings, and other celebrations. Customers selected the number of lamps they wanted and were charged by the lamp-hour. On the day of the celebration an engineer arrived from the battery company with a group of workmen, strings of light bulbs, and a set of charged batteries. The EPS sent its electric catering crews from one end of Britain to the other to provide illumination for a single event.[48]

Electric catering was a small, but steady business for the early companies. For example, at the April 1884 shareholder's meeting of the EPS, the chairman, describing the past year's business, remarked: "Another small matter in itself was the business commencing in the temporary lighting of ballrooms, evening parties, lectures, etc. It pays well in itself, but is of still greater use in bringing the value of electric light well before the right people."[49]

---

[47] *JIEE* 60 (28 February 1922): 421.
[48] *The Electrician* (London), 29 May 1885, p. vii.
[49] Ibid., 26 April 1884, p. 572.

The "right people," of course, were the rich and well-to-do
who attended such affairs. In order to more properly appreciate the advertising value for the battery companies of im-
pressing the "right people," consider the *timing* of this practice. EPS was formed in March 1882. By August the Lighting
Act was passed, all but destroying the EPS's prospects of selling batteries for use in Pearl Street-type central station distribution systems. Consequently the company's only hope for
a market in incandescent lighting was in the field of the private
lighting plant; dynamos and engines which served a single
building, or at most a very few contiguous buildings. Until the
reform of the Lighting Act in 1888, almost all British incandescent engineering was restricted to small plants for hotels, office
buildings, and the homes of the "right people."

The advertising value of electric catering, therefore, was
considerable. It insured that the very people who had the
money to afford private lighting plants were the ones who
would be exposed to the advertising. Such exposure was important, since it was only thus that the potential customers
could get a realistic impression of one of the chief advantages
of the bulb over gas—its greater brilliance.

The permanent installation of some early private lighting
plants was also often carried out by battery companies more
for their advertisement value than for the chance to make a
profit. In the early days of electric lighting, the storage battery
companies not only manufactured batteries, but also undertook to design and install complete lighting plants. This was a
crucial element in the infant industry's survival; it is unlikely
that the industry would have survived by restricting itself only
to manufacturing the batteries, given the resistance of many
early engineers to using batteries. Mr. H. Hirst, a British electrical engineer who began his career with EPS in 1882 or 83,
reminisced in 1922 on the first days of the company:[50]

> At the Head Office it seemed to be nobody's business to look
> after the securing and execution of contracts. The principal inst-
> allations which were carried out—such as the Grand Hotel and
> the Gaiety Theatre—were for the purpose of advertisement.
> They were demonstrations intended to inspire newspaper arti-
> cles and so create a movement in shares or assist negotiations for
> the sale of foreign patents. . . . While the management was
> preoccupied in such directions, it was left to subordinates like

---

[50] *JIEE* 60 (28 February 1922): 468.

myself to receive ordinary mortals who wanted to buy a few cells.

The Grand Hotel installation mentioned above was originally contracted by FEA, but upon which nothing had been done by that unfortunate firm. The project was taken over by EPS in 1883. This 400 light plant was a technical success and it provided a good deal of publicity for the firm. It was not a paying proposition, however. In fact, in 1884 the chairman of EPS admitted that the Grand Hotel project had been undertaken for its publicity value.[51]

The most important of the 1883 publicity projects of the EPS was that at the Bank of England.[52] This was its largest installation, and, like a similar installation at the Law Courts, the EPS could confidentally expect the newspapers and technical journals to report the details. Moreover, the "right people" were sure to see them.

Despite the resistance of many electrical engineers towards batteries, EPS did not have to do all its own installation of lighting outfits in the early years. Beginning in 1883, it was able to sell batteries to a number of contractors, who, though they didn't make batteries themselves, believed in their efficacy. The Anglo-American Brush Company was supplied, as was the Swan Company, Siemens Brothers, and R.E. Crompton, who used EPS cells in his famous pioneering project at the Vienna Opera.[53]

The EPS also sold cells to smaller lighting contractors with whom it had established contacts. Such a small contractor was F.E. Gripper, who specialized in installing private illuminating gas plants in the 1870s in large English country houses. In 1879, however, Gripper found that his prospective customers were becoming reluctant to install gas plants because of the publicity about the Swan and Edison bulbs.[54] Not wanting to be left behind, he learned to install the new technology and put in one of the earliest British private plants in 1881. Gripper encountered problems with its operation however: "As at this time no storage batteries were available, the lighting had to be shut off during the night until storage batteries made by my

---

[51] *The Electrician* (London), 5 January 1884, p. 185.

[52] Ibid., 19 April 1884, p. 552.

[53] Ibid., 24 March 1883, pp. 454-455.

[54] *JIEE* 60 (28 February 1922): 465-466.

friend Bernard Drake (chief engineer of the EPS) were available for practical use, and added to every country house installation."

Moreover, Gripper discovered that the irregular, unbattery-buffered output of the early generators caused the bulb filaments to pop with "very depressing" frequency.

In 1883, however, he was able to obtain storage batteries of good capacity (although undoubtedly of poor durability), as well as much cheaper bulbs and better generators. From that time on, he never received another order for a gas plant. Gripper was one of many such small contractors who entered the field of electric lighting in Europe, Britain, and the United States in the 1880s. The EPS was successful in establishing contacts with these contractors and in convincing most of them that storage batteries were essential to the proper operation of a private plant. Consequently, the installation of batteries at private plants became standard British practice. As we will see, this process did not take place in the United States, where almost all early private plants were installed without batteries. This retarded the evolution of a native American storage battery industry.

The British practice was to run the generators during the day to charge up the batteries. A servant would be trained to do this and special switchboards were developed to help them carry out proper charging. The generator was shut off at night and the lights run from the batteries alone. Bulb manufacturers provided special low voltage bulbs—usually 40 volt—to reduce the number of cells needed.

This practice became standard in Britain by the later 1880s, but it was difficult for the EPS to create this market. Until 1886 the company lost money at the rate of $75,000 a year, chiefly due to the return of failed batteries.[55] By 1884 the company's engineers had made significant progress in improving the battery durability by design changes, but it was still found that the durability of the batteries in actual service was often hardly better than that of the original Faure cells. Furthermore, the failure of these batteries did not seem to be due to design problems themselves, since the cells at some installations all held up very well—for example, those at the Bank of England, which gave no trouble for two and a half years.[56] The problem, therefore, appeared to be caused by operating conditions.

---

[55] Ibid., p. 435-437.
[56] The Electrician & Electrical Engineer, 11/1886, pp. 418-420.

The man who solved this problem and put the EPS on the road to solvency was Bernard Drake (1858-1931). Bernard Mervyn Drake, like a number of early lighting pioneers, began as a mechanical engineer, serving his apprenticeship at Whitworths.[57] In 1881 he joined Anglo-American Brush and obtained there an intimate, on-the-job education in the problems of private lighting plant engineering, from both the manufacturing and installation-operation perspectives. He was first put to work in the factory, being shifted from one phase of the manufacturing to the next until he had learned it by heart. Drake then installed a number of large Brush plants and helped establish a reputation for the Brush interests.

In 1883 or 1884 he struck out on his own, going to the United States where he obtained the British rights to the Thomson-Houston system of lighting.[58] Returning to Britain, he found no one interested in the system, British reluctance probably an offshoot of the crippling 1882 Act. Consequently, he accepted a job as plant manager of the EPS's factory at Millwall, in the east of London, in 1884.

When Drake entered EPS, its fortunes could not have been lower. The English electrical engineering journal *The Electrician*, summing up the results of 1883, made the following comment about the EPS:[59]

> . . . . the year just past has been altogether unfortunate. The batteries that were so full of promise 12 months since (when EPS began manufacture) have disappointed their best friends. The Faure-Sellon-Volckmar battery can hardly be said to be anything but a doubtful success. . . . The defects of the battery mentioned are well known, and as the battery has been withdrawn pending the introduction of a modified or a new form we must assume that the defects are insurmountable.

Drake withdrew the cells from the market and instituted a research program to find the solution. No cells were sold until the problem was solved. All failed cells were bought back from the customers and carefully examined. New cells were tested under a variety of conditions in the factory.

Drake found that these cells failed for three chief reasons: 1. the destruction of the lead grids; 2. the buckling or warping of the grids; 3. the active materials falling out of the perforations

---

[57] *JIEE* 71 (1932): 984.
[58] Ibid., 60 (2/28/1922): 435-438.
[59] 5 January 1884, p. 182.

in the grid.[60] The reason for these failures puzzled Drake until one of Pasteur's "accidents which favor the prepared mind" took place. It was the practice at this time for each workman in the "forming" room to tend only his own batch of plates. Consequently, if one of the workmen was absent, his supply of partially-"formed" plates would sit in the "forming" tanks untouched. One of the workmen became ill and his plates lay idle for some time. Drake noted that these plates began to swell and buckle all by themselves. He took these and charged them. The swelling stopped.

Drake noted, however, that if these "formed" plates were again left idle, either in the charged or discharged state, their active materials also tended to swell and bend the grids. Moreover, the longer the plates were left idle, the harder it was to recharge them. If discharged plates were allowed to stand long enough, they became highly resistant to both current flow and recharging, having become essentially inert chemically. Moreover, they displayed the three characteristic breakdown problems mentioned above.

When Drake examined the failed cells, he invariably discovered that the faces of the plates had become covered with a "hard, white enamel," which he found to be lead sulphate. This led Drake to make a most important, although quite unexpected discovery about lead-acid battery operation. This was the problem of *oversulphation;* a battery can be destroyed if too much sulphate forms in the active material. This was an unexpected phenomenon since it is normal for lead sulphate to form during discharge; the current is generated by the conversion of Pb and PbO_2 to $PbSO_4$. During the early 1880s, therefore, it was assumed that sulphate formation was a good thing. Indeed, Faure and some of the earliest battery manufacturers chose $PbSO_4$ as the raw material for the active material paste, although by 1884 EPS was using PbO and Pb_3O_4. But Drake discovered that overdischarge or improper discharge of a cell can produce an enamel-like deposit of the sulphate. This deposit both resists current flow, being a poor conductor, and resists penetration by sulphuric acid, since it is non-porous. Because sulphate formation occurs most rapidly at the interface between sulphuric acid and the active material, the "enamel" tends to form on the outside of the plate, and the more rapid

---

[60] Contemporary sources, following EPS practice, called these cells Faure-Sellon-Volckmar, or, more usually, FVS batteries.

the rate of discharge, the greater is the tendency for the sulphate to form only on the surface, since the electrolyte is not given time to penetrate. Once the surface deposit is formed, it becomes increasingly difficult for the sulphuric acid to penetrate. Thus, there occurred the *apparent* conversion of the active material into an inert substance.

In addition, lead sulphate takes up more room than either sponge lead or lead peroxide. Therefore, the discharge and charge of plates is mechanically analogous to the freezing and thawing of ice in cracks of rocks. If the process of enamel formation is allowed to proceed, more and more active material in a plate becomes inert and the plate swells and bends. The swelling pushes the pellets of active material free, so that they lose electrical connection with the grids and eventually fall out entirely.

Drake also discovered that the overcharging which produces oversulphation is not a simple process. He found that oversulphation can be produced by discharging a battery too far or by discharging at too rapid a rate. But he also discovered that batteries which had been used in this way often resisted oversulphation and recharged nicely.[61] Drake found that such treatment, in itself, did not produce oversulphation. A much more insidious cause of oversulphation was the practice, which was common at the time, of partially or completely discharging a cell and letting it stand in this condition for long periods.

If this is done, local action begins to convert any remaining PbO_2 to the sulphate. In addition, the metallic grids themselves will be slowly converted to sulphate. Thus, the lead grids are slowly eaten away, lose their rigidity, and can collapse.

Drake concluded that the formation of excess sulphate was the major reason for the lack of durability of contemporary battery plates. He reasoned further that a reverse procedure should preserve plates—that is, a mode of operation which would keep sulphate formation to a minimum. He tested this hypothesis by deliberately overcharging cells and by recharging discharged batteries immediately after discharge. This procedure worked well; plates so treated did not buckle, the active materials did not fall out of the perforations, and the grids did not disintegrate.

---

[61] Wade, *Secondary Batteries*, p. 328.

These results were completely unexpected. Up to this time, electrical engineers, including the EPS people, had feared *overcharging* of batteries, and not *overdischarge*. This was for a number of reasons. First, as a battery nears the completion of its charging cycle, less and less sulphate remains to be converted back to lead peroxide and sponge lead. Consequently, the charging current begins to decompose water, with a copious evolution of gas bubbles on the plates. There is a consequent sharp rise in the EMF as this electrochemical transition occurs. Early storage battery designers feared that this gassing would build up pressure inside the active material and disintegrate it. Moreover, they feared that when all the $PbSO_4$ on the cathode was oxidized back to PbO_2, the charging current might act on the metallic lead of the grids, "forming" them to PbO_2 also. Accordingly, the EPS had carefully warned all its customers against overcharging their batteries and had even recommended, just to be on the safe side, that cells should not be brought up to full charge.[62] As a result, conscientious customers tended to leave their cells with as low a level of charge as possible when they were sitting idle; in fact, the EPS was unknowingly encouraging its customers to destroy their batteries.

Drake found that some overcharging is good for batteries and that excess gassing is only damaging if carried out at very high rates.[63] Moreover, he discovered that occasional overcharging will not cause further "forming" of the grids, since the thick coat of peroxide on the cathode seemed to protect the metallic grids from this conversion—another reversal of what had been expected. Drake, therefore, issued a: " . . . .radical change in the instructions issued to users of the EPS batteries. Whereas, hitherto, they had been specially cautioned against overcharging, they are now urgently required to overcharge new cells, and to charge incessantly."[64]

The customers were also urged to overcharge any cells, new or old, which had lain idle for a time.

From this point on the fortunes of the EPS company improved. They received an increasing amount of praise in the pages of the technical press about the durability of their cells.[65] At the beginning of 1885, the London *Electrician*, whose

---

[62] *JIEE* 60 (28 February 1922): 435-8; *The Electrician & Electrical Engineer*, 11/1886, p. 418.

[63] Wade, *Secondary Batteries*, p. 332.

[64] *JIEE* 60 (28 February 1922): 435-438.

[65] *Proc. Inst. Mech. Eng.* 36 (1885): 395; *Electrical Engineer* (London), 4/1886, p. 142.

gloomy 1883 report on EPS was quoted earlier, reported that the company had made good improvements during 1884 and that they hoped for financial success soon.[66] These hopes were fulfilled, when the *Electrician* reported in January 1886:[67]

> During the past year probably few companies have improved their commercial position more than this company. Great hopes were entertained a year ago that the practical difficulties in connection with their secondary batteries had been surmounted, and the results have more than verified these expectations.

The success of EPS, in contrast with the failure of FEA and Force et Lumiere, illustrates a number of points about the evolution of commercial technologies. First, it illustrates the wisdom of the EPS's approach to the exploitation of their new technology and the unfortunate consequences of that used by Philippart. From the beginning, the EPS committed itself to a centralized method of organization, that is, all functions of the battery business were carried out by the EPS itself—design, manufacture, sales, installation, maintenance—and all these operations were directed from a central location—the Millwall factory.

By contrast, the Philippart's idea was to decentralize in order to reap quick profits. Thus, the central firm—Force et Lumiere—tried to make its profits chiefly by selling or licensing its patent rights to a series of companies who would then do the actual manufacturing. The central firm would use these profits to buy more patents and sell or license them in turn. This is not to say that the Philipparts saw Force et Lumiere purely as a holding company. On the contrary, Force et Lumiere did carry out a program of research, but it was for the purpose of generating patents which would then be licensed to the manufacturing companies. Thus, in the Philippart scheme, there was a distinct separation between the technology-generating and technology-controlling parts of the industry and the technology-using (manufacturing) parts.

The success of Bernard Drake, however, as well as a number of other successes of the EPS which we will discuss later, was due to the centralized methods of the company. Drake formed a link between battery consumers, manufacturers, and designers; thus a continual flow of information took place. For example, Drake was able to relate the bad experiences of many consumers with the "accidental" discovery in the EPS "form-

---

[66] 10 January 1885, p. 186.
[67] 22 January 1886, p. 211.

ing" room. His prior background as both a manufacturer and an installer of electrical equipment in the Brush company prepared his mind for this role.

A second point about the evolution of commerical technology is more subtle. The development of commercial technologies is usually connected with the patenting process. Yet notice that what Bernard Drake developed was a *technology*, but not a *patent*. No patent was ever taken out on the new technique of charging. The reason for this is that it was unpatentable—the technique pertained only to the use to which the consumer placed a pre-existing invention and it required no change either in the manufacturing methods or in the design of the invention itself. The process could not even be kept secret, since it only had value if everybody knew about it. Consequently, Bernard Drake published his results as a scientific paper, not as a patent; he claimed the technology as intellectual, not economic, property.

Here we see clearly the limitation of the Philippart scheme in its ability to develop the technology of the storage battery. The only technology of value to Force et Lumiere was *patentable* technology. Philippart's idea was to control the economics of the battery industry by controlling its patent base. He made this strategy quite explicit in an FEA shareholder's meeting in November 1882:[68]

> Our company must, henceforward, cease to be an industrial company manufacturing accumulators for sale or for its own use, and must become a company for the formation and constitution of affiliated companies; to which companies we shall sell licenses for cash, and for cash and fully-paid shares, and we shall also sell licenses to private individuals and already existing companies. . . . Therefore, gentlemen, we should purchase, resell and farm the French and Continental patents, and this is what we will do.

Simon Philippart confused patents with technology. As the future history of the storage battery was to show, a great deal of its technology proved to be unpatentable. Consequently, his plan would have proved increasingly unworkable as time went on. The EPS approach, combining control of technology with manufacturing, made non-patentable technology as valuable as patentable.

---

[68] *The Electrician* (London), 11 November 1882, p. 614.

3. The Early Evolution of Storage Battery Design

The Electric Power Storage Company was not the only manufacturer of storage batteries during the early 1880s. It was, however, the most significant company, both from the perspective of size as well as on its influence on the future of technology. In Britain, EPS had an early, albeit small-sized competitor in the firm of Elwell-Parker, which also began manufacture in 1882. In France, a series of small workshops produced storage cells throughout the 1880s. Except for Force et Lumiere, however, these were small operations—often scientific instrument makers diversifying into battery making—and little is known about these operations.

In the United States considerable activity went on during the early 1880s but the only firms to achieve any success were those which derived their technology from the EPS. German storage battery activity, on the other hand, did not become significant until the later 1880s.

Yet despite the differences mentioned above, the pattern of physical storage battery design during the 1880s was quite similar among the various designers and companies. Storage battery design during the early '80s was characterized by a mechanical approach. Towards the later '80s and early '90s, battery design evolved into a mixed mechanical and chemical approach; as a result, it became characterized by greater subtlety and sophistication of design. We will now examine how and why these changes took place.

The first significant improvement in the evolution of the pasted-plate storage battery electrode was the replacement of Faure's flat plates by perforated plates. The Faure plates not only lacked durability, but also had relatively little storage capacity. The reason for this was the small area of contact between the active material and support, as well as the poor quality of their contact, which prevented much of the active material from properly taking part in cell reactions. The improvements in storage capacity made possible by the use of perforated plates can be seen from the results of laboratory tests made in July 1885, when a current model EPS cell was found to yield 2.7 times the storage capacity of an original 1881 model Faure cell.[1]

---

[1] 48,000 ft·lbs/lb, as against 18,000 ft·lbs/lb. *Electrical Engineer* (London) 7/1885, pp. 19-19.

Yet the invention of the perforated or "grid" plate was only
the beginning of the story. The earliest grid plates were simple, with straight holes cast in the plates from one side to the
other, of which the 1881 grids of Swan and Sellon were typical.
In these early designs, the cross-section of the holes through
the plates was either parallel to the plate surface, or else with
the sides flaring outward from the middle, permitting both
ease of casting and pasting. In practice, however, such plates
did not work well. As Bernard Drake discovered, one of two
things was likely to occur; either the active material would
crumble out of the holes as a fine, incoherent powder, or it
would pop out as a hard, rock-like pellet.[2] Drake associated the
powdering with the presence of too little sulphate and the hard
pellet with too much.

This problem has been noted even before Drake, and there
were numerous attempts since 1882 to prevent the active materials from falling out of their holes. For example, since the early
EPS grids had outward-flaring holes, John Sellon in June 1882
reversed the direction of the flaring, seeking to hold the active
materials in more firmly:[3]

The casting of such grids was, of course, more difficult than
the outward-flaring pattern, forcing EPS to develop more complex and expensive two-piece grids in 1883:[4]

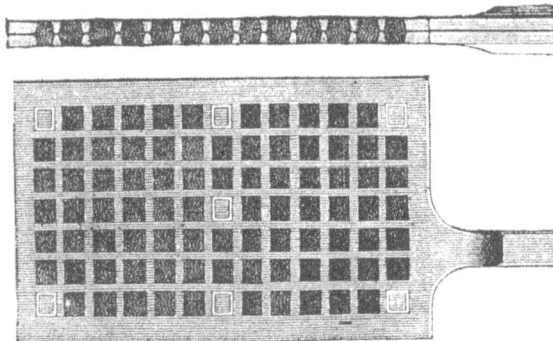

---

[2] *The Electrician & Electrical Engineer,* 11/1886, pp. 418-419.

[3] Wade, *Secondary Batteries,* p. 22.

[4] Wade, *Secondary Batteries,* p. 25.

Since the casting and filling of such two-piece grids presented technical problems, other solutions were proposed to the sealing-in of the active material masses. For example, a number of inventors proposed single-piece, "rolled over" grids. The concept here was to begin with a plate whose perforations were wide open to facilitate ease of casting and filling, and then to pass the pasted plate between rollers to squeeze the openings of the holes together. The Charles Gibson plate of 1888 was typical:[5]

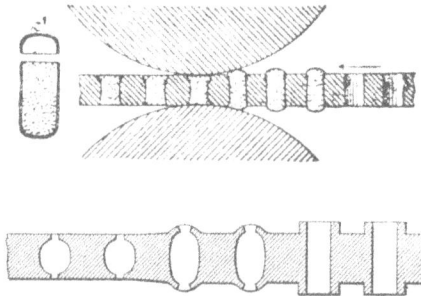

As grid designs became more complex and delicate in the late 1880s and '90s, the rolled-over grid was replaced by a lighter variant in which the grids were pressed between special dies after pasting. The pressing caused some bars of the grid to be bent over the paste. The 1890 Peral grid was typical:[6]

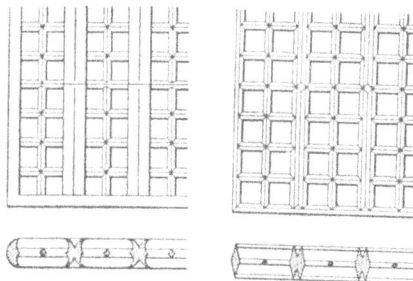

Here the diamond cross-section bars were curled over the active material masses by the pressing.

---

[5] Ibid., p. 23; German patent #45,992—20 March 1888.

[6] Wade, *Secondary Batteries*, p. 24.

The "rolled-over" or "pressed" grid designs were popular in the storage battery patent literature of the 1880s. Few such designs seem to have reached the manufacturing stage, however; this was probably because of the difficult mechanical problems involved in manufacture.

The foregoing sample of grid designs is typical of what might be called the "brute force" approach to battery manufacture; the basic idea of this approach was to mechanically wedge the active materials between lead walls so that as little room as possible was left for the falling-away of these materials. Grids of the "brute force" school were typically massive; often there was more solid metallic lead present in the grids than active material.

The reason for the reliance of early battery designers on mechanical solutions to the problem of active material durability lay in the designer's ignorance of the chemistry of the active materials. A symptom of this ignorance can be read in the wording of several of the earliest storage battery patents, where it is stated that the active material paste can be made from "any salt of lead."[7] While it is true that it makes little difference what lead compounds one begins with, since the "forming" current will produce the PbO_2 and lead sponge from a variety of substances, the structural stability of the finished plate is highly dependent on the proper choice of starting materials.

In modern manufacturing practice, the paste for battery plates is made from a mixture of dilute sulphuric acid and litharge (PbO), although red lead (Pb_3O_4) is sometimes used. These two lead compounds, when mixed with sulphuric and treated properly, will behave much like cement. The mass hardens, releases heat, and will even do so under water. The hardening should be accompanied by drying, but it is important that the rate of drying be slow and carefully controlled. If this is done, the setting process will produce long, fibrous, needle-like crystals of lead sulphate throughout the active material mass. This fibrous structure will be retained even after the sulphate has been converted to sponge and peroxide, thus giving the "formed" active material good coherence.

Also analogous to cement technology is the mixing of the lead compound paste. The paste must not be too wet. It should have a damp, greasy-crunchy feel just before it is pushed into

---

[7] See for example: U.S. patent #318,828—25 May 1885 (Swan); #382,420—8 May 1888 (Epstein).

the grids. It should shear well, but must not flow or be sticky.[8] The sulphuric acid and litharge must be vigorously mixed and the grids must be pasted immediately after mixing.

Lead grids coated with paste prepared in this manner possess a high degree of toughness and resistance to disintegration. But if some other raw material is used for the paste—say lead sulphate or lead peroxide or finely divided metallic lead—then the setting of the paste will be either completely absent or much diminished. That is, the durability difference between a plate made from litharge and one made from lead peroxide is analogous to the durability difference between a concrete building block and a mud brick.

Moreover, even if litharge or red lead are used, but poorly treated, the durability of the plate suffers. Thus, if the mix is too wet, or is allowed to dry too rapidly, or is allowed to set too long before being pasted, the final coherence of the mass is reduced. Again, this is analogous to the poor results obtained when cement or concrete is not "cured" properly.

Such detailed knowledge about the intricacies of paste handling was lacking during the early years of the storage battery industry. A result of this lack of knowledge was the "brute-force" design approach, with the massive lead walls of the grid making up for the active material's incoherence. Another result, which can also be grouped under the general rubric of "brute force," was the extensive use of binding agents mixed with the lead compounds. Binders such as gelatin, glue, plaster of Paris, and glycerine were all popular.[9]

Nevertheless, a body of craft knowledge about the behavior of active material pastes was slowly evolving during the 1880s but since it was nonpatentable and nonscientific, it is almost impossible to document the evolution of this particular phase of battery technology. These craft skills were developed in the paste workrooms of the manufacturers, where each "pasteman" worked a small pile of litharge, red lead, and sulphuric acid with a wooden trowel on top of his workbench. When the paste reached the right consistency, the pasteman pushed it vigorously into the grids, struck off any excess with a straight-edge, and set the plates aside to dry.

Like the techniques developed by Drake for charging batteries, these paste preparation methods were unpatentable. Nonetheless, a number of patents did appear during the '80s

---

[8] Nels E. Hehner, *Storage Battery Manufacturing Manual #2*.

[9] For example, Wade, *Secondary Batteries, passim*.

which contain evidence of the evolution of this craft knowledge. For example, an 1885 patent by two Hungarian mining engineers, Stephen Farbaky and Stefen Schenek, specified the use of a paste of 95 percent litharge and 5 percent pumice stone for the negative plate.[10] The patentees observed that the paste must be "a little moist, and of no plasticity." The paste plates were to be cured by keeping them moist for several days before "formation." If this was done, said the patentees, the active material "stiffens like cement."

Such references are rare, largely because they could not be part of the patent claims *per se*. Nevertheless, they do indicate a steady advance in empirical knowledge of paste chemistry during the 1880s. This progress, coupled with that of Bernard Drake in learning how to charge pasted-plate batteries, resulted in the development of the first reasonably durable storage batteries soon after the middle of the 1880s.[11]

A symptom of the evolution of paste-making and curing technology was the change in the physical designs of the grids in the latter '80s and '90s. The massive lead supports gave way to much lighter, more open, and delicate grids. The more general use of antimony as an alloying agent also contributed to this change in grid design. An early example of these new grids is this rather attractive system of interlinked lead-antimony rings:[12]

---

[10] U.S. patent #344,957—6 July 1886; #348,625—7 September 1886; German patent #37,012—8 November 1885.

[11] By "reasonably durable" is meant a battery whose positive plates would last about three to four years and whose negatives would last eight to ten years, if given proper treatment.

[12] Wade, *Secondary Batteries*, p. 27.

In 1886 Farbaky and Schenek patented this grid to go with their new paste. In the older grids, the massiveness served not only to hold the active materials in place, but also to resist the warping action of the expansion of the active materials. This problem was most serious on the positive electrode. With the adoption of lighter grids, however, there was less metallic lead to resist the warping. The system of interlinked rings shown above was one attempt at a solution—only the hexagonal spaces were filled with paste; the lentil-shaped interstices were left open to allow space for expansion.

In 1892, Frank King and E. Clark, two engineers working for EPS, developed the so-called "claw grid":[13]

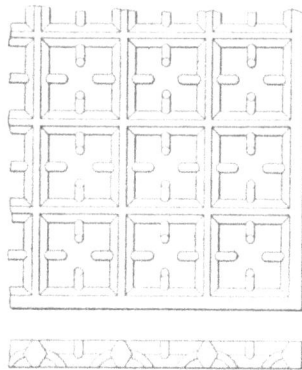

Here, the open structure and the outward-flaring walls of the grids were intended to permit easy expansion of the active material without damaging the grid itself. The curving lead pins, one pair facing outward and the other inward, were designed to grip the active material all the more firmly as it contracted or expanded. Although grids of this kind were similar to the "brute force" designs of the early 1880s in the sense that they were intended to *grip* the active materials, the method of holding the active materials was quite different. The older grids were designed to grip the active materials by surrounding them on all sides, whereas the claw grid relied on the inherent coherence of the block of active material which was formed within each small, box-shaped hole. If each of these small blocks did not possess internal coherence, the small lead claws would have accomplished nothing.

Grids with such an open, delicate structure have been popu-

---

[13] Ibid., p. 31.

lar with battery engineers ever since. They save weight as well as cost, are easily pasted, and permit the electrolyte easy access to the interior of the active mass. The favorite pattern of delicate, lattice-work grid was the "double-grid." Double grids emerged in the late '80s and remain to this day the most popular plate design. Sellon patented one of the earliest double-grids in 1889:[14]

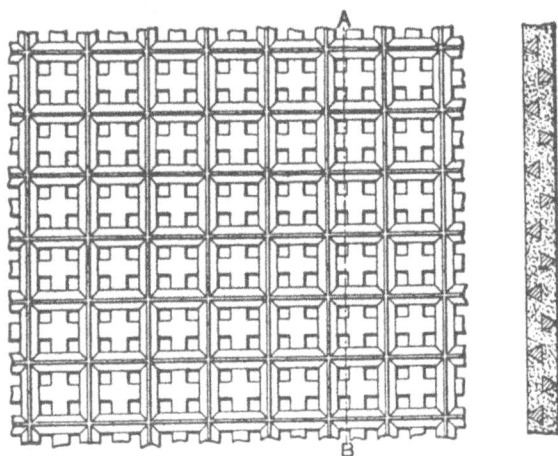

These were used commercially by the EPS; they were the first grids used by this company which both held the active materials well and could be easily cast as a single piece.

Present day double grids are different in detail from those of Sellon, but they retain the same basic design idea—rows of thin, triangular cross-section grid bars staggered on alternate sides of the grid. In grids of this design, the function of the bars is not to grip or surround the active material, but rather to provide the lightest possible skeleton around which the paste can set into a solid block. In fact, in such plates it is the paste which surrounds the grids, and not, as in the early 1880s designs, the metallic grids which formed an invertebrate-type shell around the paste.

Ever since the 1890s storage battery designers have spent much effort developing minor variations in the shape of double grids. Much of this effort has centered in the problem of matching the design of the grid skeleton to the coherence,

---

[14] Ibid.

expansion, porosity, and other physical properties of particular paste mixes. One obvious problem connected with the use of one-piece double grids is the difficulty of casting them cheaply and rapidly. It was the German storage battery industry which made many of the early important contributions to double-grid casting technology.

The German industry began to develop at the same time that the first double-grid designs appeared; the late '80s, so it began to use such designs from the first. The first German company to begin manufacture of storage batteries on a significant scale was Gottfried Hagen of Cologne in 1884.[15] He entered the storage battery business by a process which is unique in the history of the industry. His firm had been formed in 1827 to smelt lead, and by the 1880s was one of Germany's largest lead works.[16] In 1884 it engaged in some vertical expansion by licensing the battery patents of A. de Khotinsky, a Rotterdam inventor.

The de Khotinsky grids are a good example of the older, brute force approach:[17]

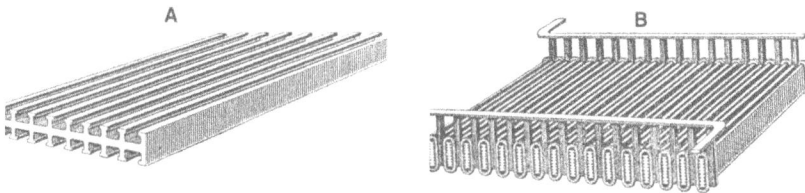

A shows an electrode strip formed by forcing solidified lead through extrusion dies under hydraulic pressure. *B* shows the strips inserted into oblong glass tubes and assembled into complete electrodes. This kind of grid was necessary, since the raw material for the active paste was finely-divided metallic lead, which does not form a coherent mass.

---

[15] Albrecht, "Die geschichtliche Entwicklung des elektrischen Akkumulators," *Verhandlungen des Vereins zur Beförderung des Gewerbfleisses* (Berlin, 1912) 91: 547.

[16] *Electrical World*, 1896: 278.

[17] Wade, *Secondary Batteries*, p. 36; German patent #35,396—18 July 1885; #30,041—12 May 1884.

In 1889, however, G. Hagen and C. Beyer patented some
double grids:[18]

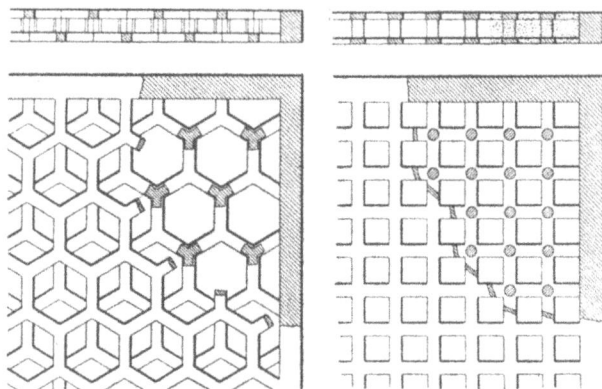

At first, these grids were expensive, since they were cast with
sand cores. However, in 1894, Hagen developed special molds
which did away with the need for separate cores and allowed
complex double and even triple grids to be cast from mechanical metal molds.[19] The Gottfried Hagen company continued to
make the Khotinsky battery at a branch factory it established
in 1890, but the larger Cologne works were converted to the
new double-grid system.[20]

Gottfried Hagen was the first significant German manufacturer, but it eventually was to be eclipsed and absorbed by its
younger rival; Accumulatoren Fabrik Aktiengesellschaft, Hagen, or AFA, as it was known. This company was formed in
1887 in the town of Hagen, Germany. As a result, the products
of both companies—Gottfried Hagen and AFA Hagen—came
to be known as Hagen accumulators; a fact which was to cause
considerable confusion among British and American customers when German storage batteries became popular in these
countries in the late 1890s.

AFA's early development was not unlike that of EPS. The
founder, Adolph Müller (1852-1928), like Bernard Drake, obtained his electrical engineering education on the job while
employed by Spiecker & Company, a Cologne electric lighting

---

[18] Wade, *Secondary Batteries*, p. 33; German patent #52,880—19 July 1889.

[19] Wade, *Secondary Batteries*, p. 32; German patent #77,492—21 July 1894.

[20] Elektricitäts-Gesellschaft of Gelnhausen, which continued to make the Khotinsky
plates into the twentieth century. Ibid., p. 451.

plant contractor.[21] While employed there, he came to realize the value of storage batteries as adjuncts to electric lighting, but he knew that no satisfactory ones existed in Germany. Then, in 1885, Müller heard about a new storage battery invented by a Luxembourg engineer named Henri Tudor.

Henri Owen Tudor (1859-1928)[22] was born in Rosport, Luxembourg, where he spent most of his life. He graduated from the Brussels Polytechnic Institute in 1881 and went on for post-graduate study in electricity at the Sorbonne. There he met both Planté and Faure, and became interested in storage batteries and their applications in electric lighting.[23] During the early '80s Henri and his brother Hubert installed a number of private lighting plants and developed the design for a new storage battery. In 1885 a small factory for the manufacture of these cells was set up in Rosport, but it was ephemeral and apparently never reached the commercial stage.

Tudor's approach to battery design was original, although a number of later inventors copied the idea. Perhaps because of his association with both Planté and Faure, Tudor developed an electrode-making process which combined the techniques of both these men. First, Tudor cast a grid with horizontal ribs:[24]

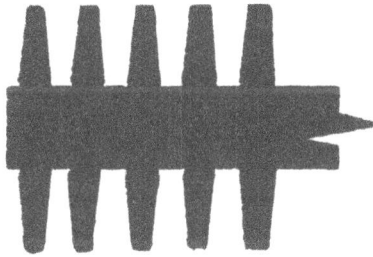

and then "formed" these by the traditional Planté technique until a thin coating of active materials had been produced.[25] After this, a paste of red lead and sulphuric acid was applied to the grids and "formed" according to the Faure method.

---

[21] Burkhard Nadolny & Wilhelm Treue, *Varta—Ein Unternehmen der Quandt Gruppe, 1888-1963* (Munich, Varta, 1964) p. 39.

[22] The family claimed descent from the founder of the English dynasty.

[23] *Luxemburger Welt,* 7 August 1976; *La Revue Technique Luxembourgeoise,* May/June 1928.

[24] H. Beckmann, "Zur Geschichte des Akkumulators und der Accumulatoren-Fabrik Aktiengesellschaft," *Beiträge zur Geschichte Technik und Industrie* 14 (1924): 250.

[25] Wade, *Secondary Batteries,* p. 85.

This process may seem an unneccessary and expensive du-
plication of effort. Yet there was a substantial reason for it, and
this reason relates to the poor understanding which early battery designers had of paste technology. Because of their inability to produce coherent pastes, at least with any reasonable
degree of reproducibility, early battery manufacturers did not
expect the paste to remain long in the grids. Henri Tudor,
instead of adopting the brute force techniques to hold the
paste in the grids, accepted the rapid decay of the paste. In his
plates, the pasted active material was expected to provide storage capacity for only a few months of service, and then fall
away. At the same time that the pasted materials were flaking
away, however, the Planté-"formed" layer of active material
on the grids themselves would be deepening. Consequently,
these second layers would gradually take the place of the
pasted layers. Of course, this meant that the ribbed grids had
to be made rather heavy, since they served both as supports
and as future active material.

In 1885 Adolph Müller purchased the patent rights for most
of Europe from Tudor,[26] and spent the next two years experimenting with the system. Finally, in 1887, Müller decided to
commit himself fully to the business of battery manufacture
and in December 1887 set up the Firm of Büsche and Müller in
Hagen; Büsche being one of the local financiers who helped
capitalize the firm. The next year the firm's name was changed
to Müller and Einbeck; Johannes Einbeck was a mechanical
engineer who joined the company that year.[27]

The crucial change in the company's fortunes came in 1890.
Müller and Einbeck was a small, local partnership, typical of
many firms in all the branches of electrical engineering business which sprang up in the early years of electric lighting. The
overwhelming majority of such firms, whether in battery
manufacture or some other branch of the electrical business,
remained small, and either went bankrupt or were absorbed
by larger competitors. For every Westinghouse or General
Electric, there were scores of such mushroom companies.

At the start of 1890 Müller & Einbeck appeared to face such a
mushroom future, since the two largest German manufacturers
of electrical equipment—AEG and Siemens—were developing
storage batteries of their own. Had these firms gone through

---

[26] Nadolny & Treue, *Varta*, p. 39.
[27] Beckmann, "Geschichte," p. 250.

with their plans, the future of the Tudor battery would have been grim indeed. Müller, however, entered into talks with AEG and Siemens. On 1 January 1890 Müller & Einbeck was reorganized as AFA, with a board of directors consisting of men from Siemens, AEG (Emil Rathenau), and several large Berlin banks. The firm also received considerable recapitalization for its expansion. Siemens and AEG agreed to give up all battery manufacture themselves and to exclusively purchase AFA batteries for their own needs for ten years.[28]

The history of AFA is one of the success stories in the evolution of the storage battery; today, under its new name VARTA, it is the largest manufacturer on the continent. Such success stories, however, are rare. The vast majority of all storage battery firms in all countries soon failed. Of these companies which succeeded, the pattern of economic development outlined above for AFA was almost universal; that is, that these successful companies were all heavily financed or otherwise supported by more powerful industrial interests. As we saw, the EPS arose as the top early English storage battery manufacturer by its connections with Anglo-American Brush. This pattern was repeated for the leading American and French companies as well. Those firms which did not obtain such support either went out of business, or were eventually amalgamated into the leading firm in its country. The linkage between AFA (or EPS) and the leading electrical manufacturers provided the firm with an outlet for its batteries, as well as technical assistance and funds to be used to improve its technology.

The improvement of its technology was something AFA needed very much at this time. During the late 1880s and 1890s AFA engineers put the original Tudor process through many changes. This process of innovation was of importance not only for AFA, but also for the entire future development of the lead-acid storage battery, since the AFA plants were to become a center for the worldwide diffusion of storage battery technology in the 1890s.

During the early 1890s, the AFA designers did away with Tudor's compromise between the Faure and Planté processes. In the original Tudor process, the ribbed plates were "formed" electrolytically at the factory until they developed about one-fourth of their rated capacity. Then the red lead paste was added to contribute the remaining three-fourths. During ac-

[28] Nadolny & Treue, *Varta*, p. 48.

tual service, the one-fourth gradually grew, and the three-fourths fell. It fell quite literally as a continual flaking of PbO_2 and lead sponge particles to an empty space below the grids. In about a year this process was supposed to be complete.

Technically, there was little wrong with this system; commercially, however, it was undesirable. First, the labor and capital costs were high, since separate sets of tools, workmen, and working spaces were needed for the two kinds of "formation." Second, the dual function of the grids placed unreasonable demands upon them. Grids which are pasted should be alloyed with antimony, to give them strength. On the other hand, grids which are to be given Planté formation should not be alloyed, since the addition of antimony causes the lead to resist conversion to lead peroxide.[29]

Consequently, Tudor grids could not be alloyed with antimony, but in order to solve the strength problem, the non-alloyed grids had to be made thick and heavy. Yet thick and heavy grids cannot have great surface area per unit weight, and great surface area is essential in a commercial Planté-type battery, since, as we saw in the last chapter, Planté "formation" is a slow process. Only great surface area can provide good storage capacity in a Planté cell.

In the early 1890s, AFA engineers did something about these self-contradictory problems in the original Tudor design. They chose a solution which was the opposite of the one chosen by EPS—that is, instead of improving the adherence and cohesion of the pastes, the AFA designers did away with the pastes entirely and developed a pure Planté plate. Because of this change, the new AFA grids could be made more delicate and with greater surface area than the original plates, since there was no initial weight of paste to bear, nor did the grids have to resist the expansion of the paste.

In the early 1890s, AFA used such Planté-"formed" plates for both the positive and negative electrodes. After a time it was discovered, however, that for the negative plate, Planté-"formation" produced a plate of limited durability.[30] The reason for this is not clear. Therefore, about 1895 AFA introduced a pasted, double-grid electrode for its negative, and retained the Planté plate for the positive. This combination of Planté positive and pasted negative was copied by a number of other battery companies, and remains to this day one of the chief means of maximizing durability and storage capacity.

---

[29] Antimony addition raises the EMF required to "form" the lead.

[30] Beckmann, "Geschichte," p. 253.

Although AFA developed a highly successful Planté-type battery in the 1890s, a number of other firms preceded them in attempts to commercialize this form of storage cell. Indeed, throughout the 1880s, Planté-"formation" had its defenders just as did the Faure method. This was particularly the case in France, where small workshops and instrument makers who made the batteries favored this approach.

Faure-"formation" held the spotlight during the year or so immediately after Thomson's letter to the *Times*, since the speed, cheapness, and apparent simplicity of the new technique caught the imagination of inventors, and seemed to make the Planté method obsolete. But then, as the early pasted batteries revealed their poor durability, a reaction in favor of the older technique took over, and it seemed to many that the Faure technique was doomed. The British storage battery authority E.J. Wade vividly recalled his memories of the anti-Faure reaction of the early 1880s:[31] "Those were the dark days of the storage battery, when the first uncritical enthusiasm over its introduction had died away and it seemed likely to prove incapable of satisfying even the very moderate demands of the engineering practice of the period." Indeed, a number of authorities at the time suggested that if storage batteries were ever to be made practical, the pasting idea would have to be abandoned and Planté-"formation" used instead.

The chief durability problem with these early batteries was with the positive (PbO_2) electrode. In fact, the positive plate has been the main problem with battery engineers ever since, because it is inherently more liable to breakdown. The reason for this is that the material on the negative plate—lead sponge—is softer and less dense than the peroxide on the positive. Under pressure from the expanding lead sulphate during discharge the sponge tends to give and therefore remains in place. Lead peroxide, on the other hand, is hard and brittle and tends to crumble under pressure. Therefore, as E.J. Wade described the emphasis of battery engineers during the 1880s and early 1890s:[32]

. . . . the design of the positive (plate) so as to prevent its rapid deterioration monopolized attention for some years, and the poor negative came to be regarded as little more than an incidental adjunct of the cell. . . . Every other consideration was ignored in the futile and misdirected effort either to abolish the positive support altogether or else to make it strong and rigid

[31] Wade, *Secondary Batteries*, pp. 255-256.
[32] Ibid.

and of such a construction that it could forcibly detain the active material and prevent its expansion.

We saw in the early part of this section how this latter objective of "forcibly detaining" the active materials was developed in the "brute force" school of battery design. The other approach which Wade mentions—that of "abolishing the positive support altogether"—was the use of Planté-"formation." Although the main durability problem was with the positive electrode, those inventors who adopted the Planté method in the 1880s used it for both plates.

One of the first inventors to manufacture Planté-type batteries for the new electric lighting field was the Frenchman, Nicolas de Kabath. De Kabath prepared long strips of lead foil 1/250th of an inch in thickness, and packed these into perforated lead boxes, each of which formed an electrode:[33]

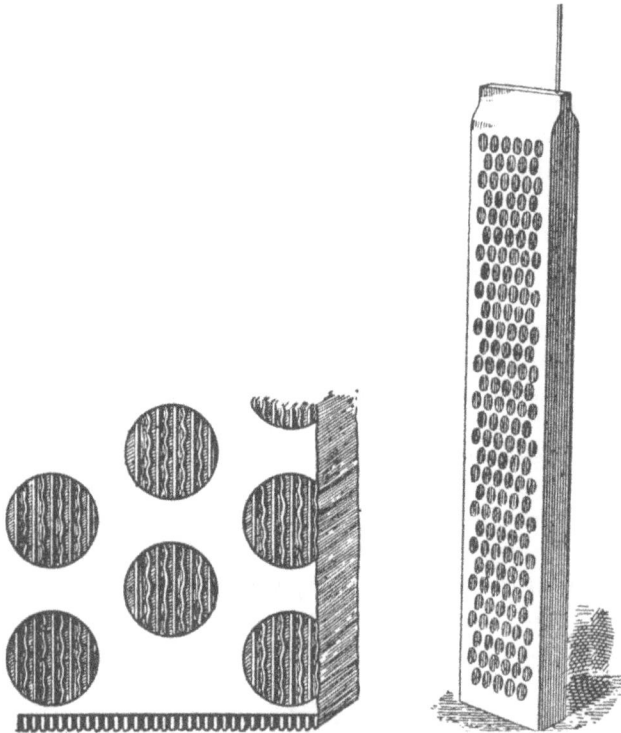

About 200 of these strips, alternately corrugated and flat, were held in the box electrode.[34] De Kabath patented these in 1881 and manufactured them in France throughout the 1880s.

[33] Ibid., p. 78.
[34] Albrecht, "Entwicklung," p. 445.

Planté-type electrodes made with lead foil or strips appeared regularly in the early patent literature and several such designs were manufactured commercially. Such fine materials provided excellent surface area/weight ratios and the technical details of manufacture were simple; various forms of lead wires, shavings, shot, chips, were packed, wound, or fastened by some other simple mechanical means to a lead support. Thomas Edison, for example, suggested a number of schemes in this vein in the early '80s, although he reduced none of the patents to practice. Typical of these is patent #274,292, filed August 1882, in which Edison suggested pouring molten lead into water to form various "arborescent shapes." These delicate lead pieces could then be removed from the water, dried and pressed into the surface of a molten lead support.

Although Edison's designs were never commercialized, a variant on them was manufactured in London during the early '80s. A workman at the top of a ladder poured molten lead into a large tub of water below. This technique, which may have been suggested by the ancient shot-tower, produced sponge lead. This material was then packed into annular rings in a cylindrical battery. An inner cylinder of sponge lead was used for the negative, surrounded by an insulating cylinder, surrounded by a final lead sponge cylinder for the positive.[35] It took about eight weeks to "form" this sponge.

Similar to these designs were the various forms of "felted plate" batteries popular in the 1880s. The Simmen-Reynier design of 1882 was typical:[36]

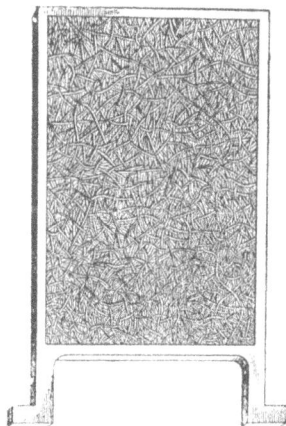

---

[35] *The Electrician* (London), 25 May 1928: 583.
[36] Wade, *Secondary Batteries*, p. 102.

Lead was poured through a collander and allowed to solidify
into crumbled threads, like masses of spaghetti. These masses
were then pressed into sheet lead containing frames.

Charles Smith in 1883 suggested stringing split shot on
wires and packing them inside lead boxes:[37]

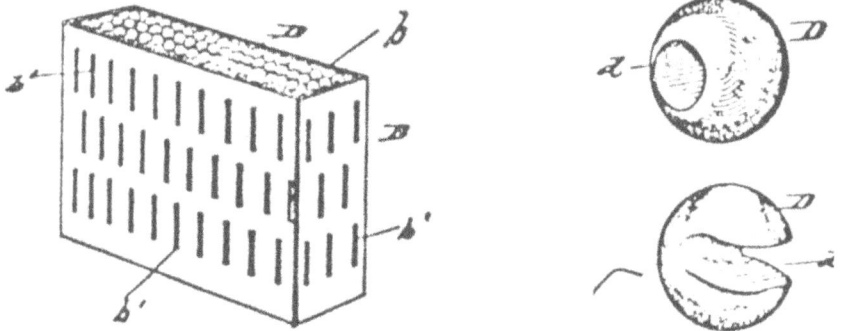

The French inventor Emile Reynier experimented with a
good number of storage battery designs during the '80s. He
began his work with Faure, but shifted his interest to methods
of making improved Planté plates in 1883. Reynier developed a
series of lead foil electrodes similar to De Kabath's. These were
made of single, large sheets of lead, folded back and forth many
times to form a pleated curtain, and then fixed into an electrode support:[38]

---

[37] U.S. patent #292,142—15 January 1884—filed 7 March 1883.
[38] Wade, *Secondary Batteries*, p. 90.

Reynier manufactured these commercially in Paris in the 1880s.

By the early 1890s Planté electrodes made according to the crude designs illustrated above largely disappeared from the market and the patent literature. The chief reason for this was that these plates, like the early pasted-plate electrodes, lacked durability. The lead foil cells, for example, were easy and cheap to make, since the great surface area made them easy to "form." The weakness of the lead foil, however, led to early disintegration of such electrodes, particularly since the thin sheets were soon completely converted into active material.

The same problems were experienced with plates made of lead shavings and chips, and the "felted plates." Although such plates are easy to make, requiring little or no special equipment or skills, the irregular nature of the manufacturing processes prevented quality control—no two plates ever performed in the same way. As a result, during the later 1880s and 1890s, designers of Planté-type batteries turned their attention towards thicker, more regular electrode designs, and away from the haphazard approach typified by Edison's "arborescent shapes."

The AFA-Planté plates discussed earlier are a good example of this more professional, quality-conscious approach to battery design. In contrast to the simplicity of making the felted electrodes, the casting techniques used in making the ribbed grids were difficult to develop, but the AFA engineers realized that such techniques were essential if Planté-type batteries were to be manufactured with guaranteeable properties. By the later 1880s and early 1890s storage battery manufacturers came to realize that one of the best ways to encourage consumer confidence in batteries was to insure him of precisely defined and guaranteed product capabilities.

An early example of the regular approach to Planté plate design was that patented in 1885 by C.P. Eliesen in London. Small ribbons of lead were rolled into spirals with strips of asbestos in between. The spirals, each an inch in diameter, were packed into holes in a grid. During "formation," the expansion of the materials locked the spirals in their holes: [39]

---

[39] Ibid., p. 79; British patent #9522—10 August 1885; U.S. patent #358,341—22 February 1887.

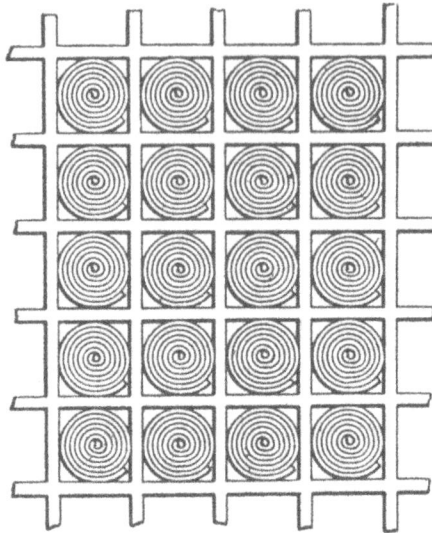

Although this was a more expensive and labor-intensive way of making Planté plates than those described above, it also made more durable plates, and therefore it has been used ever since in one or another variant. In 1891, for example, J.H. Walter cheapened the process by corrugating the lead ribbons before rolling; this eliminated the need for the asbestos separators.[40] This idea was then used as the basis for an 1895 patent of J. Rhodin for a new plate made by the Chloride Electrical Storage Syndicate of Britain.[41] The story of Chloride belongs to the next chapter, but it is sufficient to note that the technological experience of Chloride in the early 1890s was not unlike that of AFA; Chloride designers recognized the utility of using pasted plates for the negative and Planté plates for the positive electrode. Chloride began in the late 1880s using pasted grids for both electrodes, but in the mid 1890s it adopted the modified Eliesen pattern of Rhodin for its positives and continued to use them for many decades for its stationary batteries.

Chaim Eliesen was a Frenchman who came to England to work for EPS, where his interests centered chiefly on the use of batteries for self-propelled streetcars.[42] He left EPS before 1885 to found his own company; the Electric Locomotive and Power

---

[40] Wade, *Secondary Batteries*, p. 81.

[41] Ibid., p. 82

[42] *JIEE* 60 (28 February 1928): 468.

Company, which later became the Lamina Accumulator Syndicate. Eliesen spent his career developing variants on the Planté method. For example, the ladder and water tub method mentioned earlier was one of his early designs. Eliesen developed the spiral-grid design, however, as he came to realize the value of heavier and more regular construction for Planté plates. The following remarks, which Eliesen made in 1888, reflect not only the evolution of his own thinking, but that of most contemporary battery designers:[43]

> In accumulators the same evil principles have been followed (as in making electric motors too light and flimsy) and an accumulator of large storage capacity and minimum of weight is still the idea of the day. It is astonishing, considering the world-wide, and I may say indispensable, application of accumulators, that electricians have not yet found out what should constitute a good and reliable accumulator. In my experience of the same I should advise, before all, strength, or in other words, duration of life.

A number of British storage battery firms were created during the 1880s to manufacture Planté-type batteries. Perhaps the most influential of these was the D.P. Battery Company, established in 1889. D.P. is another example of a battery firm created by a larger electrical equipment manufacturer; in this case, Johnson and Phillips, who were a large cable manufacturer.[44] In 1889 Johnson and Phillips were collaborating with Drake and Gorham for the installation of electric lighting plants. Drake and Gorham was a firm set up in 1886 when Bernard Drake and his assistant John Gorham left the employment of EPS to create their own firm to contract for the installation of private lighting plants.[45] Johnson and Phillips were already making batteries of their own in a small way at their cable plant. This plant, however, was becoming overcrowded, and it was consequently decided to set up a separate company to handle the battery manufacture. Drake and Gorham were chosen to do this.

The initials *D.P.* stood for "Dujardin-Planté". P.J.R. Dujardin was a French inventor who had patented a series of Planté battery modifications in 1886.[46] None of these designs were

---

[43] *TJER* 22 (1888): 8.
[44] Barak, "History," p. 3.
[45] Barak, "History," p. B-10; *JIEE* 60 (28 February 1922): 437.
[46] Wade, *Secondary Batteries*, pp. 17, 96, 124.

very original, but Johnston and Phillips nonetheless purchased the British rights and used the Dujardin patents as the basis of their manufacture in 1889. The original Dujardin plates were built up of a series of lead strips held in a lead frame, like that shown below:[47]

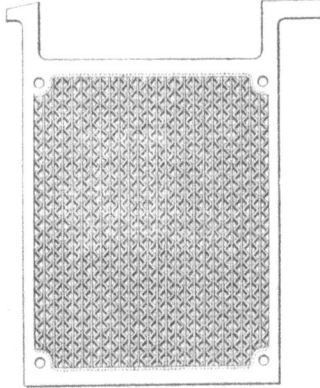

By the mid 1890s, however, the D.P. Company had given up these strip electrodes. They had been simple to manufacture, but were not durable. Consequently, by the mid 1890s the D.P. company had adopted a mixed-system approach; it used a pasted grid for the negative and a Planté-type plate for the positive. In the positive, the "ribs" were casted integrally on the plate itself. The pasted negative was a typical double-grid:[48]

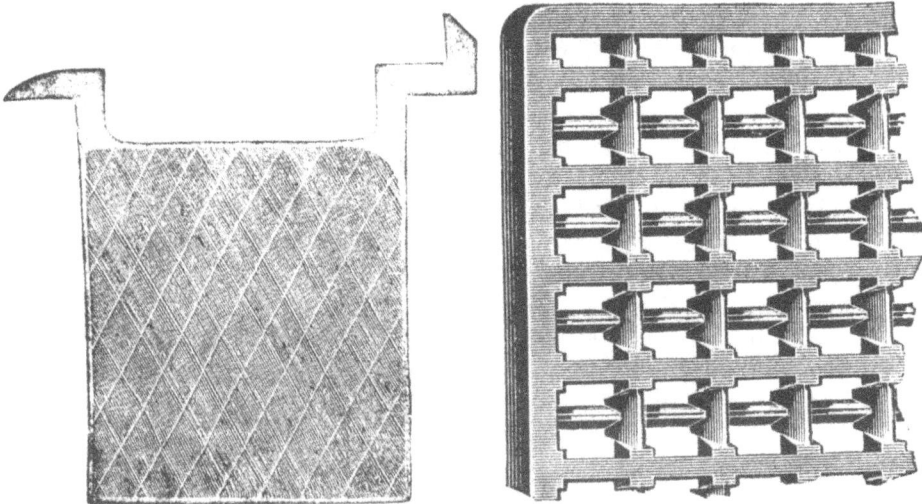

---

[47] Ibid., p. 95.
[48] Ibid., p. 402.

This is another example of the convergence of storage battery design ideas in the 1890s; the D.P. mixed system was very similar to that of AFA in Germany—the use of a pasted, double-grid negative and a "ribbed" Planté-type positive.

The first company to manufacture Planté batteries in England was the Wolverhampton Electric Light, Storage, and Engineering Company. Wolverhampton was created as a subsidiary of the Elwell-Parker Company, which was a smaller version of Anglo-American Brush; a firm for manufacturing electric lighting equipment such as dynamos, and for contracting the installation of complete plants. Again we see the pattern of the successful storage battery firm spun-off from the larger electrical equipment manufacturer and contractor. The founders of Elwell-Parker were two engineers; Paul B. Elwell and Thomas Parker, of whom Parker was the most significant.

Thomas Parker was a mechanical engineer who began his career with the Coalbrookdale Iron Company.[49] His early years were spent designing traditional mechanical engineering devices[50], but in 1876 Parker was given charge of the Coalbrookdale firm's electroplating operations. Here he got his first education in electricity, designing a dynamo to carry out the plating. Then, when the first practical incandescent bulbs appeared, Parker turned his attention in the direction of electric lighting.

In 1882 Parker and Elwell took out their first patents for batteries. The inventors appear to have been uncertain at first about the method to be used, since the patents contain ideas for both Planté and pasted-type batteries.[51] Nevertheless, when the company actually began manufacture, they chose to concentrate on the pure Planté form. The cells consisted of a nested set of concentric lead cylinders, each cylinder being alternately plus and minus.[52]

Before being given Planté "formation," the cylinders were soaked in a solution of dilute nitric and sulphuric acid for twenty-four hours. This soaking had two important functions: it dissolved metallic impurities out of the plate's surface and also roughened the plate microscopically, thus making it more

---

[49] *Electrical Engineer* (London), 22 December 1893: 581; *Electrician Electrical Trades Director,* 1898 ed., p. lxxv.

[50] In 1877, for example, he designed a steam pump which the Coalbrookdale firm manufactured.

[51] British patent #2917—20 June 1882 (Faure type); #3710—4 August 1882 (Planté type).

[52] *The Electrician* (London), 23 June 1883: 131.

susceptible to Planté "formation." The use of such chemical preparation of Planté plates to speed their "formation" became popular with commercial battery makers after 1881. Planté himself never used the technique until the 1880s, but it rapidly became a "state of the art" method in the industry with the development of commercial Planté forms. The most popular method was the nitric-sulphuric method described above, but several other oxidyzing agents were also proposed.[53] Although the development of large surface area electrodes was of greater importance in the commercialization of the Planté technique, the chemical preparation methods were also important.

It would appear, however, that the original Elwell-Parker cells were not very good. Elwell-Parker and EPS began a low-key legal battle in 1884, with John Sellon claiming that a clause in his 10 September 1881 patent anticipated the Elwell-Parker preparation technique.[54] In the manner typical of the British storage battery industry, the matter was settled by a drawing-together of the two firms. In 1886, Elwell-Parker agreed to give up its original design and license the EPS technology.[55] Two years later the Wolverhampton works was producing a battery essentially identical to that of EPS, with pasted grids for both electrodes.

In 1889 the two firms became more closely linked when a new company—the Electric Construction Company—was formed by Parker and a group of London promoters to buy control of EPS and Elwell-Parker.[56] ECC was organized to exploit the electrical products produced by its two subsidiaries. Both the Wolverhampton and Millwall factories continued to manufacture batteries, but from this time on they were, for all practical purposes, a single operation, with identical technologies.

Elwell-Parker's abandonment of their original battery design in favor of the technology licensed from EPS illustrates an important point about the general evolutionary process of storage battery technology. No company which did not specialize in battery manufacture was ever successful in producing a practi-

---

[53] In addition to Elwell-Parker, Planté himself patented the process in France in 1882 and van Depoele and C.A. Smythe patented it in the U.S. in 1883. Clarke took out another British patent on the same process in 1882. Other techniques patented in these years were to heat the plates in contact with sulphur or boil them in acetic acid. However, only the nitric acid method was used commercially.

[54] *TJER*, 27 December 1884: 515.

[55] *The Electrician* (London), 25 March 1887: 450.

[56] Although owned by ECC, these two firms preserved their independent status.

cal storage battery on its own. As we saw in the work of Bernard Drake, the complexity of battery technology requires full-time attention if satisfactory products are to be produced. EPS devoted its full attention to battery design; Elwell-Parker did not. The latter firm made a diversity of electrical goods and Parker himself was chiefly interested in dynamo design. Parker, who was the chief engineer and manager of ECC, seems to have realized this reality of battery engineering in 1889 when ECC was created—he restricted himself to dynamo design and installation from that time on and battery development was left to the EPS engineers.[57]

Despite the abandonment of the Planté plate by Elwell-Parker, this method of preparing commercial storage batteries achieved a secure position in the overall technology of storage battery manufacturing by the late '80s. Manufacturers used both the Planté and Faure methods for making plates and learned that the different characteristics of the two types could be adapted for different applications. This mature technological outlook, which emerged in the early 1890s, replaced the simplistic debates of the 1880s, in which the exponents of the two methods had each tried to prove their method superior in an absolute sense. The use of the mixed system approach, (pasted negatives and Planté positives) by such companies as AFA and D.P., is one example of this maturation.

Moreover, Planté battery advocates learned, just as their pasted-plate colleagues did, that the parameters of battery construction must be carefully controlled. Plates of inherently random structure, as for example the "felted" plates, were found to possess inherently random durability, capacity, and other properties. Since late nineteenth century electrical engineering was developing on quantitative, mathematical lines, with precision as its goal, this kind of imprecision and unpredictability was unacceptable. Therefore, those companies which made the random structure type Planté plates in the 1830s either went out of business or switched their technology.

An example of this evolution is illustrated by the history of the Crompton-Howell Electrical Storage Company. This company was created by R.E.B. Crompton, one of the leading British pioneers of electric lighting and one of the strongest exponents of the use of storage batteries as incandescent lighting adjuncts. Crompton had been involved in electrical

---

[57] *The Electrician Electrical Trades Directory*, 1898 ed., p. lxxvi.

equipment manufacturing since 1878, first in arcs and then incandescents. At first, Crompton used EPS batteries in his installation. In 1883, however, Crompton became interested in a storage battery patent taken out by F.T. Williams and J.C. Howell that same year.[58]

In this process, molten lead was allowed to cool slowly. As it cooled it was stirred. When the first crystals formed, they were ladled out with a strainer and placed in flat molds. When these molds were full, molten lead was sprinkled over the mass of crystals to produce connecting links. The resultant mass was a sintered sheet of coherent but porous crystals of metallic lead. These thin sheets were then sawn into blocks of standard size and frames of lead-antimony alloy cast around the edges to hold them in place in the battery box.[59] Following assembly, the plates were given Planté-"formation."

Crompton established a factory at Llanelly in South Wales to manufacture the cells and appointed Howell as manager.[60] This was a wise decision, since Howell was able to concentrate his entire attention on battery production. The Crompton-Howell batteries figured prominently in many of the first British central stations set up after the reform of the Lighting Act in 1888. Many of the plates in these batteries survived into the twentieth century, despite the fact that manufacture ended in the mid to late 1890s.[61] The Crompton-Howell batteries were generally given high marks by technical writers in the 1880s and 1890s.[62]

In 1902, however, it was reported that the manufacture of these cells had been discontinued for some years, the reason being the unreliability of their performance.[63] Some plates performed superbly, while others collapsed fairly soon. From the description of the manufacturing processes, it is easy to imagine the problems of maintaining proper quality control.

A comparison between some English Planté plates commercially manufactured at the turn of the century and the then

---

[58] Wade, *Secondary Batteries*, p. 102; David Salomons, *The Management of Accumulators* 1: 55.

[59] *The Crompton-Howell Electrical Storage Co.*, (advertising pamphlet) (London, 1892).

[60] R.E.B. Crompton, *Reminiscences* (London, Constable, 1930) p. 148.

[61] Wade, *Secondary Batteries*, p. 400.

[62] For example, *Electrical Engineer*, 27 March 1891: 314-315.

[63] Wade, *Secondary Batteries*, p. 400.

defunct Crompton-Howell plates illustrates the evolution towards greater control of reproducibility to maintain quality
control. The following section of the Pritchett and Gold positive plate is an example:[64]

Pritchett and Gold was an English firm which began manufacture of storage batteries in 1889 when it bought the assets of
the then bankrupt Eliesen Company.[65] By the end of the century, Pritchett and Gold was manufacturing batteries under
the patents of Dr. A. Lehmann, which it was licensing from the
Berliner Accumulatoren und Elektricitäts Gesellschaft.[66] A
number of English and American companies began importing
German storage battery technology after the mid 1890s. For
example, AFA licensed the creation of a British branch in the
late 1890s called the English Tudor Company. The following
view of this firm's positive plate at the turn of the century

---

[64] Ibid., p. 418.

[65] Barak, "History," p. T2-3.

[66] Wade, *Secondary Batteries*, p. 417: German patent #70,708—11 January 1893;
#72,199—30 May 1893; #74,068—14 April 1893.

again shows the movement towards more regular Planté plate
design:[67]

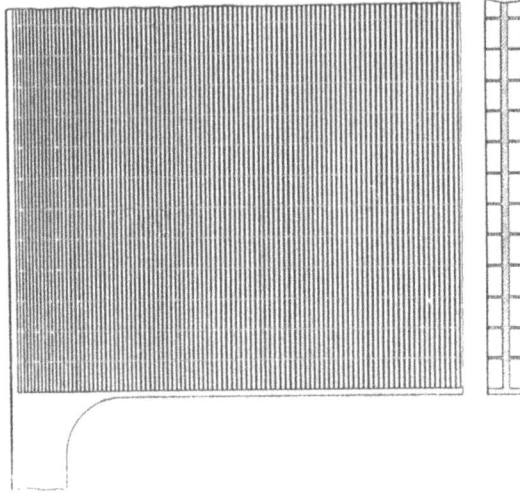

[67] Wade, *Secondary Batteries*, p. 411.

4. The First American Storage Batteries

Interest in storage battery design was not a European
monopoly during the 1880s; American inventors were equally
fascinated by the possibilities of this device. Despite this, the
early evolution of the storage battery was almost exclusively a
European achievement. Not until the mid 1890s did Americans
make serious contributions to this technology. This American
lag in storage battery design is in contrast to the history of
almost all other phases of early electric lighting and power
technology, in which American contributions were either on a
par with or ahead of those of the Europeans, for example,
dynamos, arc lights, incandescent lamps, transformers,
motors. In this section and in the following chapter we will
examine the causes for this lag.

Briefly stated, the reasons for the American lag were
economic. Until the early 1890s the only economically significant use for storage batteries was as adjuncts to electric lighting systems, and until the early 1890s American resistance to
the use of batteries for this purpose was strong. In contrast to
the situation in Britain and Germany, where many of the lead-

ing promoters of incandescent lighting supported the battery, and where a number of the leading electrical equipment manufacturing firms financed battery companies and used their cells in the lighting installations they designed, the leading American incandescent lighting people regarded batteries as an unnecessary nuisance. Consequently, the American battery promoters and companies had little market.

Unfortunately, it is difficult to provide precise documentation for the relative failure of American battery manufacturing during the 1880s; industries which fail leave few records behind. Nevertheless, we do possess numerous statements by contemporaries commenting on the strong contrast between the European and American battery business in the 1880s and 1890s. For example, when the British Post Office Telegraph's head and electrical authority W.H. Preece took a cook's tour of the United States in 1884, he reported back to the London Institution of Electrical Engineers:[1] "The use of secondary batteries has not received so much attention in the States as on this side of the water."

When he made a return visit nine years later in 1893, Preece had similar comments to make: "The accumulator in America has been a failure for electric lighting, and a very strong prejudice exists against its use. It has led to much litigation and the best forms are unknown there."[2]

These comments are borne out by Professor George Forbes, who toured the United States in 1888, and reported that:[3]

> It is a curious thing, that in America there is no work hardly being done with accumulators. . . . I was struck with the fact in 1884 and still more in the present year, that accumulators do not seem to work in the American atmosphere; at any rate, people have been importing numbers of accumulators and yet they are not recognized as the satisfactory apparatus we have come to regard them as (in Britain). . . .[4]

Percy Sellon[5], responding to Forbes's comments, observed that:[6]

---

[1] *The Electrician* (London), 13 December 1884: 95.

[2] *JIEE* 23 (1894): 79.

[3] *Electrical Engineer* (New York), 11/1888: 530-531.

[4] This statement, incidentally, gives further evidence of the great strides made by EPS since 1882.

[5] Robert Percy Sellon. British electrical engineer, 1864-1928. Early member of Anglo-American Brush.

[6] *J. Soc. Teleg. Eng. & Elec.* 17 (1888): 449.

. . . . I do not attach importance to the fact that in America the transformer system is the only one which has made any way (i.e. the A.C. system), for the Americans have practically only one company who are devoting themselves to making storage batteries, and they came over to England only a year or two back to learn how to use them. If the same amount of brains and time had been expended on batteries in America as on transformers, we would no doubt have heard from Professor Forbes of the hundreds of tons of batteries he saw, as well as the large quantities of transformers.

Sellon was wrong in placing the blame for the failure of the American storage battery on the inventors and the battery companies themselves. A graph of the number of patents granted for storage batteries, parts thereof, and equipment, shows that American enthusiasm for batteries began about the same time as it did in Britain and about as strongly:[7]

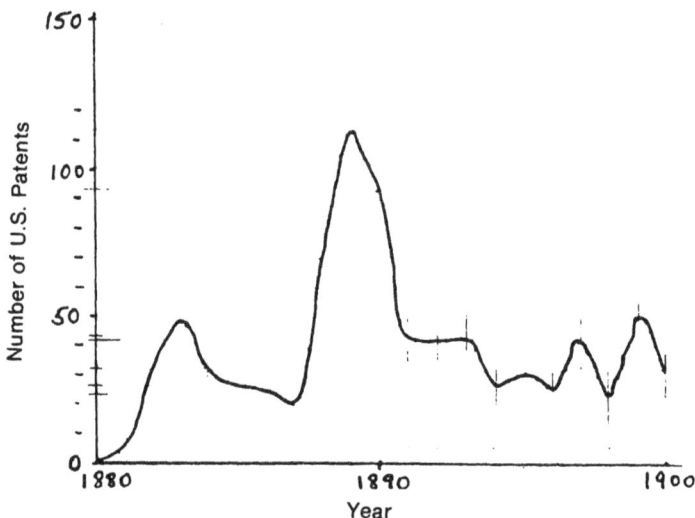

The earliest storage battery patent by Charles Brush was filed 12 June 1881 and the first Camille Faure application in the United States was filed 20 April 1881. A good number of other early American battery enthusiasts filed for patents soon thereafter. On 5 July 1881 James Maloney filed an application for an almost perfect copy of the original Planté battery.[8] Eli

---

[7] U.S. Patent Indices, 1880-1900.
[8] U.S. Patent #247,934—4 October 1883.

Starr, an indefatigable although unsuccessful early storage battery promoter, filed his first application 5 August 1881.[9] Charles Van Depoele, Nathaniel Keith, W.A. Shaw, and Alfred Haid were other American inventors who filed applications in 1882 or early 1883.[10]

Although the records are unclear, it appears that a good many of these early American patents were reduced to practice and the batteries sold commercially. The scale of such economic activity, however, was small. Several of the abovementioned patentees, for example, created companies to commercialize their patents, they were small and short-lived. James Maloney created the American Electrophore Company in Washington D.C. in 1881, but it soon disappeared. Eli T. Starr and his son E. Eugene set up the Starr Electric Storage Company in Camden, New Jersey, in 1882.[11] W.H. Preece observed Starr batteries in use in some private lighting plants in 1884 and they were given some good marks for quality by contemporaries. Nonetheless, the company disappeared sometime after the mid 1880s.

Alfred Haid was an Edison employee whose earliest storage battery patents were assigned to the Edison Electric Light Company.[12] He left Edison in late 1882 or 1883 to create Haid's Electrical Storage Company in Newark for the purpose of working his various imaginative patents.[13] Although Haid continued for some time to design and manufacture primary batteries, he gave up on storage batteries about 1885.

Nathaniel Keith created the American Electric Storage Company in Newark in 1883.[14] It does not seem to have survived 1884, perhaps because Charles Brush was given priority over Keith in patent interference litigation, but more probably because Keith was involved with other affairs after 1883.[15]

None of the companies mentioned above was linked with any larger electrical manufacturing concerns, nor, as far as can

---

[9] U.S. Patent #250,764—13 December 1881.

[10] U.S. Patents: #282,414—31 July 1883 (van Depoele); #285,529—25 September 1883 (van Depoele); #266,262—17 October 1882 (W.A. Shaw); #273,855—13 March 1883 (N.S. Keith).

[11] U.S. Patent #268,308—28 November 1882.

[12] U.S. Patent #271,628—6 February 1883—filed 13 September 1882.

[13] U.S. Patents: #294,463—4 March 1884—filed 12 September 1883; #294,464—4 March 1884—filed 17 February 1883; #296,164—1 April 1884—filed 29 December 1883.

[14] *The Electrician* (New York), 1/1883: 185; *The Electrician* (London), 7 April 1883: 481.

[15] Keith went west in 1884 to develop electrical machinery for mining.

be determined, was there any agreement between any of these companies and the contractors of electric lighting plants for the inclusion of their batteries in installations. They doubtlessly possessed very little capital for research and development and were therefore poorly placed to overcome the prejudice of American electrical engineers against storage batteries.

This argument is supported by the history of the first American battery company to achieve some degree of commercial and technological success—Electrical Accumulator Company. In May 1882, agents of Force et Lumiere arrived in this country to sell the American rights to the Faure patent and brought sixty cells of Faure batteries along with them. These were of the perforated-plate design, but still retained the old felt wrappers.[16] The agents held similar public demonstrations to those that had been conducted in Europe—running motors and operating incandescent bulbs from the cells. As in England, their demonstrations were accepted enthusiastically, and some half dozen American licensees of the Faure patents were created in various parts of the country.

The organizational structure of the companies descended from the Faure patents was complex. Force et Lumiere originally sold the rights for its American patents *for the whole United States* to a New York company called the Force and Light Company. Force and Light then sub-licensed manufacturing firms in the various sections of the country.[17] In the summer 1882, when the EPS and FEA agreed to share each other's patents and divide up the world market, EPS was given the United States.[18] Essentially, this meant the EPS received four-fifths of the stock of Force & Light. Force & Light benefited considerably from this change, since from this point on it was controlled by a well-capitalized, well-led, and innovative firm. At first, Force & Light did not see things quite this way, and when the chairman of EPS, J. Irving Courtenay, arrived in New York in 1883, Force & Light refused to acknowledge its control.[19]

---

[16] *Electrical World*, 5 June 1886: 262; *The Electrician* (New York), 3 February 1883.

[17] For example, the Western States Electric Storage Company and the Marine Electric Light and Storage Company.

[18] *The Electrician* (London), 24 March 1883: 454.

[19] The confusion about control may have been the result of the Philippart's actions. During the pre-summer patent fight between EPS and FEA, the Philipparts had spread a good deal of anti-EPS propaganda and Courtenay found that one of his chief problems in the United States was overcoming the residual influence of this propaganda. *The Electrician* (London), 26 April 1884: 573.

Courtenay, however, was able to bring pressure on Force & Light, so that by the time he returned to Britain in 1884, control was vested in EPS. The name was changed to Electrical Accumulator Company (hereinafter: EAC) and Courtenay was able to accomplish an important task: "to bring into the business some of the most substantial and shrewd business men in the United States, *with a considerable amount of American capital.*" (my ital.), as he reported to the EPS shareholders.[20]

From this point until its absorption into a new American battery firm in the 1890s, EAC prospered under the guidance of EPS. The advances in battery design made by EPS were at the disposal of the American company, so that by the late '80s cells made by EAC were not significantly different from those of EPS.[21] The relationship between the two companies was close. Courtenay secured the services of Theodore N. Vail as president. Vail was a skilled administrator in the electrical field, having been general manager of Bell Telephone from 1878 and one of the men most responsible for the organizational structure of the company.[22]

Of the five or six other licensee companies spun off the Force & Light Company, only one survived for any length of time. The reason for its survival was also linked to its alliance with the EPS. In 1882 the Viaduct Manufacturing Company of Baltimore licensed the Baltimore-area rights to the Faure patents from Force & Light. A subsidiary company called the Electric Storage Company was created to carry out the actual manufacture.[23] The Viaduct Company was the successor firm to Davis & Watts, a telegraphic equipment supply house. Viaduct continued to manufacture telegraphic and telephonic equipment, although its new president, Augustus G. Davis, oriented the firm towards the production of electric lighting equipment also.[24] Davis's acquisition of the Faure patent rights was part of this reorientation. This pattern of development was not unlike that of such firms as Siemens in Europe, although Viaduct was by no means as large.

Along with its acquisition of patent rights, the Electric Storage Company received a batch of the Faure cells which had been brought from France by the agents of Force et Lumiere.

[20] Ibid.

[21] *Electrical World,* 11 December 1886: 285.

[22] John E. Kingsbury, *The Telephone & Telephone Exchanges* (New York, Arno Press, 1972) pp. 183, 520.

[23] *Electrical World,* 4 April 1885: 135-136.

[24] *Electrical Review* (New York), 20 September 1883: 11.

Although these were perforated-plate cells, they proved no more durable than the original flat-plate Faure batteries. Mr. A.H. Bauer, the chief engineer of Electric Storage, described the performance: "These batteries gave satisfactory service for about six months, when they failed rapidly—first one, and then another, would refuse to store, until nearly all had become useless."[25]

Bauer took the problem in hand and soon introduced a few simple improvements. The felt separators were replaced by more durable strips of pine veneer. The perforated plates, whose holes had been rather crudely produced by means of punches, were not cast in iron molds.[26] Battery performance was improved as a result of these changes, but Bauer was still not able to produce a commercially viable cell. This came only after a visit in 1884 to the Millwall factory of EPS where he learned their techniques. This visit was made possible by an agreement reached between Courtenay and Davis in 1883. On returning to Baltimore, Bauer thoroughly changed the designs of his batteries. The new design "showed a great improvement over the first made. It was found to be thoroughly reliable. It would retain its charge for months without practical loss."[27]

Most of the early American electric lighting pioneers were wary of the battery and did not include it in their systems. The one exception was Charles Brush. He, like R.E.B. Crompton in England, saw the battery as a valuable adjunct to a lighting system, and, also like Crompton, he decided to manufacture his own batteries for use in conjunction with his other lighting equipment. Unlike Crompton, however, Brush did not conceive of the battery purely in terms of its value for incandescent lighting.

Most of the early batteries were conceived for making incandescent lighting more practical, or for driving self-propelled streetcars or boats. Brush, however, was one of the pioneers in the commercialization of arc lights, and his idea was to use the battery as a way of making that form of electric lighting more economical. When the first successful incandescent lamps appeared, Brush thought of combining an arc with an incandescent system, operating both from the same dynamos, and thus achieving considerable economy.

---

[25] *Electrical World*, 5 June 1886: 262.

[26] Ibid.

[27] Ibid. We must, however, allow for some exaggeration in Bauer's statement. No storage battery of the nineteenth century could possibly have stood for months without significant, if not total, self-discharge.

It was thought that a mixed lighting system would raise the profits by increasing the earnings/capital cost ratio of an electric power station. Since arc lights were only used at night, the steam engines, boilers, dynamos, at the power station lay idle during the hours of daylight. If, however, incandescent circuits were added to the power house, then current could be generated during the daytime to charge up banks of storage batteries. At night, the dynamos would be switched from the batteries to the arc circuits, and the batteries would run the incandescents. Thus the generating equipment could be used to produce earnings during the entire 24-hour cycle.

Charles Brush was not the only inventor to experiment with this tempting idea, but his experiments were by far the most extensive. Although the mixed lighting system eventually failed, it excited a considerable degree of attention during the early 1880s. Thomas Edison, for example, worried that the mixed system threatened his own system and excoriated the storage battery in remarks quoted earlier. In 1883 Edison began a campaign against the Brush Illuminating Company of New York, denouncing their system as unsafe because of the high voltages used.[28]

Edison had good reason for his fear of the mixed system, for although he had effective control of the American incandescent patents, he did not hold the arc light patents. Until the development of the tungsten filament bulb in 1910, incandescent bulbs could not compete with arcs for street lighting. Since city streets had to be wired for arc circuits anyway, the adoption of a single mixed system was more attractive than the installation of separate indoor-outdoor systems on the same street.

Brush introduced his system to the public at a demonstration on 19 December 1882 in New York, with a large number of men present who were interested in the future of electric lighting.[29] Brush showed off his new batteries, as well as an automatic switch he had designed which was supposed to switch the cells to the charging circuit automatically when the voltage fell below a minimum value. Such a device, Brush claimed, would make it unnecessary to have human attendants control-

---

[28] *Electrical Review* (New York), 22 November 1883. In 1883 Brush Illuminating began using its 1000 volt arc circuits to charge batteries of 500 cells of storage batteries in series at sub-stations. The batteries were discharged in parallel in groups of 50 each. E.H. Johnson of the New York Edison Company, acting as Edison's mouthpiece, denounced the 500 battery series as "deadly." Edison was to use the same arguments years later in his more famous anti-A.C. fight.

[29] *New York Times*, 20 December 1882: 8.

ling the charge/discharge functions. He also indicated that batteries were "practically indestructible."

During the next two years about half a dozen commercial installations of the Brush mixed system were carried out. In 1883 the Willamantic Thread Company of Connecticut installed a set of 250 Swan lamps[30] and Brush storage batteries which were to be operated from the Brush arc circuits which had been installed five years previously.[31] The Brush people were particularly proud of this system, since the thread mill had originally put in an Edison private incandescent plant which was removed to install the Brush set.

The most famous of the Brush installations, however, and the one which became the test case for the system, was that in Cheyenne, Wyoming. The Brush Swan Electric Light Company of Cheyenne was incorporated in the fall of 1882 and the system began operating its dynamos, arcs, incandescents, and batteries in early 1883.[32] By mid 1883 the system covered a district of about a mile square and supplied 1,500 incandescents. A small, wood-frame house serving as a substation was located on every block of the district. This contained the batteries and the automatic battery switches for all the lamps on that block.[33]

The system soon ran into trouble. A chief problem was the automatic charge/discharge switch. These switches were of a simple design which performed only one function; the batteries were either switched to the dynamos for charging or to the lamp circuits for discharge. Storage batteries, however, are complex electrochemical systems whose voltages continually change both on charge and discharge. Accordingly, any switching system must "tune" the changing characteristics of the battery system to the unchanging characteristics of lamps and generators. A simple on-off switching system is therefore insufficient.[34]

Charles Brush was not the only electric inventor to propose the automatic control of storage battery systems by means of such simple, on-off switches. Numerous American and Euro-

---

[30] As part of this system Brush had an American license on the Swan lamp patents.

[31] *The Electrician* (New York) 2 (1883): 347; *Electrical Review* (New York), 17 January 1884: 2.

[32] *Electrical World*, 25 July 1885: 35.

[33] Ibid.

[34] U.S. Patent #281,175—10 July 1883—filed 19 June 1882; *Electrical Review* (New York), 10 January 1884: 4.

pean patents appeared during the 1880s which played variations on this theme.[35] The simplicity of these devices shows that many electrical engineers did not yet realize the complexities inherent in electric power engineering. Most of these engineers had gained their experience in telegraphic engineering, where such problems as the control of primary batteries were much less complicated than they were becoming in the field of incandescent lighting and power house operation.

Since the Brush automatic switches were not satisfactory, the Cheyenne company resorted to manual control of the batteries. Each substation had an attendant to watch the dials and make the proper adjustments in battery operation. This was uneconomical, since one man was required to serve only a handful of consumers on each block and in 1885 the manager of the system admitted that this use of the storage battery was impractical.

Not only were the control devices impractical, but the storage batteries themselves were also very poor. In itself, this is not surprising, since all commercial storage batteries made in the early '80s deteriorated rapidly. Brush, however, never succeeded in producing a practical storage battery, even though he continued to be involved with the storage battery industry until the early 1890s.

The technological reason for the failure of the Brush storage batteries is clear from an examination of his patents. Brush designed a process whereby perforated lead plates would be packed with lead granules. These granules would first be given a surface oxidation and then would be subject to intense hydraulic pressure to fuse them together and into the walls of the grids. Such a process will not produce coherent plates.[36] Brush manufactured his batteries for a number of years, selling them for a variety of purposes including the lighting of railroad cars. They were invariably given poor marks, however, and by the late 1880s Brush battery installations were being replaced by the superior products of the EAC. The reason for Brush's failure to develop a quality battery was the failure of his mixed lighting system itself. During the brief period of enthusiasm for this system, Brush invested much time and energy on battery design. He took out nine storage battery patents in 1882 and six more in 1883. Then, in 1884 when the system encoun-

---

[35] For example, U.S. patents: #393,147—20 November 1888; #394,642—18 December 1888; #418,748—7 January 1890.

[36] U.S. patents: #262,533—8 August 1882—filed 27 May 1882; #266,762—31 October 1882—filed 20 July 1882; #275,986—4/17/1883—filed 5/27/1882.

tered setbacks, only two patents appeared. There were none in 1885; two in 1886, and no more until a final one in 1891.[37]

Storage battery engineering is an inherently demanding skill, requiring full-time specialization to produce a commercial product. All those companies which succeeded in the field were specialists. With Brush, on the other hand, storage batteries were only one of his electrical products. Unlike Crompton, he did not turn the design of batteries over to a subsidiary company, but designed and manufactured them himself.

The history of the Brush battery would be an unimportant part of our story were it not for the unfortunate fact that Charles Brush held the fundamental American patent for pasted, perforated-plate storage batteries. In an almost continuous series of patent interference cases held throughout the 1880s and early 1890s, the basic Brush patents were awarded priority over all other American patents. Unlike the situation in Britain, where such conflict was usually settled out of court by amalgamation or purchase of patents, the American industry wasted much capital on these endless legal proceedings. The uncertainty which was caused by these infringement fights also had a retarding effect on the willingness of American engineers to invest in batteries.

Electrical engineering journals often blamed the poor showing of the early American storage battery industry on this legal warfare. It is true that these problems had a depressing influence on the industry and a number of instances can be cited of the cessation of storage battery projects as the result of the unfavorable outcome of a trial. On the other hand, it is uncertain if these legal problems were a major factor in the slow evolution of the American industry during the 1880s. Other parts of the contemporary electrical industry equally plagued by legal fighting advanced rapidly nevertheless; the incandescent bulb industry is perhaps the best example of this. A number of self-propelled storage battery streetcar experiments were terminated after unfavorable court proceedings, but since these experiments were commercial failures anyway, the legal proceedings could merely have been an excuse to stop.

In 1889, the Brush company was sold to Thomson-Houston, who thus obtained the patent rights to the Brush battery. They

---

[37] *1882* (#262,533; #260,653; #260,654; #261,512; #261,995; #264,211; #266,090; #263,756; #266,089) *1883* (#274,082; #275,985; #275,986; #276,155; #274,905; #276,348) *1884* (#293,709; #293,710) *1886* (#337,298; #337,299) *1891* (#445,422).

appear to have done nothing with the patents, except to continue infringement proceedings. The Brush patents were an undoubted nuisance for the early American industry, but had there been a substantial demand for batteries in the United States during the 1880s, it is likely that a means to avoid this roadblock would have been found.

During most of the 1880s, however, there was no substantial demand for batteries. The rapidly-falling curve of the graph on page 112 indicates a sharp decline in American interest in storage batteries after 1883. This is borne out by other indicators; a number of the companies created before 1883 disappeared afterwards, and during the period from the end of 1883 until the end of 1886 no new American storage battery companies were formed. At the beginning of 1887, however, a sudden upsurge in the patenting rate occurred, accompanied by the creation of new companies and the expansion of some old ones.

It is tempting to speculate that the downturn in the battery business after 1883 was due to the bad reputation established by the early commercial cells. Such an explanation does not explain, however, the much sharper decline in the patenting activity after 1889, by which time a number of relatively good American cells were on the market, nor does it explain the sharp upturn in 1887. The explanation appears to lie rather with the state of the entire American economy. The 1883 and 1889 peaks on the graph both occurred about a year after major peaks in the American economy; the lag was due to the time between filing and granting of patents.

As Jacob Schmookler pointed out in his *Invention and Economic Growth*, the rate of innovative activity, as measured by the patenting rate, is a direct function of the economic demand for the technology. Storage batteries, as used in those days, were capital goods, and it is characteristic for capital goods spending to fall most sharply in time of economic downturn. It is significant to note that the late 1880s were the most intense period of storage battery interference fights, and yet these years also saw a strong period of patent application and company creation. The patent litigation, therefore, although disruptive of individual firms, does not seem to have had a strongly negative influence on the entire industry.

The company which grew most rapidly after 1886 was EAC. In September 1886 the company erected a large, new factory in Newark, New Jersey. According to the press release, this was the first American factory devoted exclusively to storage battery manufacture.[38] At the same time that it was erecting this factory, EAC began the creation of a network of franchised dealerships throughout the country. These dealers were electric lighting contractors who installed complete incandescent plants in the same manner as the EPS in Britain. Each dealer bought the batteries and associated equipment from EAC, and the dynamos, wiring, from other manufacturers. Between January 1887 and December 1888, the following EAC dealerships were created:[39]

Date	Company	Territory
1/87	Elec. Storage, Light & Power Co. East St. Louis, Ill.	West (undefined)
1/87	L.C. Kinsey Co., Penna.	Penna & surrounding states
4/87	Pacific Accumulator Co., S.F.	Pacific Coast
5/87	New York Isolated Accumulator Co.	New York
3/88	Northwestern Accumulator Co., Chicago	Northwest
4/88	Atlanta Accumulator Co.	South
11/88	Maine Accumulator Co.	Maine
12/88	Boston Accumulator Co.	Mass. & surrounding

In addition to these franchised firms of EAC, a series of smaller, independent storage battery companies were also formed in this period, for example, Macraeon Storage Battery Company of New York (10/88); Johnson Storage Battery Company of Boston (2/89); Gibson Storage Battery Company of New York (1886); River & Rail Electric Light Company of West Virginia (1887); Electric Storage Battery Company of Philadelphia (1888); Woodward Electrical Company of Detroit (7/88); Bradbury-Stone Storage Battery Company of Lowell, Mass. (88 or 89).[40] This list is not exhaustive.

---

[38] *The Electrician & Electrical Engineer* (New York), 9/1886, p. 353. The health safeguards in these early lead factories were minimal. Workmen complained about the "lead cholic," but rather than ask for safeguards, they demanded more pay as recompense.

[39] *Electrical World*, 15 December 1888: 318; 24 March 1888: 156; 14 May 1887: 234; 25 June 1887: 308; *The Electrician & Electrical Engineer* (New York), 4/1887, p. 157.

[40] *Electrical World*, 4 May 1889: 263; 23 February 1889: 110; 27 October 1888: 232; 21 July 1888: 33.

Unfortunately, no statistics of storage battery manufacture or sales survive from the 1880s. It is apparent, however, that American sales of batteries increased substantially in late 1887 and 1888. In the electrical engineering journals, reports of sales of batteries before this time were very sparse. Beginning in late 1887, however, many American battery companies reported excellent sales. All of the franchise firms of EAC, for example, reported excellent business in 1888. The Maine branch at Bar Harbor reported closing contracts for five small lighting plants batteries; the Boston office reported three contracts; and New York Isolated was working on a yacht and was beginning to enter the catered electricity field—they provided the light for several balls, including a massive one for Cornelius Vanderbilt which required 100 large batteries.[41]

The president of EAC reported good sales throughout the East in 1888. Things were equally good in the West; the Chicago office had contracts for two central station batteries and the St. Louis franchise was installing a plant for a large office building.[42] Moreover, it is clear from internal correspondence between L.C. Kinsey Company and the EAC that the optimistic press releases in the technical journals were not an exaggeration; Kinsey reported more business at this time than it could handle.[43]

In 1889, however, economic downturn began. Most of the new firms created during 1886-1888 disappeared. The EAC weathered the storm, but most, if not all of the franchises disappeared. By the early 1890s, the American storage battery industry was little better off than it had been during the mid 1880s. It had made significant progress technologically, but its economic base was almost non-existent. In 1892, the president of the Consolidated Electric Storage Company of New York gave the following appraisal of the current status of the business:[44]

The aggregate demand for storage batteries throughout the whole country is small at best; and it would require all the

---

[41] *Electrical World,* 26 January 1889: 49; 11 August 1888: 73; 24 November 1888: 281; 3 November 1888: 244; 14 July 1888: 22.

[42] Ibid., 1 September 1888: 111; 2 March 1889: 137; 31 March 1888: 169.

[43] Bell Telephone Company Letterbooks, Pennsylvania Historical & Museum Commission, Harrisburg, Pa., *passim*.

[44] Testimony of Charles Bracken, Brush Electric Co. & Consolidated Electric Storage Company, vs. Milford & Hopedale Street Railway Co., etc., 1892, United States Circuit Court, District of Mass.

customers for storage batteries in the United States to keep a factory of moderate size in full operation if indeed that would be sufficient to do so.

It was not until the mid '90s that the American storage battery industry entered a period of sustained and stable growth.

5. Storage Battery Transportation

During the past hundred years, the idea of using storage batteries to propel electric vehicles has gone through several cycles of enthusiasm and neglect. As this is being written in the late 1970s, the increasing cost of gasoline is stimulating renewed enthusiasm after many decades of neglect. The first high point for the idea of battery transport came in the early 1880s. At that time, cities were expanding rapidly and it was clear that new methods of urban transport were needed and would become increasingly more necessary in the years ahead.

As soon as Faure and Brush inventions appeared, the idea of using storage batteries to power self-propelled vehicles stimulated the imaginations of many European and American inventors. This idea was more obvious than that of using batteries as incandescent lighting adjuncts. This stemmed from the simplicity of the idea—the easy-to-manufacture pasted-plate cells would simply be slipped under the seats of conventional horsecars equipped with electric motors. William Thomson had publicized the great amount of energy that pasted-plate batteries could store; inventors calculated that battery-equipped electric streetcars could be made to run an entire day with only two or three changes of batteries.

The earliest battery streetcar experiments were conducted with Faure batteries in Paris. In 1881 the Compagnie generale des Omnibus, which owned almost all the Paris horsecar lines, provided Force et Lumiere with a car or two for its tests. The first trial was conducted in May 1881 when a car was driven with a battery of 225 cells weighing over two tons.[1] This was too heavy a load for the horsecar; horsecars were built lightly so as not to strain the animals. Therefore, when Force et Lumiere resumed the tests two months later they reduced the battery to only 160 cells.[2]

---

[1] T. Illingworth, *Battery Traction on Tramways & Railways* (Locomotion Paper #14).

[2] *TJER*, 15 July 1881: 269.

In the early spring of 1882, Force et Lumiere conducted the first experiment with in-plant industrial use of battery transport. The large Duchesne-Fournet bleaching ground wanted to mechanize the laying and take-up of cloth, but the use of a small steam locomotive was precluded, because of the smoke and cinders. Therefore, Force et Lumiere built a small electric locomotive, weighting 2,500 pounds, driven by a Siemens dynamo run backwards as a motor, and carrying 1,400 pounds of batteries in a small tender. The locomotive ran up and down the bleaching ground on small rails and drew in the cloth by attaching the motor to a small winch which drew in 10,000 square meters at a time:[3]

---

[3] The Electrician (London), 27 May 1882: 28-29.

These experiments by Force et Lumiere and by its successor company, the French Electrical Power Storage Company, continued into 1883. The results heralded an unhappy pattern repeated frequently by dozens of inventors during the next decade and a half. The batteries worked splendidly for a few runs, giving the motors plenty of power, but after a few charges, the electrodes started to collapse and refused to receive further charging. Soon, the active material coatings lay in flakes at the bottom of the cell box.

Rather than discouraging further experimentation in self-propelled streetcars, however, the Force et Lumiere work appears to have encouraged further experimentation. Like the automobile during the 1890s, the electric streetcar was a powerful, stimulating motif for inventors of the 1880s and early 1890s. The rapid collapse of the first propulsion batteries was of far less importance in the eyes of these men than their initial high power output, which seemed to prove that battery drive was at least potentially practical. As a result, battery streetcar enthusiasts throughout the next fifteen years emphasized the short-term economics of their system and downplayed durability and maintenance costs.

Several dozen inventors experimented with battery streetcars in the 1880s and early 1890s. Rather than give a catalog of the work of these various men, I have chosen to describe the work of the two most professional and successful of these inventors, since the history of their projects is typical of the general pattern—initial high hopes and favorable results, followed by inevitable failure. These two engineers were the Englishman, Anthony Reckenzaun and the Belgian, Edmond Julien.

Reckenzaun was originally a mechanical engineer. Like such early electrical engineers as R.E.B. Crompton, Reckenzaun did his first work in steam engine design, working on marine engines when he arrived in England in 1872.[4] The event which turned Reckenzaun's attention towards electrical engineering, as it did to Camille Faure, was the Paris Exhibition of 1878. Reckenzaun became interested in storage batteries and in 1881 went to work for Force et Lumiere, but later transferred to EPS.

The Millwall factory of EPS provided Reckenzaun with the facilities needed for his experiments. A circular track was installed for the streetcar work, and the location of the factory on the Isle of Dogs provided Reckenzaun with excellent opportunities for his electric boat work. The first experiments were unsuccessful, but by March 1883 Reckenzaun had a five-ton car, capable of carrying 46 people, running experimentally on a London horsecar line. It ran for seven hours on a charge, but went at a top speed of only six miles per hour—no faster than a horsecar. Two of its five tons were accounted for by batteries, which took up over thirty cubic feet of space. They were slipped into compartments below the passenger's seats; this was typical practice on almost all the storage battery streetcars.[5]

The EPS was encouraged by the initial results of these tests. They estimated the cost of a day's run by batteries and horses at $1.50 and $6.25 respectively.[6] These calculations are significant since they represented the first published figures "proving" the economic superiority of the battery streetcar system over horses. Many more sets of such calculations were published during the next two decades by various promoters, each showing the battery car to be cheaper than the horse or cable,

[4] Obituaries: *The Electrician* (London), 17 November 1893: 66; *Electrical Engineer* (London), 17 November 1893: 469.

[5] Thomas C. Martin & Joseph Wetzler, *The Electric Motor & Its Applications*, pp. 100-101; *The Electrician* (London), 17 March 1883: 430-431.

[6] Statement of John Sellon, *The Electrician* (London), 17 March 1883, p. 431.

and at least as cheap as the overhead conductor system. Such published data kept interest in battery cars alive.

The EPS's figures, giving batteries a four-fold plus advantage over horses, look so good that we might expect exaggeration. This does not seem to have been the case, however. The horsecar statistics were supposedly obtained from London streetcar companies and are, if anything, much too low. They work out at seven cents per mile. In 1893 Newcastle on Tyne horsecars were operating at an overall cost of 23¢ per car mile.[7] Birmingham cars operated at 20¢ in 1892; in Paris in 1893 at 18¢, and in the United States in 1890 at an average of 10¢ per car mile.[8]

In actual practice, however, battery streetcars never operated this favorably. Over the long term, when all expenses were figured in, battery cars were either as expensive as horsecars, or only slightly less so. The EPS, making the same mistake as numerous contemporaries, prepared its calculations on the basis of a few test runs and grossly underestimated deterioration and maintenance—EPS knew about the poor durability of the early traction cells, but they also knew that in only three years much progress had been made in storage battery design. Future progress, it seemed likely, would be just as rapid, enabling battery streetcars to operate as economically in the long-run as they did for just a few trial runs. Unfortunately, these sanguine projections were not to prove accurate.

Reckenzaun continued his experiments. By the end of 1884 EPS announced a new experimental car, whose battery weighed only 1.25 tons. In exchange for the weight advantage, however, the cruising time of the car was reduced from seven to two hours, thus increasing the labor cost of shifting the batteries. From the reported figures, it is clear that the company had learned much by this time about the actual running costs of battery cars. They reported an overall cost of operation of 3.5 pence per car mile, or $6.56 per day.[9] They also reported an annual depreciation rate for the batteries of 50 percent as compared with only 15 percent for motors, generating machinery and running gear. The top speed was still six miles per hour.[10]

---

[7] *Electrical Engineer* (London), 8 June 1894, p. 677.

[8] Philip Dawson, "Electric Railway & Tramways," *Electrical World*, 18 October 1890: 282-283.

[9] Calculated at 6 mph for 15 hours/day.

[10] *The Electrician* (London), 20 December 1884: 113; Martin & Wetzler, *Electric Motor*, pp. 101-105.

In 1885 Reckenzaun left EPS to set up his own electric streetcar and boat company.[11] He continued to use EPS batteries; EPS used some of Reckenzaun's patents to make special motive power cells, thus combining the advantages of the EPS experience and Reckenzaun's. The first public trial of a Reckenzaun car took place in Battersea in April 1885 on the tracks of the South London Tramway Company.[12] The technical press reported that the trial "proved that the scheme of working a tramcar by accumulators is perfectly feasible and practicable." This kind of press comment followed almost all initial battery streetcar experiments. At the same time, Reckenzaun began expanding his operations abroad. He had a number of cars running on three Berlin lines.[13] In 1887 he took his first trip to the United States and interested a number of American horsecar lines in his system. By the summer Reckenzaun had a car running in Philadelphia and another was operating in Sacramento by November 1888.[14] A month later EPS was even completing a Reckenzaun car for Melbourne.[15]

Reckenzaun's career ended shortly after this. He was suffering from tuberculosis and died in late 1893 at the age of 43. At the time of his death his cars were still being run experimentally, but no streetcar line ever placed them on regular service.

Somewhat more successful in commercializing his system was the Belgian engineer, Edmond Julien. He, unlike Reckenzaun, was originally connected with S.A. Force et Lumiere; he was director of Compagnie Belge et Hollandaise d'Électricité, which operated under Force et Lumiere patents. The earliest report of his battery streetcar work dates from June 1881, when Julien demonstrated a car in Brussels which carried one and one-third tons of his batteries.[16]

Julien exhibited his cars at the 1885 Antwerp Exhibition, where they ran for six months against competing steam and compressed air-driven streetcars. The Julien cars received the prize.[17] The lack of competition from electric cars using the

[11] Due to space limitations, the subject of the electric boat has regrettably been omitted from this study. For information on this subject, see, for example: Thomas C. Martin & Joseph Sachs, *Electric Boats & Navigation* ; Maurice Gaget, *La Navigation sous Marine*.

[12] *The Electrician* (London), 4 April 1885: 435.

[13] Ibid., 14 March 1885: 361; Martin & Wetzler, *Electric Motor*, pp. 103-104.

[14] *Electrical Engineer* (London), 8/1887: 375.

[15] *The Electrician* (London), 14 December 1888: 157.

[16] Martin & Wetzler, *Electric Motor*, p. 202; *The Electrician* (London), 7 June 1884: 94.

[17] *The Electrician* (London), 24 December 1886: 150-151; Martin & Wetzler, *Electric Motor*, p. 105.

overhead conductor system was significant in earning Julien favorable publicity. At the same time, the Compagnie Belge et Hollandaise operated cars on a Brussels streetcar line which had been using horses.[18] The company claimed a 30 percent saving in overall costs as against horses. These figures were probably accurate, since they are similar to other horse-battery comparisons made in the 1880s and 1890s. In any case, the streetcar company decided to retain the battery cars for regular service. They were only used on this small line, however, which had only three cars, rather than on any other Brussels line. The line was run on a half experimental, half commerical basis until 1890, when the company reinstated horsecars. They found that the long-term costs of battery deterioration more than offset the higher daily costs of horse operation.[19]

Encouraged by his initial successes and by the publicity his cells had received at the Antwerp Exhibition, Julien in late 1885 changed the name of Compagnie Belge et Hollandaise to the Julien Electric Company and began looking farther afield for places to market his inventions. A branch of the firm was opened in the United States in 1886 and this American firm soon overshadowed its European parent in its contributions to storage battery transporation. The Julien Electric Company was one of the very few electric traction concerns ever to install a fully commercial, non-experimental storage battery line anywhere in the world.

The American Julien Company was one of the large number of new storage battery companies which sprang up in the United States during the late 1880s. A number of American cities tried Julien cars experimentally in the late 1880s, but the most significant of these installations began in Manhattan in 1886.[20] In that year the Fourth Avenue Streetcar Company, a horsecar line, began running a nine ton Julien car on its Madison Avenue route. This carried about two and one-third tons of batteries and could go thirty-six miles on a charge.[21]

The Fourth Avenue Line continued the experiments for two years and then analyzed the comparative costs of horses and batteries, as well as cable costs from New York cable car com-

---

[18] *The Electrician* (London), 14 March 1885: 362.

[19] The Brussels cars had apparently worked well, but cost 4 centimes per car mile *more* than horses. This may have been because the Brussels line had steep gradients, which placed severe strains on the batteries. *Electrical Engineer* (London), 4 January 1889: 5; *Electrical Review* 41 (12 November 1897): 633.

[20] For example, in St. Louis in July 1887. *Electrical Engineer* (London), 7/1887: 354.

[21] Ibid., 15 January 1889: 138-140; 27 February 1891: 210-211.

panies. The overhead system was not considered, because of
municipal opposition to this system. The analysis led the company to fully commit itself to battery operation. The findings of
the analysis were as below:[22]

	Battery	Horse	Cable
Cost of Cars	1	.54[23]	.81[23]
Motive Power Cost	1	1.45	1.06
Roadway Construction Cost	1	.53[24]	2.09
Depreciation & Labor Cost	1	1.47	2.04
Operating Expenses	1	3.38	1.71
TOTAL	5	7.37	7.74

In September 1888 ten 18-foot cars were installed on the line,
together with a powerhouse for charging the cells midway on
the Madison Avenue line.[25] The design of this powerhouse
represented a major change in the operation of battery cars.
Hitherto, the cells were kept as individual units, with one or
two in compartments under each passenger's seat. They were
withdrawn one at a time through panels on the outside of the
car. This represented a high labor cost, considering the removal of 100 to 200, 30 to 40 pound acid-covered boxes and
their replacement with an equal number of such boxes. This
method of operation also probably shortened the life of the
batteries, since the bouncing of the cells in and out of the cars
was a source of deterioration.

The Julien cars, however, were not merely modified horsecars, but streetcars built specifically for battery service. The
batteries were held on special trays which could be lowered
from the bottom of the cars as complete units. The car was
positioned over pits in the powerhouse, where an electric
elevator removed the discharged trays and replaced them with
fresh ones, all within three minutes.[26] In 1889 the Julien company reported the results of its Madison Avenue operation.
The costs per car mile were suspiciously low—four and five-
tenths of a cent; almost certainly an exaggeration. These figures, however, as well as the speed of the cars—nine to ten miles

[22] *The Electrician* (London), 29 June 1888: 230.

[23] These lower costs were due to the fact that horse and cable cars, as well as the track
for these systems could be made lighter than those for batteries, since they do not need
to bear the weight of the lead.

[24] Ibid.

[25] *Electrical World*, 22 September 1888: 161.

[26] *Electrical Engineer* (London), 13 September 1889: 209.

per hour on the level—created increased interest in the Julien system in Europe.[27] An agreement between the EPS, the Frank Sprague Company and the Julien Company in Britain made possible the creation of a trial battery line in Birmingham in October 1889.[28]

The Birmingham line, however, was not a total commitment to one system like that of Madison Avenue. The Birmingham Tramway Company tried four systems on its various lines in the late 1880s and early 1890s and compared them—cable, horse, steam and battery.[29] Again, as in New York, overhead conductors were not tried because of municipal opposition. The Birmingham battery line continued to run throughout the 1890s, but was never a success. *Electrical World* reported in April 1897 that the Birmingham battery line was "financially unsuccessful."[30] Eventually, Birmingham changed all its cars to the direct conduction system, using the underground, or conduit system of conduction.

The Julien cars on the Madison Avenue Line were discontinued even earlier, in 1892.[31] The reason for this was the failure of the batteries; contemporary reports indicate that they could not bear heavy discharges. If heavy discharge were taken, the positive plate active material fell out of the grids.

Edmond Julien died in 1894, within a few months of Reckenzaun's death. The work of these two men was the most extensive of any battery streetcar work carried out before the mid 1890s. Nevertheless, dozens of other inventors, in the United States and in Europe, worked on such ideas during the 1880s and early 1890s. Unlike Reckenzaun's and Julien's work, most of these were ephemeral experiments, with one or two reports at most appearing in the technical press. Little can therefore be said about this work. The reports do indicate, however, the high level of contemporary interest in this technology.

The earliest American work on storage battery streetcars was conducted by A.H. Bauer of the Viaduct Company. When he visited the EPS's Millwall factory, Bauer was stimulated by

---

[27] *The Electrician* (London), 18 October 1889: 610.

[28] *Electrical Engineer* (London), 25 October 1889: 322-323.

[29] *Electrical World*, 17 April 1897: 536.

[30] Ibid.

[31] Accumulator Company vs. New York & Harlem Railroad Company, U.S. Circuit Court, Southern District of New York, 1892; *Electrical Engineer*, 22 August 1890: 147.

Reckenzaun's work. Returning to the United States, Bauer collaborated with Leo Daft, mounting his batteries in a Daft car in July 1885. The Daft car ran on some lines which the Daft company had established in Baltimore.[32] As with most such early cars, the battery was very heavy—the full car weighed six tons, of which almost half (5,400 lbs.) was battery. Moreover, the speed was no better than a horsecar—six miles per hour. Bauer, however, claimed that he could reduce the weight of the battery, and reduce operating costs of the cars to almost half that of horsecars.[33] As was typical of most such early promoters, Bauer greatly underestimated the rate of depreciation of a traction battery—he based his cost calculations on the depreciation of a *stationary* lighting battery he had set up in the cellar of the Baltimore Academy of Music two years previously.[34] The Daft-Bauer experiments ended sometime in 1886 or 1887.[35]

Storage battery traction schemes boomed at the end of the 1880s, after remaining at a low level of activity during mid decade. In February 1889 the Woodward Electrical Company of Brooklyn introduced a storage battery car, which ran experimentally on a Detroit line.[36] It received some excellent initial reviews but soon disappeared.[37] Elias E. Ries described a combined conduit-battery car in 1887, thus combining the highest maintenance cost system with the highest capital cost system.[38] It is unlikely that the system was ever tried in practice.

In 1889 the River & Rail Electric Company of West Virginia introduced a car and storage battery invented by William Main; a prolific patentee of battery designs. The car could go twelve miles per hour and supposedly had a successful run in Washington, DC. in March.[39] The Union Electric Car Company

---

[32] Martin & Wetzler, *Electric Motor*, p. 108-109; *New York Review of the Telegraph & Telephone*, 1885; *Electrician & Electrical Engineer* 6/1886: p. 215-216; 9/1885: p. 359.

[33] Bauer calculated $4 per car day for horsecars and $2.21 for battery cars. *Electrical World*, 15 May 1886: 225.

[34] *Electrical Review* (New York), 25 July 1885: 9.

[35] The association of Leo Daft with this battery scheme is interesting, since he is usually known only for his work with the direct-conduction system.

[36] *Electrical Engineer* (London), 22 February 1889: 147.

[37] The company guaranteed to keep the battery in repair for a premium of 10 percent per annum for 17 years, which must have been a losing proposition. Ibid., 19 April 1889: 307.

[38] *The Ries Electric Railway System* (advertising pamphlet) (Baltimore, 1887).

[39] *Electrical Engineer* (London), 27 March 1889: 227.

of Boston, a firm which also operated under Main's patents, tried his cars on a three-mile long Massachusetts line in October 1888.[40] Experiments with the Main battery cars continued into the early 1890s.[41] No record of their eventual outcome survives, but it is safe to say that all the experiments must have failed—Main's battery was a variant on the usual lead-acid system in which the lead sponge anode was replaced by zinc anode. To this day, no one has succeeded in making a successful storage battery with zinc electrodes.

During 1890 a number of short notes appeared in the technical press reporting the trial runs of a battery car built by the J.F. McLoughlin Electric Company. These tests appear to have been short and were restricted to the Philadelphia area.[42] Galveston was reported considering battery cars in late 1889, but eventually dropped the idea.[43] At the same time a car designed by the Electric Car Company of America was being tried experimentally on Broadway; the reason for these tests was a ban on overhead conductors in Manhattan.[44] The president of the Electric Car Company was William Wharton, who had originally been an assistant of Reckenzaun. In 1889 and 1890 the company also tried out one of its cars in Philadelphia.[45]

The Compagnie Generale des Omnibus of Paris began to consider battery streetcars again at the end of the decade. In late 1889 Gustav Philippart fils was in charge of a car running experimentally on a Paris line.[46] The construction of the cars was much influenced by Julien's work, although he took no part in the experiments. By 1890 three more cars had been added and the Compagnie Generale soon committed itself to the full commercial operation of battery cars on three of its central Paris lines.[47] These cars continued to operate until after the turn of the century.[48]

In 1890 the Belgian engineer Jarman, working closely with the EPS and the London Tramways Company, established the

---

[40] Ibid., 1 November 1889: 356.

[41] Testimony of William Main, Brush & Consolidated Electric Companies vs. Milford & Hopedale, etc., Loc cit; Ibid., 26 July 1890: 54-55.

[42] Electrical World, 8 February 1890: 94; 4 January 1890: 15; Ibid., 7 March 1890: 182.

[43] Electrical Engineer (London), 25 October 1889: 321.

[44] Ibid., 11 October 1889: 282.

[45] Electrical World, 6 April 1889: 211; The Electrician (London), 3 January 1890: 227.

[46] Electrical World, 17 November 1888: 269; Electrical Engineer (London), 10 May 1889: 365.

[47] Electrical Engineer (London), 28 March 1890: 243.

[48] Ibid., 8 December 1893: 535-538.

Jarman Electric Tramcar Syndicate to exploit his new eight-ton cars.[49] The influence of Reckenzaun's work is evident in these cars especially in the paralleling of the motors; the use of series-parallel switching of the batteries; the removal of the batteries on complete trays. Although numerous reports of the imminent runs of the cars were reported through 1890 and 1891, no further reports appeared after 1891.[50]

The above list of late 1880s–early 1890s battery streetcar experiments is by no means exhaustive. However, the list given above gives a representative picture of this activity. One authority in October 1890 took a survey of the worldwide use of storage battery cars and found sixty-four cars then running. Compared to the overhead conductor system, therefore, this was a very small-scale effort.[51] Most of these sixty-four cars were scattered in small, experimental clusters and the history of these experiments followed an ephemeral pattern—one or two cars, usually refurbished horsecars, were run on the lines of a horsecar company for some months, sometimes for as much as a year. Only in rare instances did the experiments lead to long-term, commercial operation. To understand the reasons for this pattern of activity, it is necessary to look at the situation facing street railway companies during the 1880s.

By the end of the decade streetcar companies knew that the slow, inefficient horsecar had to be replaced. Electricity was the most logical alternative; the problem was to choose the *best* electrical system. From the beginning most electrical engineers predicted that the overhead trolley wire system would be more economical, since the battery cars needed to drag around a large mass of lead and the battery plates decayed rapidly. This became particularly evident after Frank Sprague proved the practicality of the overhead system in Richmond in 1887. Nevertheless, it was difficult to demonstrate the economic superiority of the overhead system over batteries, since each system claimed to achieve its economies in different ways. The battery system claimed a capital cost advantage, since no conductor was needed and existing horsecar tracks could be used essentially as they were. The overhead system, on the other hand, claimed lower operating and maintenance costs. It took time and extensive cost comparisons to prove which system was more economical.

---

[49] Ibid., 10 October 1890: 316-317.

[50] Ibid., 12 September 1890, p. 209; 10 October 1890, p. 306; 27 June 1890, p. 500; 20 February 1891, p. 178; *The Electrician*, 3 January 1890, p. 221.

[51] Twenty were in England, four in France and the remainder in the United States. *Electrical World*, 18 October 1890: 282-283.

Moreover, streetcar companies could experiment with battery cars less expensively and with less risk than with the overhead system. The company could provide the experimenter with a horsecar on which to install his motors and batteries, and this could be run on the same lines with the horsecars. The dynamo and charging facilities occupied a small section of the horse barns. If the experiment failed, nothing much was lost. Once a line opted for overhead conductors, however, there was little chance of turning back. The installation of the conductors and powerhouse had to be paid for, and this would be difficult if the company also had to pay the high operating costs of a horse stable.

Reliable and detailed comparative cost data on the various electric traction systems were not available until the early 1890s. Nevertheless, streetcar companies knew of the overhead system and were therefore unwilling to opt for full commercialization of the battery system until they knew more. Many lines experimented with a few battery cars, and found several systems workable, but few were willing to go any farther. Those lines which did fully commercialize the battery system did so for non-economic, non-technological reasons—in Manhattan, Paris, and Birmingham the use of overhead conductors was either prohibited or strongly discouraged.

The battery streetcar was not a failure in the sense that it didn't work. On the contrary, most of the better-designed systems operated about as economically as a horsecar line, and the continued operation of a few experimental battery cars on the lines of a horsecar company was no burden on the company. The rapid deterioration of the early traction batteries, however, quickly led most streetcar companies to see that the solution to future urban transport problems did not lie with the battery, and therefore most such experiments ended quickly.

Little is known about the detailed construction of these early traction batteries, except that they performed badly. In 1892, a series of witnesses at a patent interference trial described the leading American traction storage batteries as: "utter inadequacy" (EAC and Julien); "in every commercial sense entirely useless and deserved no attention" (Brush); "required constant attention and gave serious trouble" (EAC and Julien); "they have no practical value for traction work" (EAC); "went entirely to pieces in a few weeks' time" (Johnson); "with none

---

[52] Various testimonies, Brush & Consolidated Electric Storage Companies vs. Milford & Hopedale, etc., loc. cit.

of these did we secure satisfactory results" (EAC, Sorley, Barrett and Currie); "an absolute failure" (EAC); "were failures; would not do the work" (Herring); "very unsatisfactory" (EAC and Julien). Most of these witnesses had experimented with battery cars themselves.

One of the witnesses, for example, was Dwight Washburn, who had been involved with storage batteries since he experimented with the first Faure cells in 1881. Since 1886 he had been involved with various battery streetcar projects. Washburn's 1892 testimony, which ended on the following note, may therefore be taken as fairly authoritative:

> From my experience I do not believe that today there are any storage battery cars run by any of the above mentioned batteries, or by any battery I have ever seen, that are practical either from a financial standpoint, or satisfactory to the patrons of the road, for the reasons that the batteries are constantly giving out, the plates are always buckling or short-circuiting, the active elements are unstable, the cars stick on curves and comparatively slight grades and the enormous weight destroys the road beds.

Many of these failures were undoubtedly due to the poor treatment given these batteries on the road, such as over-discharging them and discharging them at much too fast a rate on steep grades. It is also clear, however, that these early traction cells were unmodified forms of the stationary batteries designed for powerhouse use. The Julien cell, for example, was a pasted battery with the usual open-grid design with no protection for the electrode faces; many people complained that the active material pellets on the cells were knocked out by the vibration of the road.[53]

A number of these inventors proposed the first steps towards developing vibration-resistant batteries. Reckenzaun, for example, patented in 1886 one of the first of the so-called molded-pellet electrodes. The idea behind these plates was to form small, highly coherent pellets of active material by casting them with various binders. The pellets would then be held in place by casting a molten lead plate around them:[54]

---

[53] *Electrical World*, 16 January 1892: 45.

[54] U.S. patents: #385,200—26 June 1888—filed 22 March 1888; #371,358—11 October 1887—filed 14 July 1887; Wade, *Secondary Batteries*, p. 44-45; Augustus Treadwell, *The Storage Battery*, p. 90.

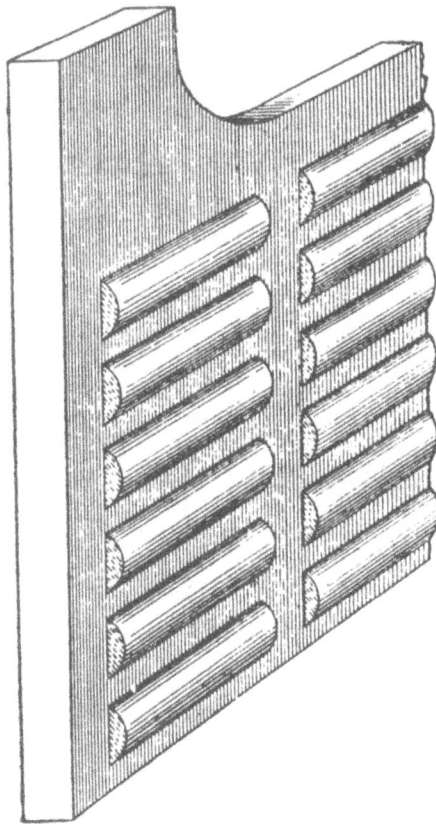

A number of other inventors interested in traction batteries patented molded pellet designs at about the same time. Although this particular approach to battery design was abandoned by the turn of the century, it stimulated a good deal of traction battery design during the 1890s. In the end, this was the significance of the 1880s experimentation with battery streetcars; it set in motion a tradition of design which eventually led to the development of the first practical transportable batteries around the turn of the century, and prepared the groundwork for the gasoline automobile and electric vehicles which have been the mainstays of the storage battery industry throughout the present century. The battery streetcar was a failure, but several of the engineers who worked on streetcar battery design later put their experience to work in the design of electric automobile cells.

III. The Maturing Industry: 1890-1900

1. The Battle of the Systems

During the later 1880s and early '90s a long and noisy debate by historians took place in the electrical engineering profession, dubbed the "Battle of the Systems." The battling systems were the alternating current method of power distribution versus the direct current method. The A.C. side had such distinguished supporters as Gisbert Kapp, Mordey, Ferranti and Westinghouse, while D.C. had the advocacy of such men as R.E.B. Crompton, Hopkinson and, of course, Edison.

Like all historical terms, however, "Battle of the Systems" gives an oversimplified picture of what in fact took place in the later 1880s and early '90s. The A.C./D.C. debate was only one sympton of the maturation of electric power engineering. This maturing process influenced more factors than simply the relative importance of A.C. and D.C. distribution and one factor which it influenced was the role of the storage battery in power distribution.

In the last chapter, our analysis of the use of batteries as incandescent lighting adjuncts concentrated on private lighting plants and the smaller central distribution systems. By the late 1880s, however, central distribution systems were growing rapidly both in size and complexity. The British revised the Lighting Act in 1888, opening the door to the erection of large central stations. In the United States, where central stations had been built earlier, but almost always without battery adjuncts, the battery slowly began to win acceptance. The Germans and the French also began constructing large central distribution systems using batteries near the end of the 1880s. In these larger and more complex systems, the storage battery was used in a more sophisticated way than it had been in private lighting plants.

The "Battle of the Systems" must be viewed against the background of this increase in the size and complexity of power distribution systems. The debate depended on electrical

engineers making a crucial assumption about their
technology; namely, that power generation from a central
generating station to a large number of consumers was more
economical than either private plants or small stations supplying a handful of consumers. To us, this assumption may appear so obvious as to need no discussion, but it was by no
means self-evident to many leading engineers of the early
1880s.

For example, in 1892 Professor George Forbes recalled that in
1885 one of the basic questions facing British electrical engineers was whether central station generation could ever be
made as economical as private generation: [1]

> Consequently there were many men at that date who did not
> think that central stations could compete against isolated plants
> for even moderate sized establishments, and who saw no pros-
> pect of small consumers being supplied, except by neighbors
> clubbing together to generate electricity for themselves.

It is beyond the scope of this study to analyze the reasons for
this belief; suffice it to say here that A.C. generation offered no
advantages for engineers who held this belief. Forbes went on
to state, however, that by 1892 the general consensus was that
central stations were economically practical and that the problem which remained was to determine which system of central
station distribution was best.

The two basic systems were A.C. and D.C., but there was
also a good deal of competition and variety between individual
systems *within* these two broad categories. In the A.C. camp,
for example, there were the exponents of small transformers,
who recommended that one transformer be placed in each consumer's cellar, and there were those who favored large transformers, which were to be placed in substations. In the D.C.
field, there were the low voltage, no-battery people, like Edison, who favored generating and distributing current at the
same voltage as that at the consumer's outlet. These men were
opposed to the high voltage battery and rotary converter D.C.
engineers, who favored the use of these devices as ersatz transformers for "stepping up" D.C. voltage for distribution.

These debates were stimulated by the growing size of distribution systems at the end of the 1880s. The earliest D.C.
systems, as designed by Edison, were called "direct" direct
current, since there was no conversion of current or voltage

---

[1] *Electrical Engineer* (London) 9 September 1892: 267.

between dynamo and lamp. As systems grew in size, however, the high currents required by such low voltage transmission began to produce increasingly expensive i^2R losses in the feeder cables.[2] Thus, engineers looked more favorably on A.C., with its ability to use transformers to raise and lower voltage and current.

Direct current cannot be used to operate conventional transformers. Nevertheless, a number of techniques were proposed for raising the voltage of D.C. so that it could be economically distributed to a distance. In 1882 John Hopkinson invented the Three-Wire System, which was soon adopted by Edison.[3] This system, only capable of doubling the transmission voltage, was limited in its ability to extend the economic range of D.C. distribution. Nevertheless, by the mid 1890s some D.C. systems were operating on main's voltages of 2,000-3,000 volts, with the voltage being "stepped down" to 110 volts at the consumer's end.

Storage battery "transformers" provided a method of "stepping down"; therefore, just as electrical engineers grew more interested in A.C. in the late 1880s because of its ability to use conventional transformers, so also their interest in storage batteries grew.

Storage batteries can be used as electrochemical transformers. As an example, let us say that an engineer wants to transmit current at 500 volts and distribute to consumers at 100:

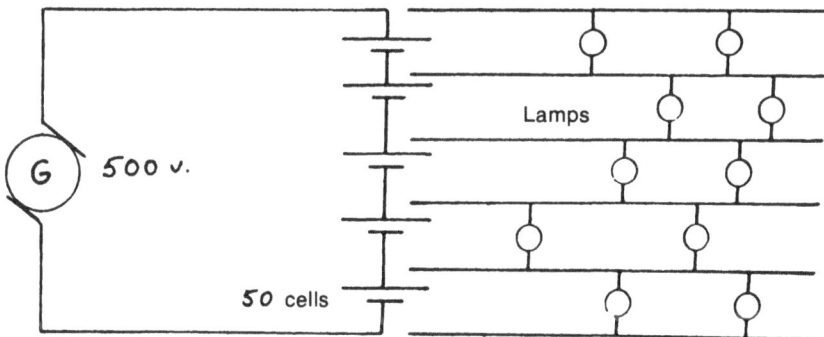

---

[2] R = resistance, i = current. I²R is the measure of heating losses in conductors. On this point, see, for example, Samuel Insull, *Central Station Electric Service* (Chicago, privately printed, 1915) *passim*.

[3] A method of transmission in which two sets of consumers receive power from a common neutral wire and either of two conductors whose potential differences from the neutral are equivalent but opposite in sign. In practice, a generator produced current at 220 volts, and transmitted this through the two extreme conductors. The neutral wire served to split this voltage, giving the consumer standard 110 volts. See next section for details.

The 500 volt D.C. generator transmits current through the mains to the ends of a 250-cell battery located in a sub-station near the consumer's premises. This battery is tapped by local feeder circuits every 50 cells. This particular application of storage batteries was conceived at the time of the original Faure and Brush patents. It was, in fact, a rather natural idea for the early lighting and power engineers to conceive of, since most of them were telegraphic engineers, and such battery configurations were standard practice at the larger telegraphic offices for many years. The only factor missing in the telegraphic use was the recharging function, since telegraphic batteries were primary cells.[4]

Because of the resistance of American electrical engineers to storage batteries during the 1880s, the high voltage D.C. battery-"transformer" system was never widely discussed in this country. In Britain, however, considerable debate centered on this system, and the A.C./D.C. Battle of the Systems was sometimes referred to as the Storage Battery System versus the Transformer System.[5]

In comparison with the A.C. transformer method of high voltage transmission, the D.C. battery system had both advantages and disadvantages, the former coming from the undeveloped state of early electrical hardware. Early generators for example, often broke down, making interruption of service common. The batteries could tide a system over a shutdown of several hours, whereas transformers could not. This was no mere theoretical advantage for early systems; the early Ferranti A.C. systems in London were placed under severe economic strain by constant equipment failure breakdowns causing disruption of service and consumer dissatisfaction.[6]

Moreover, the nonexistence of practical A.C. motors until the late 1890s gave D.C. another economic advantage over its

---

[4] For example, at the larger, urban telegraphic offices, it was impractical to maintain separate sets of primary cells for each circuit, since there could be hundreds of circuits and dozens of cells per circuit. Circuits of different lengths and those connected to different kinds of instruments required different voltages. The New York-San Francisco circuit, for example, needed far less voltage than the New York-Newark line, so the circuit diagram used on p. 141 might have been used analogously at a telegraphic office, with the left-hand side representing the San Francisco circuit, and the right-hand five circuits representing circuits going to five cities near New York like Newark. The only difference was that at telegraphic stations primary cells were used instead of storage batteries. This telegraphic practice was common by the 1870s.

[5] For the lack of American interest, see, for example, the series of papers in the New York *Electrical Engineer* during 1888 concerning the A.C./D.C. controversy.

[6] John E. Smart, *The Deptford Letterbooks* (London, The Science Museum, 1976); R.H. Parsons, *The Early Days of the Power Station Industry*, pp. 21-41.

rival. Motor load was very light in the early 1880s, but in the late 1880s it picked up with the development of the trolley. Trolley motor load was never as significant as lighting, but its importance exceeded the total kilowatt hours consumed, since most of the consumption was during daylight, light-load hours. Another, although lesser problem was the undeveloped state of A.C. metering technology.[7] In addition, the details of many early A.C. systems made them considerably less economical than modern A.C. systems. For example, Professor Forbes observed that until 1892 the standard A.C. practice was to run the high voltage mains to individual stepdown transformers right in each consumer's house.[8] This method tends to maximize energy waste, since heat-generation losses in each transformer will be the same whether or not electricity is being consumed in the house.

During the Battle of the Systems debates, A.C. advocates rarely referred to this negative aspect of their system. Rather, they pointed to the greater *full-load* efficiency of the transformers versus batteries. On an energy-in versus energy-out basis, and operating both devices at their most desirable rate, transformer efficiency was about 94-98 percent, whereas batteries averaged only 75-80 percent.[9] But, if we again refer to Professor Forbes's 1892 observations, we see that this efficiency was more apparent than real:

> When the pioneers (of A.C.) started work it was of course looked upon as a great feat to get nearly the whole of the energy transformed and supplied in the secondary circuit; and it was a great feat to do this cheaply and economically unless we used very small transformers; and the only way in which we could use very small transformers, with pressures of 100 volts, was by having rapid alternations, and, consequently, the pioneers used dynamos which gave very rapid alternations and a very small transformer. The result is that you get a very considerable consumption of energy due to the hysteresis of the transformer.

As Forbes points out there were two reasons for the small size of early transformers: a design limitation and the idea that each house should have its own transformer. No doubt these

---

[7] In 1888 O.B. Shallenberger invented the first practical A.C. current meter for Westinghouse. Before this, Westinghouse had difficulties in billing customers. Passer, *Electrical Manufacturers*, pp. 139-141.

[8] *Electrical Engineer* (London) 23 September 1892: 314.

[9] Francis B. Crocker, *Electric Lighting* (New York, Van Nostrand, 1901) (Vol. 2) p. 159.

two reasons had a feedback effect on each other and on the inefficiencies each produced. During the 1890s and afterwards, the evolution of A.C. technology favored the development of larger transformers and the gathering of these into substations serving increasingly larger groups of consumers. One of the factors favoring this change was the creation of city-wide A.C. grids; the impedance of transmission lines to the flow of A.C. current increases with the frequency.[10] This favored the use of lower frequencies, and, consequently, larger transformers.

This development began at the same time that electrical engineers started abandoning the in-house transformer system in favor of sub-stations. In sub-stations a series of transformers can be used. Human attendants or automatic devices can be used to adjust the number of these transformers under load to the current demand. Thus, hysteresis losses are reduced. In 1892 Professor Forbes spoke of these ideas, but indicated that they were largely undeveloped in practice at the time. He indicated that if sub-stations were used "we may safely look for an average efficiency of 90% during the whole of the 24 hours."[11] This shows that the much greater full-load efficiency of transformers over storage batteries was not a significant advantage in actual practice, particularly if we consider that the difference between peak and average daily consumption of electric current was much greater in the 1880s and 1890s than today.[12]

Another advantage which the high voltage D.C.-battery system had over that of A.C.-transformers was that of the paralleling of dynamos. If dynamos at a power station can be run in parallel, the current output of the station can be adjusted to the variable daily current demand. This is done by starting up a series of dynamos one by one as the demand increases. Such a procedure can save a considerable amount of energy, since the efficiency of a dynamo falls as it is called upon to produce current below its rated capacity. The paralleling of D.C. dynamos was commonplace by the mid 1880s, but not until the 1890s did engineers develop techniques for paralleling A.C.

---

[10] Passer, *Electrical Manufacturers*, p. 280.

[11] *Electrical Engineer* (London) 23 September 1892: 314.

[12] As an example of this argument, assume a 95 percent efficient transformer (full-load) operating a house which employs the device at 100 percent load for 2 hours, 50 percent for 4 hours, and 0 percent for 18. This would be a typical pattern for the time period under consideration. The actual, 24-hour efficiency would then be 82.5 percent. Arthur V. Abbott, *The Electrical Transmission of Energy* (New York, Van Nostrand, 1904) pp. 593-594.

generators.[13] For this reason, early A.C. systems were segregated into electrically-separate sections, each run by its own alternator. This meant that no alternator could be cut in or out, regardless of how the demand varied.

The chief advantage of A.C. transformers over storage battery "transformers" can best be demonstrated with a picture. The photograph shows a battery sub-station around 1898:[14]

Power system storage batteries were large devices; a twelve-cell battery might be the size of a large refrigerator, and a large station might contain a hundred or more of these. The substations designed to hold them were not only large, but had to be expensively constructed of acid-resistant materials, since sulphuric acid vapor was always present in the air; the cells were not sealed like modern batteries, but were left completely open-topped. The biggest item of expense, however, was maintenance. Usually the plate replacement cost was 10 percent of total installation cost per annum. Moreover, the cells

---

[13] This was due to the different electrical characteristics of different machines. As Percy Dunsheath remarks: "Even if two such machines could have been run in parallel the wasteful circulating currents must have been enormous. Consequently, sections of the load were generally allocated to particular alternators. . . ." Dunsheath, *History*, p. 170.

[14] J.A. Fleming, *50 Years of Electricity*, p. 256.

had to be continually cared for, or else deterioration would be more rapid; authorities advised continual inspection of the plates, cleaning of cells, removal of shed active material, refreshment of acid, and so forth.

Consequently, the storage battery method of transmitting and transforming high voltage D.C. current was never widely adopted in practice. It is important, however, because it was a much-discussed idea in the late 1880s and therefore acted as a stimulus to battery research and investment. R.E.B. Crompton, for example, favored battery transformers. This was the chief stimulus for his creation of the Crompton-Howell battery. In 1888 Crompton delivered a paper before the Institution of Civil Engineers defending the battery transformer system. The title of this interesting paper, as well as its tone, reflects the sense of battery advocates in the late 1880s that their chief competition came from the A.C.-transformer system: "Central Station Lighting: Transformers Versus Accumulators." At the end of the paper, Crompton had a heated debate with Gisbert Kapp, a former assistant of Crompton's who had deserted the D.C.-battery camp for A.C.[15]

Ironically, the reason why the battery transformer system was used in so few installations turned out to have little to do with A.C. at all. First, despite the headway made by A.C. during the '90s, D.C. held the lead. During the next twenty years the construction of new D.C. systems remained at a high level and sometimes even exceeded that of new A.C. systems. In 1891, one source reported 18 A.C. and 34 D.C. systems operating in Britain,[16] in 1892, there were 19 A.C. and 38 D.C. central stations operating on the Continent.[17] By 1899, the British mix of A.C. and D.C. stations was about equal; by 1910 D.C. had retaken the lead with nearly 75 percent of the total, and only in 1918 did it again fall to 50 percent of all stations.[18]

Since A.C. stations were usually much more powerful than D.C., A.C. took the lead from D.C. on a kilowatt basis by the late 1890s. Nevertheless, D.C. continued to control many areas in the late 1890s and early decades of the twentieth century, particularly in the central areas of large and medium sized cities. These were the places where the first incandescent systems had been installed during the 1880s; they, of course, were D.C. Another reason for the retention of D.C. in central city areas was the large use of electric motors here, both for man-

[15] *J. Soc. Teleg. Eng. & Elec.* 17 (1888): 349-419.

[16] *The Electrician* (London) 2 January 1891: 274-275.

[17] Killingworth Hedges, *Continental Electric Light Central Stations*, pp. 160-162.

[18] *Electrician Electrical Trades Directory*, 1899, 1910, 1918 editions.

ufacturing and transport.[19] In London in 1914, for example, of 35 systems operating, 17 were pure D.C. Therefore, it was not for the lack of survival of D.C. systems on which it might have been used that the battery-transformer system was never employed extensively.

The reason why the battery-transformer system was used for so few systems was that a more convenient electro-mechanical device was invented to take its place. This was the rotary transformer. These did not exist in 1888 when Crompton defended the battery-transformer against A.C., but they were available commercially by the early 1890s.[20] The rotary transformer was simply a large D.C. motor directly coupled to a D.C. dynamo; the motor's voltage was that of the high voltage mains, and the dynamo's voltage that of the consumer. Actual efficiencies for these rotating devices in the range of 90-95 percent were not unusual, making them slightly less efficient than their static A.C. counterparts.[21] The manufacture of high quality rotary transformers was not a problem for the electrical industry, since the motor and dynamo technology already existed.

As soon as commercial rotary transformers became available, the battery transformer system stood little chance of competing with it. In 1888, storage batteries were being used as transformers at two British stations, but these disappeared by the 1890s.[22] In 1892, of 57 central stations reported operating on the Continent, three used the battery transformer.[23] These were all older installations, from the pre-1890 period. One was R.E. Crompton's famous 1887 system at the Vienna Opera.[24] Electrical engineering textbooks continued until the 1920s to mention battery transformers as a possible, albeit undesirable method of supply, but the actual use of such systems appears to have ceased by the mid 1890s.

---

[19] Most of these central city D.C. systems were Three-Wire, 110-220 volt. Harry B. Gear & Paul F. Williams, *Electric Central Station Distribution Systems* (New York, Van Nostrand, 1911) p. 20.

[20] In the electrical engineering practice of the time the nomenclature of these devices was never fixed. They were also called D.C. transformers, motor-generators, dynamotors, rotary converters. In this study, we use the term rotary transformer specifically for the mechanical device which uses D.C. current for both input and output.

[21] Ninety-two percent efficiency at full-load was reported for an 1895 Elwell-Parker rotary. Wormell, *Service of Man*, p. 648. During the late 1880s A.C. transformer efficiency (full-load) was reported at about this same level. By the early '90s, as a result of the use of pure iron cores or silicon-iron alloys, and the use of laminated cores, A.C. transformer efficiencies were in the general range of 95-96 percent. J.A. Fleming, *Fifty Years of Electricity*, pp. 133-137.

[22] Killingworth Hedges, *Central Station Electric Lighting*, pp. 116-125.

[23] Ibid.

[24] Brian Bowers, *R.E.B. Crompton* (London, The Science Museum, 1969) pp. 25-26.

This failure of the battery transformer system follows a pattern characteristic of the entire history of storage batteries. Even since Professor Moritz Jacobi's 1850 telegraphic proposal, inventors had suggested uses for the storage battery based primarily on other properties than its energy-storage capacity. None of these schemes got very far. The storage battery transformer system is a case in point. Despite their illusory simplicity, storage batteries are inherently expensive devices, both in capital and maintenance costs. Electromechanical or electromagnetic devices are almost always cheaper and better than electrochemical ones, if both can be used for the same function. Only for energy-storage itself is the electrochemical solution necessary.

Rotary transformers were used in a number of very ambitious high voltage D.C. distribution systems in the late nineteenth century. One of these, built in the Rhône River valley in Switzerland, was designed to transmit 23,000 volts 35 miles from a waterpower generating site to Lausanne.[25] It is clear, therefore, that the usual stereotype of the D.C. system as restricted to short distances is not entirely accurate. Of greater importance than the pure D.C. rotary transformer, however, was the combination A.C./D.C. rotary transformer. In this device, an A.C. polyphase induction motor was directly coupled to a D.C. generator. They looked very similar to pure D.C. rotary transformers. Although these machines went under a number of names, they wi'l be referred to hereinafter as rotary converters.

Rotary converters became available as soon as the A.C. induction motor became practical. The early A.C. systems were single phase. At the end of the 1880s, two and three phase A.C. systems began to displace single phase, and one of the chief reasons for this change was that polyphase currents made possible the operation of induction motors. By the beginning of the 1890s the induction motor was sufficiently perfected to permit the commercialization of rotary converters.[26]

From this time on, a system of mixed A.C./D.C. systems began to develop in the centers of larger American and European cities. The old, three-wire D.C. systems which had been installed in these cities in the 1880s and early '90s were powered by large A.C. central stations located in the suburbs. The A.C. station fed its high voltage current to two types of substations—those in the suburbs, which contained ordinary A.C. transformers, for the distribution of low voltage A.C. to

[25] Fleming, *Fifty Years*, pp. 250-251.
[26] Ibid.

consumers; and those sub-stations in the center city, which contained rotary converters for the production of low voltage D.C. By 1910 this pattern of electric supply was common to most larger American cities, as it was also for London.[27]

The combination of A.C. distribution with the rotary converter was a powerful argument against the use of storage battery transformers in a pure D.C. system. The mixed system permitted electric companies in center cities to use most of their installed D.C. capital equipment and to profit from D.C. motor load, while at the same time taking advantage of the economy of scale of huge A.C. generating plants located on cheap suburban land where supplies of coal and water could be more cheaply obtained. In the Battle of the Systems, these were strong economic inducements for the use of A.C. distribution. On the other hand, the proponents of D.C. had the argument of load factor, and load factor became the source of the considerable economic success which storage batteries had during the late 1880s and 1890s.

Load factor was a term invented by R.E.B. Crompton in 1890.[28] It is defined as the ratio between the average current demand at a power station over a 24-hour period, and the current demand at the peak point of the day. The graph below shows the kilowatt demand on all London power stations on 20 December 1894. The average load was 4,000 kw, while the peak was 11,600 kw at 5:30 P.M. The ratio 4,000/11,600 = 34.5 percent is the load factor:[29]

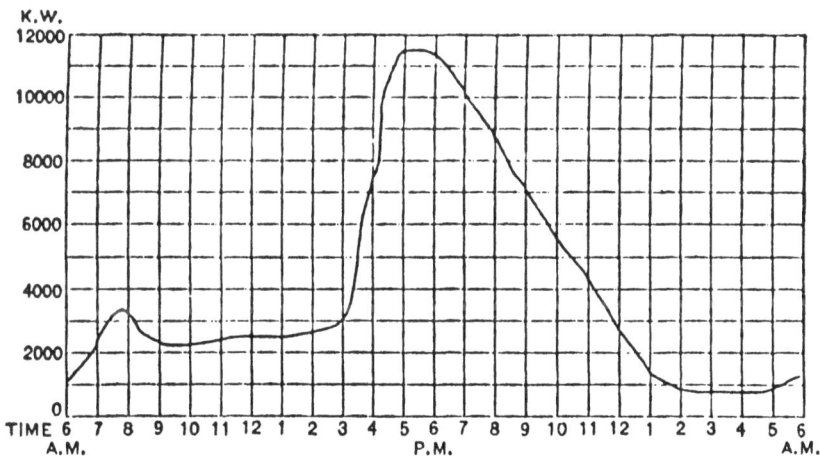

Combined Load Diagram of London Stations.

---

[27] Gear & Williams, *Electrical Central*, p. 120; *Electrician Electrical Trades Directory*, 1910 edition.

[28] R.E.B. Crompton, *Reminiscences* (London, Constable, 1930) p. 136.

[29] Crocker, *Electric Lighting* 1: 390.

Crompton invented this concept at a time when electrical engineers had solved many of the more basic problems of their craft, such as the design of reasonably efficient, constant, and durable dynamos. They had time, then, to concentrate more of their attention on some of the more sophisticated problems of electric power economics. The basic techniques of building and operating the various pieces of equipment having been solved at least to a reasonable degree, engineers could therefore pay attention to the question of how best to integrate the operations of the parts of a system for greatest economy.

Crompton invented load factor as a method of gauging energy efficiency. The generating equipment at a power station must be designed to supply the peak amount of current ever demanded of that system, even if this peak lasts only a few minutes. If the equipment is used to provide current at less than its rated capacity, the cost of each kilowatt will be proportionately greater as the output falls. Electrical engineers before 1890 were aware that the cost of each kilowatt hour increases as a plant operates below full load capacity. Before the 1890s, however, quantitative measurements of the loss of efficiency of plant equipment operating at various levels below rated capacity were not generally available. Crompton began to provide such data in 1890, and Dr. John Hopkinson expanded upon his work in 1892. In that year Hopkinson delivered an address which showed how the economics of a power station might be calculated. The following graph, prepared from his data, shows the variation with load factor of the total cost of producing a kilowatt hour at a 2500 kw D.C. station: [30]

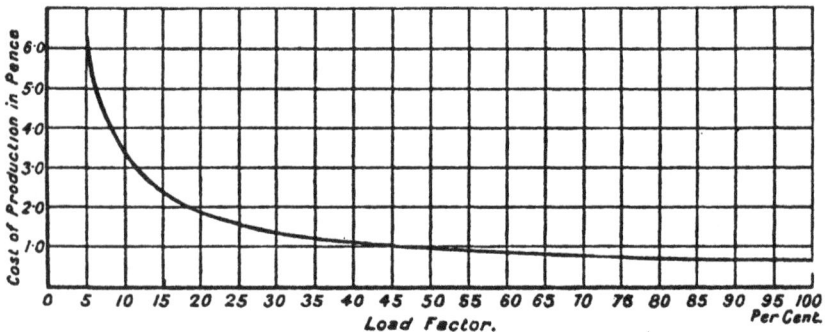

This does not appear to be so bad, except when it is considered that the load factor for a residential district in London in

[30] Charles H. Wordingham, *Central Electrical Station* (London, Charles Griffin, 1901) p. 333.

the 1890s averaged about 10 percent and might fall to 5 percent at certain times of the year.[31] In 1901, a sample of ten British systems of widely varying supply pattern had an average load factor of 15.2 the highest being 17.8 and the lowest 10.8.[32]

A number of techniques were used to raise load factor. The simplest method was the complete cessation of service between dawn and dusk. This was common practice during the early years of electric supply, when the majority of central station systems were small and resembled private plants both in size and method of operation. For larger systems, however, this was not a good solution. Systems which had offices, shops, warehouses, theaters, restaurants, and other commercial buildings in their circuits could not operate in this way. Only pure residential districts could afford such service, and these only in areas where gas service was not available, since the chief reason for paying the greater price of electric light was its greater convenience. The dusk-to-dawn method was not economically desirable either, since it required the capital investment of the plant to lie idle most of the day. This method also wasted a great deal of energy firing up and cooling the boilers each day.

By the mid 1880s, therefore, most electrical systems managements—especially those in large cities—saw the dusk-to-dawn technique of increasing load factor as basically retrograde; they realized that the future of electricity lay in *extending* the range of service, not in *diminishing* it. One of the major interests of electrical engineers in the later 1880s therefore, was the problem of *increasing* the day-load, rather than eliminating it entirely. Electrical appliances were encouraged—fans, irons, electric heaters. The electric streetcar was helping to even the load of some systems, although most car lines preferred to build their own power houses. The reason for the rapid replacement of the horsecar by the trolley was the extraordinarily high profits of the latter, and one of the reasons for this profitability was the high load factor enjoyed by many streetcar lines—often over 30 percent.[33]

A more satisfactory method of increasing load factor than the dusk-to-dawn method was the construction of powerhouses with a series of smaller dynamos and boilers instead of one

---

[31] Crompton, *Reminiscences*, p. 136.

[32] Wordingham, *Central*, p. 346.

[33] The trolley system also had peak and trough-use periods, but these were never so severe as for incandescent systems. Wordingham, *Central*, p. 334.

large set. These could be switched on and off to match the load. Thomas Edison preferred this technique. Although this was a practical method, the shape of the load curve at most power stations made it less than ideal economically. If the load varied smoothly and evenly, as in a sine wave, the matching of generating capacity to load would have been fairly easy. The real shape of most load curves, however, was by no means so smooth and even, and therefore, not as easy to match by switching on extra machinery. For example, the diagram shows a typical week's load curve for an American power station in the first decade of the twentieth century: [34]

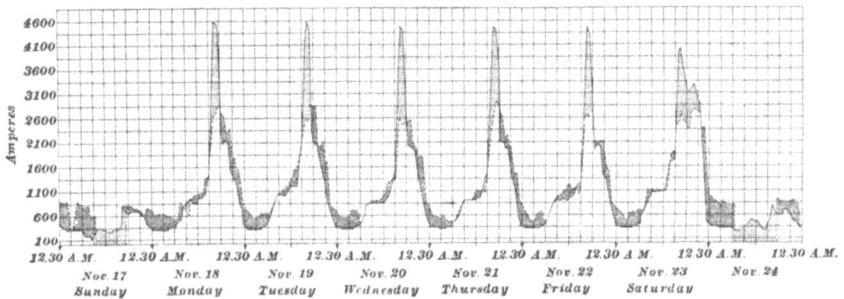

The spiky peaks of such loads required double the number of non-peak dynamos and boilers to be started up and run for two to three hours at most—a large capital expenditure for such a short time period each day.

This annoying characteristic of the lighting business long antedated the invention of electric lighting. Gas lighting engineers had to deal with this problem for decades; this is what led them to the invention of the gasholder. The gasholder is an economical way of eliminating the need for large numbers of peak-period retorts in the gas plant. The obvious analogy between the gasholder and the storage battery started electrical engineers considering the use of the battery as a load-factor improving device.[35]

Consider the diagram above. The station could be designed with a large storage battery and a single, large generator of about 1,200 amperes full-load capacity; a single, large generator is more efficient than several smaller machines. Then, between about midnight and 6:30 A.M., when not much

---

[34] *International Correspondence School Reference Libary*, p.55

[35] Early writings on the storage battery often use the analogy. Simon Philippart, for one, was fond of it.

current is needed, the generator can use its excess capacity to charge the battery. Between 6:30 A.M. and midnight, the battery can discharge into the line, making up everything over 1,200 amps. Thus, the generator would "see" a perfectly flat demand curve, and the load-factor would be 100 percent.

Such excessive attention to perfect load-factor was unnecessary and counter-productive of economy. Batteries were expensive capital investments and therefore had to contribute significantly to powerhouse economy in order to justify their expense. As the graph on p. 152 shows, load-factor losses do not begin to become serious until the factor falls below 50 percent. Consequently, batteries were only used where they would give significant cost leverage. This was by the technique of shaving off the spiky peaks. This is shown in the diagram above, where the single-hatched areas represented battery discharge, and double-hatch represents charge. With such techniques, a great deal of load-factor gain could be bought for a reasonable battery cost. The cost leverage is roughly proportional to the ratio between the height of the shaved peak and its base: H/B.

From the very beginning of incandescent lighting, engineers had conceived of storage batteries as load equalizers. However, this technique was not practically applied until the end of the 1880s, and it was only in the 1890s and the early years of the new century that it was extensively used. In 1888, of seventeen principal central stations operating in Great Britain, only 11 percent used batteries for load-leveling. Of thirty major continental stations reported, none used batteries for this purpose.[36] In 1890, of seventeen central stations in London, about half used load-levelers.[37] However, outside London in 1891, batteries were used at only 19 percent of all British stations.[38]

Despite the growing importance of A.C. throughout the '90s, the percentage of all stations with load-leveling batteries continued to increase. By the end of the '90s in Britain, of a total of 163 central stations reported, 62 percent had load-leveling batteries. Only 58 percent of these were either D.C. or mixed A.C./D.C., which means that 14 of the *pure* A.C. systems considered load-levelers to be so valuable that they went to the extra expense of internal rectification to use them.[39]

---

[36] Hedges, *Central Station*, pp. 117-125.

[37] *Electrical Engineer* (London) 6 (12 December 1890): 529-530.

[38] *The Electrician* (London) 2 January 1891: 274-275. Provincial systems, being usually built later than those of London, were usually A.C.

[39] *Electrician Electrical Trades Directory*, 1899 edition.

In 1895 in Germany, of 148 stations reported, 54 percent used batteries; of the D.C. stations only, two out of three had batteries.[40] This was already a considerable improvement over 1892, when, out of 27 German stations reported, only 30 percent used batteries as load-levelers.[41] In 1896, of 180 German stations reported operating, 57 percent had batteries; of only the D.C. stations, three out of four had load-levelers.[42] It is interesting to note that despite the growing relative importance of A.C. in Germany throughout the 1890s, storage battery use also grew in relative importance:[43]

Percentage of All German Stations:

Year	Using A.C.	Using Storage Batteries
1892	11%	30%
1895	18%	54%
1896	26%	57%

This table is not presented in order to imply a causal link between these two developments, since they were independent. Nevertheless, it must have come as something of a surprise to early electric lighting pioneers, such as Crompton, who during the '80s considered the A.C./D.C. struggle to be one of transformers versus batteries.

In 1910, 86 percent of all British central stations contained batteries. It is again significant to note that ten *pure* A.C. stations carried battery installations and that almost without exception all D.C. and mixed A.C./D.C. systems in that year had batteries.[44]

For the United States such accurate statistics are unfortunately not so plentiful. One authority, however, reported only two American central stations using batteries at the end of 1893.[45] Shortly before the turn of the century this situation changed somewhat, and a number of writers referred to the use of storage batteries in American central stations as becom-

---

[40] *The Electrician* (London) 26 April 1895: 789.

[41] The percentage may be 5 percent higher due to ambiguities in the reporting.

[42] *The Electrician* (London) 20 March 1896: 684.

[43] Ibid., 4/26/1895: 789; 20 March 1896: 684.

[44] *The Electrician Electrical Trades Directory*, 1910 edition.

[45] *The Electrical Engineer* (London) 1 December 1893: 512-515.

ing "recently more general."[46] The New York journal *Electrical World*, for example, in its review of the year 1895, observed:[47]

> Electric lighting presented no marked developments during the year, if we except the progress made in the education of central station men with respect to the utility of storage battery auxiliaries. The excellent results achieved during the year (at several stations). . . . will probably fix 1895 as the missionary year to what promises to be a wholesale conversion of central station managers to the use of the storage battery.

It is significant that this article speaks of battery use in the future tense, and that the half dozen or so American central stations then using batteries were considered as somewhat experimental at a time when such was already the standard practice across the Atlantic. In 1896, one list of 2,400 American generating stations indicates only fifteen using batteries.[48] The contrast with the above European data is dramatic. At least in Europe, however, central station use of batteries became the economic mainstay of the storage battery industry during the 1890s and remained significant for several decades.

---

[46] Crocker, *Electric Lighting* 1: 16.
[47] *Electrical World*, 4 January 1896: 3-4.
[48] Ibid., 6/12/1897: 753.

2. Change, Growth and Complexity

Load-leveling was one of the stimuli for the growing popularity of central station batteries in the late 1880s and '90s. Another important factor was the growth of the three-wire system of D.C. distribution. The earliest D.C. systems were two-wire, but the ability of the three-wire system to raise mains voltages and thus reduce copper in the cables permitted a considerable extension in the range of D.C. systems. For example, a 1905 electrical engineering textbook gives 3,500 feet as the maximum economic radius of supply for a two-wire system, but 6,000 feet as the maximum for an equivalent three-wire installation.[1] Accordingly, by the 1890s the great majority of all D.C. systems were three-wire.

---

[1] Abbott, *Electrical Transmission*, p. 523.

In the original three-wire system, two dynamos were used on either side of the neutral third wire:

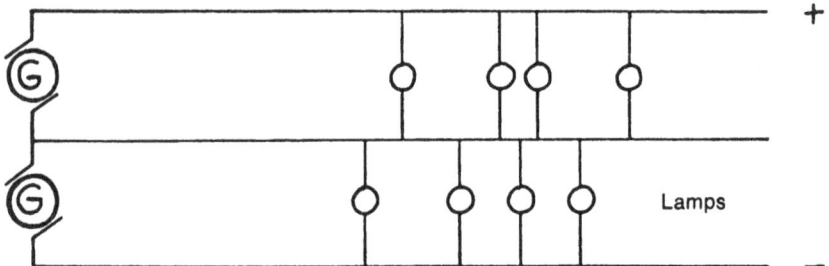

The advantage of the system is that it saves copper. If the load on either side of the neutral wire is kept equal, then this will truly be a neutral wire and no current will flow in it. Therefore, the amount of copper used in this wire can be minimized and total copper savings maximized. Moreover, the two dynamos can be made to run always near their most economical full-load (assuming the use of load-leveling battery as well).

Such a perfectly-balanced load, however, is rarely attained in practice. Consequently, allowance must be made for the neutral wire to carry a significant current and the dynamos will run above or below their most economical load as they make up for the differential in demand between the two legs.

When engineers began to install batteries as load-levelers in the late 1880s and '90s, however, they soon perceived a way to get more value out of the investment by using the batteries to improve the economics of the three-wire system. This was done by playing a variation on the old battery-transformer theme. The function of the load-levelers is to boost station *current* output. Therefore, they are wired *in parallel* with the main generators. If this is done on a three-wire system, then it becomes feasible to replace the two smaller dynamos shown in the graph above by a single, double-size machine, with the battery being used to split the voltage on either side of the neutral wire:

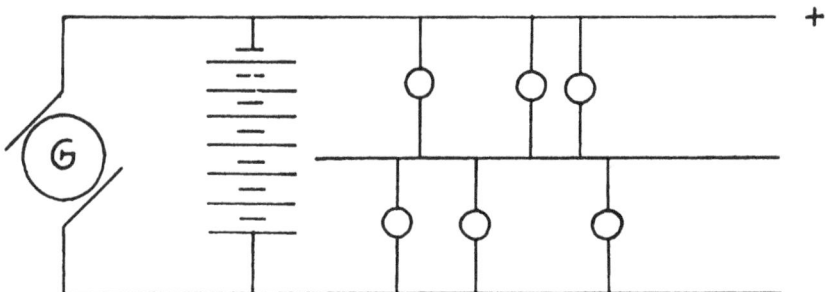

Savings may thus be achieved in several ways. The single, larger dynamo is inherently more efficient than the two smaller machines, and it may be run at the same load independently of the differential between the two circuits, which only the battery "sees." If the neutral wire is permanently fixed to the electrical center of the battery, then it will absorb the variations. One half of the battery will discharge, and the other half charge, depending upon the direction of current flow in the neutral wire. A more economically satisfactory solution, however, was for the contact of the neutral wire to move to either side of the center of the battery to balance the differential between the two consumer loads. In this way, the size of the neutral wire was kept to a minimum and the two halves of the battery were not subjected to unequal states of charge.

Here we see another reason for the growing popularity of storage battery auxiliaries as electric power stations grew in size and complexity. The resourceful station engineer used the battery for a number of different functions, and as operations grew more complex, the number of roles the *same* battery played grew accordingly. Along with its ability to balance three-wire loads and carry peak loads, the battery acted as an electrochemical buffer to smooth out the jagged output of early power generators. In addition, it could carry the full load of the station for some time if an accident occurred, or if the station was shut down for maintenance. It is necessary to describe each of the battery's functions separately in order to avoid confusion, but it should be kept in mind that when engineers in the 1890s and early years of the new century considered the installation of a battery, they evaluated the economic advantages of all these functions together. As several electrical engineering textbooks published around the turn of the century state, most central station batteries were used for all of the function mentioned above.[2]

If we look closely at both the failure of most battery electric lighting schemes of the 1880s, as well as the success of most such schemes in the '90s, we can observe something crucial about the evolution of these two technologies; that is, electric power engineering and storage batteries. During the 1880s, many inventors assumed the storage battery to be a simple device whose use could be controlled by similarly simple control equipment, such as the automatic control switch of Brush described in the last chapter. Such simplicity proved illusory.

---

[2] For example, Crocker, *Electric Lighting* 1: 395.

Likewise, the designers of central station distribution systems often assumed that a simple, uncomplicated method of design and operation would suffice. This conception also proved illusory, particularly as the economic realities of electric power generation led to the kinds of complexity described above. As the inherent complexities of these two technologies were realized, it became possible to match the reality of storage battery technology to the needs of central station operation to produce practical methods of operation.

An example of this process was the failure of many early storage battery promoters to anticipate the problems which the battery's variable output had on the practicality of their designs. As a battery discharges, the concentration of the active chemicals falls, and the voltage will therefore also fall. In addition, the voltage required for charging the battery is quite different from that obtained from the battery on discharge. In this sense, it is misleading to think of charging as the *reverse* of discharge, since the chemical and physical realities of these two processes are not mirror images of each other. The figure shows charge and discharge curves for a lead-acid cell: [3]

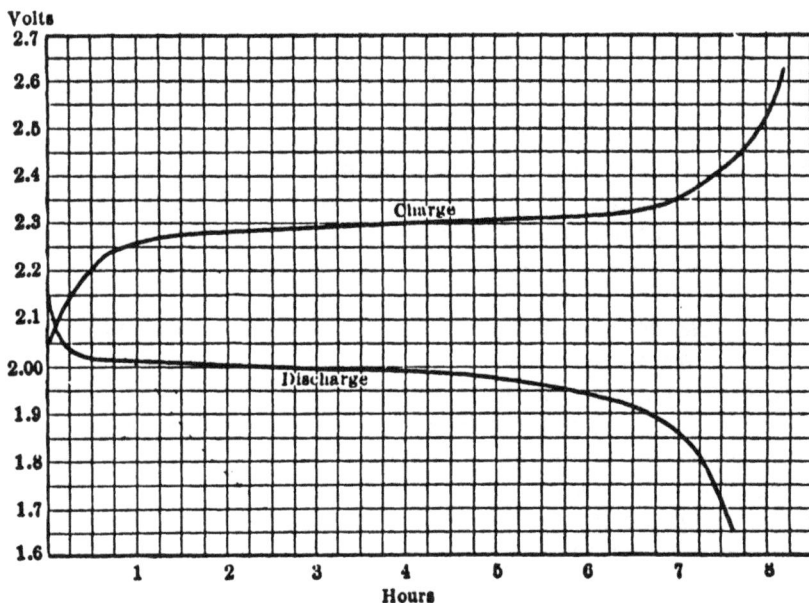

---

[3] Lamar Lyndon, *Storage Battery Engineering*, p. 14.

Although these characteristics are common knowledge to modern electrical engineers, early engineers either were ignorant of them or failed to appreciate their significance. A symptom of this lack of perception was the analogy which many of them drew between the gasholder and the storage battery; an analogy which they took as more than a vague comparison.

At the beginning of discharge, the voltage of a lead-acid cell is about 2.05 volts. Therefore, if electric lights are run from the battery at the standard 110 volts, the number of cells in series must be 53. But when the cells near exhaustion, the 53 cells will produce only $1.8 \times 53 = 95.4$ volts. Therefore, 61 cells will be needed near the end of the run. Alternatively, a station might use larger batteries, and therefore not need to discharge them all the way down to 1.8 volts. In such a case, only 60, 59 or 58 cells would be needed at the end. Whichever alternative was chosen, some means had to be provided to cut in additional batteries at intervals during the discharge, since incandescent lights are sensitive to small voltage variations. This addition of extra cells was known as the "end cell technique," and all central stations which adopted batteries as load-levelers used this method or an equivalent one.[4]

The end-cell technique, however, would have been impractical for many of the 1880s schemes of electric lighting. These systems, of which that of Brush was typical, involved the placement of batteries either in small sub-stations near the consumer's home, or in the consumer's cellar itself. The only human attendants would be at the central station; the battery installations were to run automatically. The battery would be divided into two equal halves, with one charging while the other discharged. Various forms of automatic switches were designed to control the charging cycle. Some worked on gas pressure—the production of hydrogen and oxygen gas by the storage battery as it neared the end of the charging process would fill a float which in turn operated the switch.[5] In a number of patents, the changing weight of the battery plates during charge was used to trip a delicately-balanced electric

---

[4] Since the fifty-three cells of a series string were usually lined up across the battery room floor with the end cells at the *end* of the line.

[5] U.S. Patents: #399,755—19 March 1889, filed 20 October 1888, patented England 7 December 1887; #371,893—18 October 1887, filed 15 March 1887, patented England 2 April 1886. To Frank King.

switch; one of the plates was balanced on the end of a pivoted scale:[6]

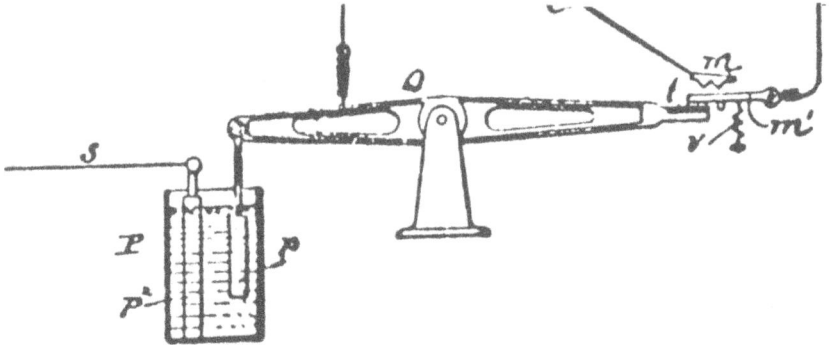

During the 1880s, dozens of patents describing similar systems were taken out in Europe and the United States. No mention is made in these early patents of the need for end-cells, yet these systems would not have worked effectively without them. On the other hand, if end-cells had been used, this would have meant filling up the consumer's cellar with his main batteries, end-cells, and a complicated electro-mechanical feedback device for cutting in the end-cells at the right time. In addition, the consumer would have required a second automatic feedback device for performing the same functions during the charging cycle; the different shapes of the charge and discharge curves would have precluded using the same device for both cycles. Finally, the automatic switches for shifting the batteries from the charge to the discharge cycle, which we do find mentioned in the early patents, would have completed the cellar installation. It is clear that by the time a consumer would have finished such an installation, he would have had a very full cellar and an empty wallet. This is not to mention the trouble which the inevitable malfunctioning of all these automatic control devices would have caused.

In private lighting installations, where a single dynamo was used to charge up a single bank of batteries, and where servants or workmen could be relied upon to operate the system, such automatic control devices were not necessary. In central station systems, however, where the consumer took no part in the operation of the dynamos, engines or batteries, automatic control would have been essential. Yet, as we have seen, the kind of automatic control designed by the early inventors was

---

[6] U.S. Patent #406,354—2 July 1889, filed 25 August 1888. To W.P. Kookogey.

woefully inadequate for the complexities of the battery. But if such complex control devices had been built, and installed in each consumer's cellar or local substation, they would have made central station lighting impractically expensive.

The success of the battery in electric lighting during the 1890s and the early decades of the new century was due in large part to the fact that a large central station had the man-power to eliminate the need for all the abovementioned automatic control devices. Unlike the battery systems of the 1880s, the load-levelers of the '90s were installed at the power house, not in cellars or substations. Attendants manually switched each end cell in or out as the battery switchboard voltameters reached certain predetermined points. Even with maximum reliance on manual control, however, the typical end-of-the-century end-cell switching gear at a large power-house was complex and expensive. The figure below shows a typical 1900 motorized end-cell switch. The sliding contact *a* could insert any of 20 end-cell combinations, and was driven by a motor started by an attendant at the switchboard. Such switches might be twenty feet long:[7]

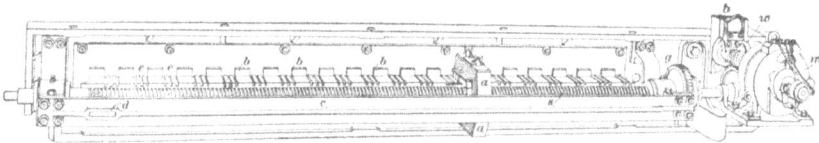

Central stations could afford the expense of these large machines only because the machines could be used for the control of some very large batteries. With regard to this point, the general evolution of powerhouse design during the late 1880s and 1890s should be kept in mind. Stations were made larger, covering increasingly larger areas with their service. Consequently, more sophisticated and expensive primary as well as auxiliary equipment should be purchased for these stations. The following remarks, made in 1888 by the experi-enced Edison-station manager L.A. Ferguson, illustrate this process:[8]

> The general design of the modern direct current central station and its equipment has been fairly well established during the

---

[7] *International Correspondence School Reference Library*, p. 66.

[8] *Electrical World*, 25 June 1898: 779.

past five years, and while it was originally the custom to erect
the generating stations as nearly as possible in the electrical
center of the city or town, it is now generally conceded that
direct current may be more economically distributed from a
condensing station, situated even a mile distant from the electrical center than from a non-condensing station located at the
electrical center of the city, and it has been further demonstrated
that the former practice of building many central generating
stations in various centers of distribution in cities is to be
supplanted by the use of one or two large condensing stations
generating direct current for distribution throughout the business district if the station be within one mile of the electrical
center of the district, and alternating current for transmission to
sub-stations located at the electrical centers of districts more
remote from the main generating station.

During the 1880s, the concept of reducing total costs by reducing labor costs was a popular theme in the thinking of
lighting system designers. Small stations operating at low
voltages would be relatively easy to operate, and the role of the
battery would be to provide current for a full 24-hour period
while being charged for only five or six hours. Thus, the attendants would be required only for this short period; during the
remaining 18 hours both they and the machinery could rest. As
we have seen, however, this concept was based upon the er-
roneous idea that a storage battery is simple to use and does
not need labor-intensive attention.

During the late 1880s and 1890s, the concept of low labor cost
operation of power stations declined as engineers realized the
need for full 24-hour attendance. A number of factors were
responsible for this, such as the larger size, power, and voltages
of stations, but perhaps the most important factor was the
campaign to lower operating costs by increasing the load-
factor. This could only be done by a continually active staff,
reading the meters and making the necessary adjustments in
operating conditions. In other words, the earlier concept of
lowering operating costs by lowering labor costs was replaced
by the newer concept of lowering fuel costs by increasing
overall efficiency.

Simplicity of operation, therefore, gave way to efficiency of
operation, and this called for more complex instrumentation
and auxiliary devices, as well as more professional human attendance. This, in turn, made the complexity of storage battery
technology integrable within the broader technology of electric
power engineering. It is ironic, therefore, that the storage bat-

tery could only be used practically in central stations once the concept of 24-hour attendance was established, whereas it had originally been conceived during the 1880s as one of the most effective means of permitting minimal human attendance in such plants. The history of the storage battery can serve as a warning to those who underestimate the difficulty of technological forecasting.

End-cells and end-cell switching machines were not the only pieces of auxiliary apparatus needed for the proper management of powerhouse batteries. This can be seen by considering the complete battery and end-cells. The 61 cells require a charging voltage of 2.65 volts each, or $61 \times 2.65 = 162$ volts. Since the generators are designed for 110 volts, 52 volts are short. This problem could be solved by reconnecting the battery halves in parallel and adding a resistance of 29 volts equivalent: $162/2 + 29 = 110$. This technique was not preferred, however, because of the difficulty of charging the two battery halves equally. The preferred technique was to employ small, auxiliary dynamos in series with the main generators on the charging circuit:[9]

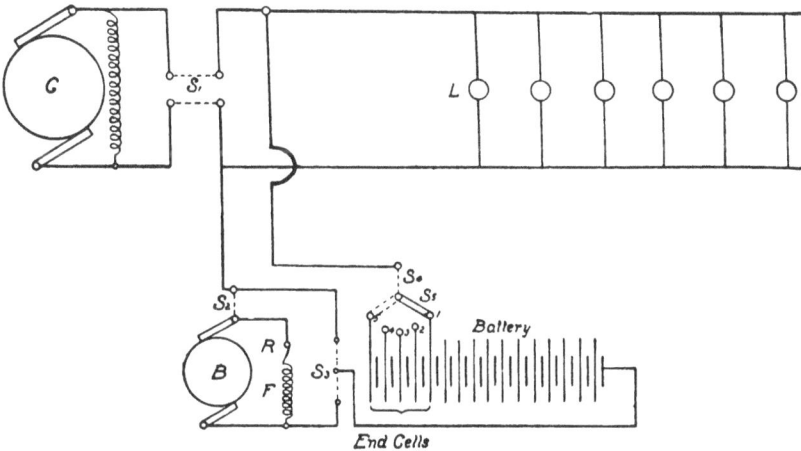

End Cells

Here B is the small dynamo, called a *booster*. The name comes from the function of the device, which is to boost the primary generator voltage from 110 to 162. Boosters were generally run from primary generator current, and they accordingly resembled rotary transformers, with the small booster directly coupled to motors which received current from the main

[9] *Cyclopedia of Applied Electricity* (Chicago, American School of Correspondence, 1905) p. 59.

dynamos. Control was manual, operated by an attendant who watched meters on a switchboard.[10]

Boosters began appearing in electrical engineering practice in the late 1880s. Their use was not limited to storage battery service,[11] but they contributed to the applicability of batteries for purposes which would have been impossible without the booster. For example, various forms of automatic, self-excited and differential boosters, which operated more rapidly than the manual boosters, made the battery a valuable adjunct in streetcar power houses during the 1890s.

Before the introduction of the booster, the design of storage battery electrical distribution systems was limited by the fact that the generators could be used either to run in incandescent bulbs directly, or to charge the batteries, but never to do both at the same time, since the voltages for the two functions were different.

---

[10] Boosters could also be used to do away with the need for end cells.

[11] For example, boosters were sometimes used in substations at the end of load feeders. When the load on the system was light, the voltage fell little from the power-house end of the feeder to the substation. During periods of high demand, however, there could be a significant voltage fall. The boosters were used to raise this voltage.

3. The First Central Station Battery Systems

In the previous sections of this chapter we examined the changing technology of storage battery use in electric lighting systems. In this section we will examine the process in greater detail by looking at the actual history of various battery central station projects in chronological perspective. The earliest central station battery systems, both in the United States and Britain, were the products of the various Brush companies. There was a considerable degree of variation between the details of design of these several systems, but, as far as the use of the battery was concerned, they were all very similar and resembled the pattern of the Brush Cheyenne system discussed in the last chapter. That is, the batteries were used as transformers, and they were used to store a large percentage, if not all, of the current produced in the dynamos. Another major similarity between all the Brush schemes is that they were all financial failures.

The English engineer Henry Edmunds was involved in the design of Brush central station systems on both sides of the Atlantic. Edmunds began his association with the Brush interests in 1882-1883, when he was in charge of the system of the Brush Illuminating Company of New York. In 1884 he returned to Britain, where he participated in the design of the Colchester system of the South Eastern Brush Electric Light Company.[1] These two systems were essentially identical. The central stations generated current at 1000 volts and transmitted this to batteries in substations near the consumers. Actually, these substations were merely the cellars of consumers who were located in the center of a small cluster of consumers, each substation served perhaps a dozen consumers on the same block. In the Colchester system there were five such substations. The similarity to the Cheyenne system is apparent.

The battery consisted of 480 cells, which received the current in full series, and transmitted it in series-parallel connection of eight groups of sixty cells each.[2] Each substation had two such batteries; these worked in tandem, one charging while the other discharged. The switching, as at the Cheyenne station, was supposed to be done automatically. This was accomplished either by the Brush automatic switch, or by the gas-pressure method, or by an interesting variant on the modern system by which gas station attendants check for the state of charge of car batteries—a hydrometer floated in one of the battery cells below a delicate micro-switch. When the battery discharged, the hydrometer rose, pushing upward on the switch, activating a motor which connected the battery to the charging mains.[3]

No end-cell switches or end-cells were provided, nor was there any other means of controlling the voltage at the lamps as the battery voltage changed. There was no means for using the generators to run the lamps and charge the batteries at the same time; accordingly, all current used to run the lamps had to pass through the batteries. Therefore, these stations suffered to a maximum extent from the low energy efficiency of the batteries—between 25-35 percent of generator-produced energy was lost in the batteries.

A number of years after the failure of the New York and Colchester systems, Henry Edmunds commented on their chief problems:[4]

---

[1] This was one of the franchise companies of the Anglo-American Brush Company.

[2] Parsons, *Early Days*, pp. 52-55.

[3] U.S. Patents: #371,893—18 October 1887; #399,755—19 March 1889.

[4] *Electrical World*, 13 October 1888: 199.

The pressure at the station or the charging mains being at least equal to the sum of all the batteries in series, was either very high or else limited the number of groups on circuit that could be charged at one time (in other words, either the voltage was too high for the lamps, or too low for the batteries—see discussion of boosters in the last section); also, the freshly-charged cells having a higher EMF than those that had been discharging caused serious fluctuation in the lights during the change from the main to the local (circuit). And as the cells automatically changed their position from the main to the local only when they had themselves arrived at a certain degree of charge, sometimes the local cells had got over-discharged, causing serious damage through buckling and deterioration generally. The result was that the Colchester system was ultimately abandoned. (the word *main* refers to the generator-to-battery circuit, and *local* to the battery-to-light circuit—these were old telegraphic engineering terms)

It is clear that without end cells or boosters, the filaments of the lights would indeed receive a shock, since there would have occurred an approximate fifteen-volt jump. In telegraphic engineering, simple control devices such as were used in the New York and Colchester Brush schemes were tolerable, since problems such as damage to filaments did not exist, and the energy-economics of primary cell operation were an insignificant part of the total cost of the telegraphic service. Moreover, since the parts of telegraphic batteries were cheap and quickly consumed, the problem of excessive deterioration through improper maintenance was of less concern than with expensive, permanent storage batteries.

One man who tried to learn the lessons of these two early failures was Henry Edmunds who worked with the Colchester system in 1886, when the plant was auctioned off as junk. He developed a set of more sophisticated control devices which were supposed to avoid the problems indicated above, but still clung to the original Brush conception of the battery-transformer-in-the-consumer's-cellar idea. Edmunds intended his control devices to be automatic, thus clinging to the older conception of power house operation whereby costs were reduced by keeping *labor* costs down.

In 1888 he had a chance to put his ideas into practice when he designed a system to light the Chelsea district of London for the Cadogan Electric Lighting Company.[5] In order to avoid the

---

[5] Parsons, *Early Days*, pp. 56-57.

variable voltage problem, each consumer was provided with four separate batteries, three of which in series gave the proper voltage for lamps. Every few minutes an electric motor-driven distributor shifted one of the batteries out of the three-battery discharge series and placed it on charge. At the same time, one battery which had been charging was shifted into the discharge series.[6] In this way, the state of charge of the three-battery series was kept constant and no noticeable voltage variation occurred.

Despite the Cadogan system's improvements over Colchester, New York, and Cheyenne, it too proved a failure. Cadogan was liquidated in February 1891. At its greatest extent only 25 houses were connected to it.[7] The basic idea behind all these systems had been to save money through automatic operation, an effect the designers hoped to achieve through the use of batteries and self-acting feedback-control devices. These devices never worked properly, however, and the deterioration and inefficiency of the batteries made the systems more expensive than the D.C. ones using no batteries. When Thomas Edison and most of the American electrical engineering community denounced the storage battery, it is important to understand that what they were denouncing was the *Brush-type system,* and not the storage battery *per se.* During the 1880s, the only storage battery system to which American engineers were exposed was the Brush system, and it was natural to associate the storage battery with it. In Britain, however, such was not the case; as we will now see.

R.E.B. Crompton was responsible for pointing the way to the practical use of batteries in electric power systems. Among British electrical engineers, Crompton was perhaps the strongest exponent of batteries.[8] As noted above he began as a defender of the battery-transformer system, and the earliest incandescent systems he erected used batteries in this way. For example, in 1885 Crompton designed his famous system for the Vienna State Opera. Because of design restrictions, Crompton had to locate the powerhouse almost a mile away

---

[6] *Electrical World,* 13 October 1888, p. 199. For distribution patent, see U.S. #393,147—20 November 1888; filed 16 July 1888.

[7] Parsons, *Early Days,* p. 58.

[8] See, for example, his statement in *Electrical World,* 6 July 1895: 20, that "he was not a low tension man at all, but he was most strongly an anti-alternating current man. Alternating currents were no doubt a splendid thing in their way for professors and schoolboys, but nasty things to deal with in central stations chiefly because they do not allow accumulators to be readily used."

from the opera.[9] To raise transmission voltage, therefore, he
generated current at 400 volts and used four sets of 54 cells each
to receive the current. Despite the early date of this installation, Crompton was an astute engineer to see the need for
end-cells and end-cell switches.[10]

The Vienna Opera system was a success, but we know
nothing of its economics, since it was a project of the Imperial
Court and not designed to make a profit. Moreover, Crompton
made no attempt to use automatic controls there. Furthermore,
in all the subsequent systems which Crompton built in England, he used batteries as load-levelers only, and not as transformers. In all these sytems, the economics of the operation
was important.

The first of Crompton's English stations was the Kensington
Court system, built in London in 1886. Kensington Court was
one of the very few central station systems built in Britain
before the 1888 revision of the Lighting Act. In fact, when the
system began in 1886, it was little more than an extended private plant, covering only a few blocks and occupying no public
land. Consequently, it was not subject to the Act.

The system began as a two-wire D.C. system, although it
was later changed to three-wire operation.[11] Crompton used a
dynamo of his own design and incorporated Crompton-
Howell batteries. These cells acted as load-levelers and carried
the full load in case of machinery failure. The specific load-
leveling technique used was minimum load carrying, in which
the battery carried the complete load during the late night and
daylight hours minimum load period, at which time the
dynamos were shut off.

Many of the details of storage battery central station operation were first investigated at Kensington Court. Crompton
built his own home adjacent to the powerhouse so he could
study its operation more intimately and at all hours of the day.
He described this pioneering work in central station
economics in his memoirs: [12]

> . . . . for a long time this question of how our accumulators were
> to be charged and kept in circuit, and how to combine in the
> best way economical working with absolute assurance against

---

[9] Hedges, *Continental*, pp. 160-162; Crompton, *Reminiscences*, pp. 117-118.

[10] Ibid.

[11] Parsons, *Early Days*, pp. 90-91.

[12] Crompton, *Reminiscences*, p. 149.

stoppage of the light, was a difficult problem, and it was some time before we were able to point to long continued periods of working without failure of any kind.

It was these studies which stimulated Crompton to invent his concept of load-factor and publish it in his famous 1890 papers. This stimulation can be better appreciated when it is considered how unfavorable were the load-factors at the Kensington station—16.5 percent at the peak in November, and below 5 percent at the trough in summer.[13]

After the 1888 revision of the Lighting Act, Crompton expanded the size of the Kensington system. In 1890 he created a nearly identical system in nearby Knightsbridge, and the two systems were combined as the Kensington & Knightsbridge Electric Lighting Company. At the same time, Crompton was building a similar battery system in Notting Hill.[14] This one opened in 1891, remained independent of the older company until 1900, when the two companies established a joint A.C. generating station at Shepherd's Bush, a 5000 volt, three-phase system, which transmitted power to rotary converter substations in the neighborhood of each company. Until 1929, these substations continued to supply customers with D.C. After 1929, the system was gradually shifted to pure A.C. and was absorbed into the all-London grid.[15]

During the early years of the systems, the batteries were used to carry the full load during most of the day, with the dynamos supplying current direct to the consumers only during four peak hours in the evening. This method of operation was used at many early and small stations, reflecting that day-load in many early systems was extremely light. It probably reflects the early tendency of powerhouse managements to reduce labor costs by shutting down most machinery most of the day, thus requiring only one full shift per day.

Later, in the 1890s, as these systems grew in size and complexity, the technique of load-leveling used changed to the peak-shaving method described in the last section. Day-load was becoming greater, augmented by electric motors and the daytime use of lights; thus, the cost leverage obtainable from the ratio of day-to-night load went down, while the leverage obtained by shaving off the spiky peaks remained high. When

---

[13] Ibid., p. 136.
[14] *Electrical Engineer* (London), 5 June 1891, pp. 548-549.
[15] Parsons, *Early Days*, pp. 93-95.

the Shepherd's Bush station was built, batteries were retained at the rotary converter substations, and they still were used for peak-shaving. Only with the coming of pure A.C. in the 1930s did the batteries disappear, although they were used in some other London stations well into the 1930s.[16]

The evolution of the Kensington, Knightsbridge, and Notting Hill systems is a microcosm of the technological development which took place in almost all larger European and American cities during the late 1880s, '90s and early years of the new century. The original, two-wire D.C. systems were very small at first, and they often shut down completely between midnight and dusk. As these systems grew, they adapted their operation to a full 24-hour load, and used devices such as storage batteries to improve load-factor. As the systems continued to expand, they met and blended at their edges, and this stimulated the abandonment of local power houses in favor of larger, A.C. stations feeding local rotary converters. In most larger cities, these mixed systems continued into the 1920s, at which time the evolution towards pure A.C. caused the replacement of the last rotary converters and batteries by transformers.

As far as the use of batteries is concerned, the evolution of American D.C. systems was somewhat different from those of Europe, since the Americans never used batteries as extensively. Nevertheless, the technical evolution of D.C. systems on both sides of the Atlantic was identical in almost all other respects. Consequently, the later adoption of batteries as load-levelers in stations not originally designed for them was no problem, and this permitted the American storage battery industry to find a market in the later '90s.

In 1889 the EPS company entered the field of central station electric distribution, choosing Chelsea as a starting point. The generating machinery and distribution network of the system was either manufactured or contracted for by the Electrical Construction Company (ECC), under the direction of their chief engineer, Thomas Parker. As indicated in the last chapter, EPS and Elwell-Parker amalgamated in 1889 and were brought under the control of the holding company ECC.

The Chelsea system was a financial success, and, like the systems set up by Crompton, it grew and developed, eventually to blend into the greater London A.C. grid in the late

---

[16] Personal communication from Dr. M. Barak.

1920s. This success is not surprising, since the Chelsea system was similar in design to Crompton's London plants. Ironically, however, the Chelsea system grew out of a failure; it was descended directly from the Colchester scheme.

The man in overall charge of the Chelsea system was Parker, but the engineer who designed the battery parts was Frank King (1854-1899), who had been hired as chief engineer of the EPS in 1886 specifically to develop equipment for the central station market. Prior to this, King had been one of the designers of the battery-charging equipment of the Colchester system.[17] When the EPS hired King, they also bought the patent rights to the Colchester system's equipment. Although the Chelsea system evolved out of Colchester, it was a more technically mature conception.

First of all, the batteries themselves were better. The cells used in the Colchester project were fragile Planté plates whose structure collapsed quickly, probably from the total "formation" of the metallic lead supports. The Chelsea batteries were the product of over five years of EPS development.[18] Regarding the system's design, the number of battery substations was reduced, and these were maintained by the company, rather than being placed unattended in consumer's cellars. One battery was located in the central station itself, with two others in substations. A variety of end-cells and end-cell switches were introduced to maintain constant voltage.[19]

At Colchester all the current used by the lamps had to be passed through the batteries; at Chelsea, however, rotary transformers were used in the substations. At times of peak load, these machines carried the load. The batteries carried the load during the day and late night hours. Gradually, the roles of battery and rotary transformer were reversed, with the rotary transformers becoming the dominant source of current, and the batteries the auxiliaries. Thus, the evolution of Chelsea was analogous to that of Kensington, Knightsbridge, and Notting Hill.

---

[17] The B.T.K. system—Beeman-Taylor-King. King was the inventor of the gas pressure, as well as the hydrometer method of controlling charging. See f.n. #3, this section.

[18] Parsons, *Early Days*, p. 62.

[19] The actual method was to use a series of counter EMF cells, whose function it was to reduce the initial high EMF of the full battery. To be precise, this is the reverse of the end-cell systems, where cells are added to increase the EMF of a partially-discharged battery. The principle of operation, however, was the same in both cases.

In his initial design for the Chelsea battery stations, Frank King retained the original Colchester idea of reducing labor costs by making the operation of the substations fully automatic, but though the automatic machinery installed at Chelsea was more sophisticated than that at Colchester, it was soon abandoned in favor of manual operation.[20]

The success of the Chelsea undertaking resulted directly from the organization of the companies which designed it. The durability and maintenance of the batteries were assured by their manufacture and care by EPS. The rotary transformers, which were some of the first ever made, were the design of Thomas Parker, who was responsible for designing dynamos more suitable for charging batteries than those used at Colchester. Another factor was that King had reoriented his original thinking about battery use in lighting systems more along the path which R.E.B. Crompton was following.

The lessons learned by EPS and ECC at Chelsea were applied two years later when they designed a lighting system for Oxford. This system, which was chiefly the work of Parker, downplayed the central role which batteries had had at Chelsea, and relegated them to a purely auxiliary, load-leveling function. At Chelsea, the batteries had been used as load-levelers *and* as transformers—receiving the current at 500 volts and transmitting at 100. Parker's rotary transformers gradually replaced the battery's transforming function at Chelsea. At Oxford, however, Parker rotaries were used from the first for the conversion of 1,000 volt transmission voltage to 100 volt consumer current. The batteries were used only as load-levelers, carrying the day-load.[21]

In addition, the concept of using automatic machinery to control the batteries was never even tried at Oxford. The battery was installed at the fully-manned central station and substations where it was constantly looked after. The Oxford system can be seen as a further evolutionary development away from the original Brush-King Colchester scheme. It was widely copied during the 1890s and became one of the prototypes for the pure D.C. high voltage transmission system. For example, the Oxford System was used in 1898 when the two towns of Bromley and Chiselhurst in Kent were connected by a mutual

---

[20] R.H. Parsons, *Early Days*, pp. 64-65.

[21] The station was located ¾ of a mile away from the electrical center of the system, and accordingly some method of voltage elevation was necessary to maintain economical operation. Parsons, *Early Days*, p. 65.

system.[22] The generating station was built at Bromley, with a substation at Chiselhurst about three miles away.[23] Both stations contained batteries for load-leveling. A nearly identical system to that of Bromley was installed at Wolverhampton in 1895. It also used transmission current of 2,000 volts to rotaries and storage batteries for the purpose of carrying day load.[24] Since Wolverhampton was the center of ECC operations, it is not surprising that the Oxford system was used there. A similar system was installed at Shoreditch in the early 1890s.[25]

As a result of the lessons learned in setting up the systems described above, as well as other systems of essentially similar design, the broad outlines of the proper use of storage batteries in central stations had become established by the early 1890s in Britain. This technology was transferred to the United States through the Electrical Accumulator Company of New York, as well as through close cooperation between European and American battery firms. Blueprints and detailed descriptions of new central stations appearing in the technical press in the early 1890s reveal that by this time the auxiliary equipment required to control batteries had reached a good degree of sophistication—end-cell switches, boosters, and special switchboards were all commercially available.

---

[22] Textbooks of the period often used this term generically for any high voltage D.C. rotary transformer system.

[23] Transmission current was 2,000 volts. *The Electrician* (London) 26 January 1900: 462-463.

[24] Parsons, *Early Days*, pp. 69-70.

[25] *Electrical World*, 19 March 1898: 351.

4. American Growth

On 25 July 1889 the engineer-in-chief of the Edison Electric Light Company of New York sent a letter to Thomas Edison recommending the use of storage batteries in the Edison lighting systems. In particular, he drew attention to the Philadelphia Edison Company.[1] Mr. Marks observed that the suburban areas around Philadelphia had no electric service

---

[1] Marks to Edison, 25 July 1889, Edison National Historical Site Archives (hereinafter: ENHSA).

yet. His plan was to run high voltage D.C. charging mains from the central city powerhouse to five rotary transformer and battery substations in the suburbs. The batteries would reduce the amount of copper needed in the mains. Suburban districts had very low load-factors, and if batteries were not used, the mains leading from the central station would have to be large enough to handle the peak load, even though this might last only two or three hours. A battery, however, could absorb current all day long through much thinner mains, and dispense it quickly through thicker but shorter local supply feeders during peak hours. Marks also stated that the battery would be useful as a load equalizer on either side of the three-wire neutral of the consumer's feeder circuits.

In the late '90s, these ideas became almost universal practice in big city Edison systems. This letter, however, was one of the first references by a representative of the Edison interests to such techniques. Thomas Edison's reply to Marks show that he had not altered his early suspicion of batteries:

Marks ---

> Would like to see a rough estimate about storage battery. I have never yet been able to learn of a 1000 horsepower of storage than can be bought as cheap as 1000 hp of boiler and dynamos.

> Edison

Thomas Edison's opposition to batteries was probably the most important factor in the delayed emergence of an American storage battery industry. During the 1880s a considerable number of firms were set up in the United States to install incandescent lighting systems. The only companies of significant size, however, were Thomas-Houston Electric Company, United States Electric Lighting Company, Westinghouse Electric Company, and Edison Electric Light Company.[2] Of these, only Edison and United States were significant for battery use, since Westinghouse concentrated on A.C. and Thomas-Houston was concerned mainly with arc lights and A.C. incandescents. The Edison company concentrated solely on D.C. incandescent lighting and dominated this field both in the area of private plants as well as central station systems. It had no significant competitors in central stations and in private plants

---

[2] After 1889, Edison General Electric. Passer, *Electrical Manufacturers*, pp. 105-128.

it far outdistanced its only serious competitor, the United States Company.

In the 1880s and early 1890s, the only market for storage batteries sizable enough to support an industry was incandescent lighting. Consequently, Edison's decision to exclude batteries from his systems doomed the American storage battery industry to a peripheral existence. The United States company also rejected the use of batteries.[3] In 1885, 448 or 449 private plants installed in the United States were the work of these two firms.[4] Although these data are doubtlessly incomplete and omit the work of small, local companies, they strongly illustrate the difference between battery company economic opportunities in the United States and Europe. In Britain, the leading private plant designers included such battery exponents as Crompton & Co., EPS, and Elwell-Parker. In Germany, AFA was begun with financing and support from electrical giants such as Siemens.

Therefore, during the 1880s in the United States, we find that the only lighting companies to use batteries were those of insignificant size and brief lifespan. One such ephemeral enterprise was L.C. Kinsey & Company; a Pennsylvania firm formed in 1886 and dissolved in 1889 or 90. Kinsey was one of the firms mentioned briefly at the end of the last chapter which were set up to install the batteries and control devices manufactured by the EAC. The company purchased the dynamos, cables, bulbs, fixtures and other non-battery equipment from other manufacturers and designed the specific lighting installations themselves.[5]

L.C. Kinsey began with a wave of enthusiasm in 1886. Responding to the general upturn in the national economy, scores of requests were received from prospective customers for estimates for private plant installations through 1886 and 1887. Although Kinsey closed few actual contracts, they had all the business they could handle.[6] By early 1888 the company had established a 130-light plant at a furniture factory, as well as

---

[3] Bell Telephone Company Letterbooks, Pennsylvania Historical & Museum Commission, Harrisburg, Pa. (hereinafter: BTCLB).

[4] Eighty by the U.S. Company, 369 by Edison. Passer, *Electrical Manufacturers*, p. 115.

[5] L.C. Kinsey to Jackson-Morgan Co., New York, 23 February 1887, p. 72 (of microfilm); L.C. Kinsey to C & C Electric Motor Co., New York, 15 March 1887, p. 132. BTCLB.

[6] Rhodes to Edward Joy, 5 June 1888, p. 154. BTCLB.

the full equipment of three small to moderate sized central stations, and the battery equipment alone for a pre-existing small central station.[7]

By the latter part of 1888, however, the Kinsey company came upon hard times. The battery at their Watsontown central station failed and this led to bad publicity for the company.[8] As one of the firm's engineers wrote in January 1889:[9] "The Watsontown Electric Company is in a precarious condition and as on account of having been so extensively advertised, its failure would be a death blow to the sale of accumulators in this section of the country."

In addition, the worsening economic conditions produced fewer contracts and the company soon found itself with no business. On 15 September 1888, L.C. Kinsey wrote a long letter to EAC, detailing his firm's problems and some possible solutions. This letter is worth examining at length, since it provides a vivid picture of the conditions faced by most American storage battery interests at this time. It begins with a survey of the general situation:[10]

> Now is the time for obtaining contracts for electric lighting; but, our present high prices preclude success. The conditions of 21 months ago (date of firm's founding) are not those of today. Then our field was large in extent and possibilities, the accumulator wasted an inappreciable amount of energy, practically lasted for years, required very little attention of the cheapest kind, could be charged easily by any machine, used in any way and, generally, was the universal panacea for all the ills of electric lighting.

Most of this last sentence was not intended as an actual description of the battery's characteristics, but as an ironic comment on the way those characteristics had been described by the EAC 21 months earlier.

Kinsey went on to describe the potential customers that his company had targeted in 1886:

---

[7] The new central stations were at Watsontown (500 lights), Bethlehem (1000 lights), and Allentown (2000 lights). The existing plant was at Phillipsburg (200 lights). All were in Pennsylvania. Although these plants seem minuscule by modern standards, they were of respectable although undramatic size compared with typical Edison central stations of the period. Of 185 Edison central stations installed by the end of 1888, for example, the average number of lamps per station was 2,080. Passer, *Electrical Manufacturers*, p. 121; Rhodes to Dey Bros., 5 June 1888, p. 155. BTCLB.

[8] Kinsey to EAC, 11 September 1888. BTCLB.

[9] Davis to D.H. Bates, 31 January 1889, p. 217. BTCLB.

[10] Kinsey to EAC, 15 September 1888. BTCLB.

1. The larger towns having no (lighting) plants.
2. Towns with arc light companies, which could run the dynamos during the daytime and store the energy in accumulators for use at night (i.e. the Brush system)
3. Small towns where the direct system (i.e. an Edison-type non-storage battery system) could not enter (i.e. where it would not pay); but where surplus power, in connection with the extraordinary economy of operations (i.e. where high load-factor could thus be achieved) admitted of profitable 500 light plants on the accumulator system. (Here he is speaking of the minimum-load load-levelling concept, as described earlier for the Kensington station).
4. Towns possessing plants which would be compelled in self-preservation to seek accumulators.

The writer then shows why each of these potentially rich markets proved illusory:

Our experience has demonstrated:
1. That there are no larger towns in Pennsylvania or New York which do not possess (lighting) plants; while those who do, use the experience of these plants as an unanswerable argument against accepting the risk of establishing a plant to compete with one already having the choice lighting, without enough margin for unestimated but actual expenses.

Kinsey's free-floating syntax gives this sentence a number of possible meanings, but he apparently refers to the fact, discussed in the last chapter, that the early central stations were established in the wealthiest areas, but that even they were not very profitable. How, therefore, could a new group of entrepreneurs expect to make a profit by installing a station in a less lucrative area and with the added expense of a storage battery, which was not in the original station?

2. That the use of ordinary arc machines (i.e. dynamos) is not practicable (for charging batteries) and that the installation of an accumulator incandescent system in connection with an arc-light plant is not cheapened thereby.

As electrical power engineering grew more sophisticated, engineers learned that all dynamos were not the same.[11]

---

[11] For battery use, shunt-wound dynamos were preferred, whereas series-wound machines were usually preferred for arc lights, thus, the combination of arc lights and storage batteries, although not absolutely precluded because of the difference in preferred generators, was definitely a negative factor. C.C. Hawkins & F. Wallis, *The Dynamo* (London, Whittaker, 1903) pp. 500-510.

3. That as the batteries will supply on 83% of the number of
   lamps and less than 90% of the energy originally calculated
   (i.e. that their energy efficiency is low; and lower than 90%
   too), and as their life is very limited, as an extraordinary
   number of high priced extras are necessary (boosters, end-
   cells, etc), as the use of ordinary surplus power is not satisfactory (see below), the first cost as well as operating expenses, render impossible a profitable 500 or even 1000 light
   plant.

The reference to surplus power refers to the idea that factories could use part of the energy generated by their pre-
existing water wheels or steam engines to drive the charging
dynamos, instead of only driving power looms, lathes, and
other machinery. The difficulty of making such systems pay
was apparently no exaggeration, if we compare this with
R.E.B. Crompton's statement about his own difficulties at the
Kensington station in 1886-1887: "4. That motors, in connection with a direct system, transform the accumulator from a
necessity into an expensive luxury."

This is a reference to the improvement of load-factor due to
the increase of day-load by electric motor service. This comment was probably more an anticipation of future developments than a description of existing conditions, since the use
of electric motors was not extensive in late 1888. Nevertheless,
it was an accurate and perceptive bit of forecasting.

The letter then goes on to show why the battery system of
lighting is at an economic disadvantage in comparison with
"direct" D.C.:

The actual first cost of installation of an accumulator system is
from 33 to over 100% greater than that of the direct[12] and in
consequence, we have lost four contracts, which have been
taken by the latter (i.e. Edison, Westinghouse or United States
contractors), have failed to close six others and now see possible
chances drifting to the direct people.

The writer then goes on to show that if EAC was to provide
the equipment to Kinsey for a plant of equal lamp capacity to
one which Westinghouse contractors just installed in Kinsey's
territory, the EAC would have to sell the equipment below cost
in order to meet the Westinghouse price. In order to reduce
costs, he suggests:

---

[12] The wide variation here was caused by the higher percentage costs for smaller
plants. The costs for auxiliaries didn't vary much with size, and because a minimum of
52 cells must be provided, regardless of the number of lights, makes economy of scale
most important for the success of battery lighting systems.

. . . . turning the attention of your electrical experts to devising means for doing without some of the material now considered necessary and of your mechanical experts cheapening the cost of production of that material (yet as we have seen, the evolution of electrical engineering practice with regard to batteries was in the other direction). That some such policy is needed seems evident when it is remembered that we have lost contracts and are losing them today by the United States, Edison and Westinghouse people installing plants at prices with ours impress the public as closely approximating cost.

While we regard it as an experiment, something must be done when after two years of trial the Westinghouse (company) can point to a 16% dividend paying plant near an accumulator plant, accepted only as a compromise over 30% discount on first cost. The presence of the lighting season requires immediate action—the Edison people are now working on a plant which we should be able to get—and as we cannot exist on nothing, if the accumulator system can be installed in competition with others we want to go ahead, but, if it must be regarded as a commercial failure, the sooner we begin at something else, the sooner may we hope to retrieve past misfortune.

A similar letter, written seven months later, further clarifies Kinsey's precarious position: [13]

Within our territory, a town of less than 10,000 inhabitants will not, as a rule, support even a 1000 light plant. All towns of that or larger size are already supplied with incandescent light from direct (i.e. non-battery) system plants, less than a dozen of which are paying dividends and none where the number of lights are less than 1500.

Capital has discovered this fact and refuses to invest in a competing system, which, whatever its claims, is in an experimental state and insists upon increased first cost.

The experience of our plants, in failing to equal the original expectations, has destroyed confidence in the accumulator system and today, it is impossible to install a competing plant without the aid of foreign capital.

This letter emphasizes an important, although often ignored fact about early electric utility service—its low profitability. The larger city systems often paid handsome dividends, but those in smaller towns and cities were frequently marginal. Yet even if we grant the point made by the author of this letter, the fact remains that these economic conditions must have been as

---

[13] Davis (Kinsey) to Bates (EAC), 5 March 1889, p. 259. BTCLB.

true for Britain as for the United States, and yet in the former country the use of storage batteries as lighting adjuncts made significant progress in the mid 1880s.

The crucial head-start of Britain lies in the attitude of her top electrical engineers towards the battery. Many of these men assumed, without much economic data, that batteries were essential to the proper working of a D.C. system. During the early 1880s, they saw to it that a large portion of the early private plants were battery-equipped, and, after 1888, that most British D.C. central stations were also equipped with batteries. In the United States, on the other hand, the men who dominated the early lighting industry came to the opposite conclusion and therefore almost all the early American private plants and central stations had no batteries.

Neither the pro-battery position of the British, nor the anti-battery position of the Americans was correct *per se*. Looked at from different perspectives, both positions can be considered either "correct" or "incorrect." Those engineers who simplistically took either a totally anti or pro battery position failed to realize that the individual operating characteristics of particular D.C. stations would determine whether or not the battery could save money for that system. During the 1890s electrical engineers in both the United States and Europe were to come to this more sophisticated conclusion and were to use the data then becoming available to prove it, as we will see in the next section. During the 1880s, however, the opposite *a priori* conclusions about battery usefulness in the United States and Britain determined the different technological evolution of the two countries.

Once such a pattern of technological evolution becomes started, it becomes hard to break. The two Kinsey letters quoted above reveal this. The salesmen of that company found themselves underbid by the private lighting interests of the Edison companies. Much of this disadvantage was no doubt due to the dominant position of the Edison interests in the diversity and scale of equipment manufacture. The Kinsey company, for example, had trouble obtaining properly designed shunt-wound dynamos for battery charging.[14] Private plants installed by the Edison company had all their important equipment manufactured by Edison. Moreover, the Edison private plant company guaranteed the performance of many of its components.[15]

---

[14] Kinsey & Co. to Jackson-Morgan Co., New York, 23 February 1887, p. 72. BTCLB.

[15] Passer, *Electrical Manufacturers*, pp. 116, 118.

In this country, Charles Brush was the only electric lighting pioneer who possessed sufficient importance in the developing industry to have created an early American tradition in storage battery lighting. All the other American battery enthusiasts were very small fry compared to Edison, Westinghouse, et al. Whatever potential Brush had, however, he failed to develop, and the attempt by the Kinsey company to challenge the established economic and technological lead of the Edison companies in the late 1880s was doomed.

The only way that an American battery lighting tradition could have started would have been for the Edison stations to have adopted batteries themselves. To compete against the developed Edison position was impossible, but cooperation might have succeeded. In fact, on a number of occasions the L.C. Kinsey Company tried to get Edison stations to try the battery, but they would have none of it.[16] Eventually, the acceptance of the battery by Edison stations was the way that batteries became integrated into American power engineering practice, but that was not to be until the mid 1890s. One reason for this later acceptance may have been due to the decline in Edison's influence and that of his closest associates after 1892 in the control of the Edison companies; in 1892 the creation of General Electric was accomplished with the top management composed of Thomson-Houston, and not Edison G.E. people.[17]

There is no direct proof that the removal of Edison from control of the Edison systems helped in the introduction of batteries into Edison central stations. Nevertheless, the amalgamations which took place in the 1890s in the electrical industry, of which the formation of General Electric was the most important, had as one of their chief functions the rationalization of electric power technology by the pooling of all existing patents. The commercial success of the Welsbach mantle in the early 1890s significantly cheapened gas lighting and made the search for greater economy in electric lighting all the more crucial.[18] The detailed discussions of the economics of central station operation which American electrical engineering groups began to conduct in the early 1890s were part of this general process. The Battle of the Systems gave way to the Battle of the Bookkeepers, as we will see in the next section.

---

[16] L.C. Kinsey to EAC, 27 April 1887, p. 231. BTCLB.
[17] Passer, *Electrical Manufacturers*, pp. 321-329.
[18] Arthur A. Bright, *The Electric Lamp Industry*, pp. 126-127, 168, 171, 212.

One result of this was that more and more Edison central station managers began publishing their station records in defense of storage battery economy.

Harold Passer, in his pioneering study: *The Electrical Manufacturers, 1875-1900*,[19] makes a fundamental comment about the establishment of patterns of technological evolution. He notes that new electrical systems excited imitation only if they were carried out on a large, expensive scale. Systems which were tried only on a small scale, even if successful, failed to excite great public interest, and did not lead to a rapid burgeoning of their technology. The two examples Passer cites are A.C. power transmission and the overhead conductor electric streetcar:[20]

> It is important to see that use of an innovation on a small scale in a few scattered installations is not of much significance. Almost no one is encouraged to follow and adopt the innovation. The Van Depoele railway installations did not convince Whitney that electric traction was the most profitable kind of motive power.[21] In contrast, Sprague's Richmond installation, because it possessed the technical features which were necessary for economical and reliable operation on a large scale, led Whitney to abandon completely his plans for a cable road. Similarly, the Westinghouse hydro-electric-power installation at Telluridé or the several General Electric polyphase installation in New England did not establish alternating current as the universal technique of long-distance power transmission. But the Niagra [sic] Falls electrification did. . . . When the Niagra Falls Power Company chose alternating current as its method of power transmission, and when alternating current performed satisfactorily, the future of polyphase alternating current for central-station power was assured. Because a firm that had a multimillion dollar investment at stake chose polyphase alternating current, the hundreds of central station owners and the entrepreneurs interested in water-power development were no longer reluctant to invest in polyphase power.

The failure of the American storage battery industry during the 1880s is another instance of Passer's rule. The American Electric Storage Company of Nathaniel Keith; the Pumpelly Storage Battery Company; the Macraeon Storage Battery Company; the Gibson Electric Company; the Anglo-American Elec-

---

[19] P. 339.
[20] Ibid.
[21] See section 8, this chapter.

tric Light Company; the Omaha and Grant Smelting Company—all these firms and all the dozens of similar minis- cule and evanescent American storage battery companies of the 1880s were too small to make the kind of technological "splash" Passer describes above when speaking of the Niagara Falls power project or Frank Sprague's Richmond trolleys. Even the EAC, which was the largest American manufacturer, never created an installation which had publicity-value. Those installations which it did create through the intermediary of L.C. Kinsey, or the other franchised firms mentioned earlier, were not large enough to excite imitation. None were ever reported upon in the technical press, except as very small news notes without detail. In Britain, on the other hand, companies such as EPS sought from the first for maximum publicity and maximum size for their installations.

5. Bookkeeping and the Battery

A number of factors were responsible for the upsurge of interest in battery load-leveling in the 1890s. The factor which was probably of the greatest importance, however, was the availability of detailed design data. By the 1890s, engineers possessed sufficient data about batteries and other pieces of electrical generation equipment to allow them to calculate the costs and benefits of installing a battery in dollars and cents. This factor was particularly important for a piece of apparatus like the battery which is not essential to the operation of the plant. Dynamos, engines, and boilers had to be bought, re- gardless of their cost effectiveness, but a battery had to prove it could save rather than consume money.

During the 1880s, the lack of data created a sense of uncer- tainty which retarded investment in electrical equipment. This is well illustrated in a comment made by W.H. Preece in 1888 during the Crompton-Kapp A.C./D.C. debate discussed earlier: [1]

> There is no question that occupies us, as electricians, at the present moment of more paramount interest than that of the proper distribution of electricity over areas that want to be lighted. There are over this country at the present moment some thirty or forty corporations (i.e. municipalities, not companies)

---

[1] *J. Soc. Teleg. Eng. & Elec.* 16 (1888): 382.

and communities who are anxious to introduce the electric light. Some of them are guided and controlled by consulting engineers, and these consulting engineers are in the unpleasant position of having to postpone their fees by advising their clients that the time has not yet arrived when they can say to them, "Go ahead; I advise you to adopt this system, or I advise you to do that." I am in that position, and have held back for the past two or three years some very large installations because it has been impossible to decide whether the best system to adopt is that of secondary batteries (i.e. D.C.) or that of secondary generators (i.e. A.C.); and I think I shall fulfill the purpose of Mr. Crompton if I give some of the reasons that kept me in the balance like a seesaw between secondary generators (transformers) one day and secondary batteries the next.

Here, of course, Preece refers to the battery-transformer, and not the battery load-leveler system.

One of the most serious uncertainties about batteries was their rate of depreciation. The extremely short life of the first Faure cells had been corrected rapidly, so that by the mid 1880s cells of economically acceptable durability were being manufactured. Nevertheless, great variability still existed between individual plates of the same manufacturer, and these problems of quality control were not solved until the end of the 1880s. It was not until the beginning of the new decade that battery companies felt confident enough to guarantee the life of their products. This was done by offering the purchaser a maintenance contract, which was usually calculated at between 6 and 15 percent per annum of the total purchase price. Although this rate of depreciation was higher than that for mechanical pieces of generating equipment, it was not much higher and was at least a calculable, predictable cost of operation which the station designer could use to predict the amount of savings obtainable from a battery's use.

In order to perform these calculations, however, more data were needed than only the battery's durability. But these additional data were also beginning to appear by the early 1890s. For example, the load curves shown in the previous section were being printed in the technical press. These curves, of course, were different for each station and for different times of year, yet by the early 1890s enough had been learned about their general characteristics for them to be used in preparing estimates of the probable load-factor conditions for planned electric lighting systems.

One other important class of data which became available by the early 1890s was the efficiency of storage batteries—

efficiency defined as the ratio between the amount of energy put in during charging and withdrawn during discharge. Efficiency figures had been given by manufacturers throughout the 1880s, but their credibility was low. Quotations in the 95 percent range were common, although some in the 40 percent range were also given. Most of these figures were advertising gimmicks with little experimentation to back them up. Another of the chief problems connected with early battery efficiency figures is that they were improperly defined. The energy efficiency must be calculated on a watt-hour basis, but manufacturers often used the ampere-hour efficiency instead, since it is higher. Amp-hour efficiencies are often in the 90-95 percent range, with watt-hour figures being about 10-15 percent less.[2]

By the 1890s, engineering journals were educating their readership in this difference. Instead of being quoted in terms of pure numbers, manufacturers began properly to define the units used in efficiency reports.[3] Consequently, electrical engineers grew to place more trust in such quotations and learned to expect energy efficiencies from quality batteries in the 75-85 percent range.

By the early 1890s, therefore, the technical debate over storage batteries in electric lighting had left behind it the simplistic "Battle of the Systems" stage, and had entered a more mature, cost-accounting period. During the earlier period, the exponents of the battery, such as Crompton, had tried to have it fulfill a more grandiose role than its inherent limitations could support. The battery was supposed to have played a *central* role in electric distribution schemes—receiving all the current generated by the dynamos, transforming it, and then feeding it to the consumers. The battery was seen as a direct competitor with the transformer and A.C. dynamo. This was an exaggerated role for such an expensive, bulky, heavy, maintenance-intensive, inefficient device, and the 1890s saw the battery assume its more modest, auxiliary and cost-saving role as a load-leveler.

The Battle of the Systems debates ended during the early 1890s with an unspoken compromise between A.C. and D.C. D.C. was assumed to be the preferred system in central city

---

[2] The product of watts and hours is a measure of energy and differs from the ampere hour efficiency in that the watt is the product of volts and amps. Watt hour efficiency is lower than amp hour efficiency because of the fact that a battery's charging voltage is always greater than discharge voltage.

[3] For example, *Electrical World*, 30 November 1895: 585-586.

areas—areas densely populated and therefore requiring relatively short transmission lines and filled with applications for electric motors. A.C. was the preferred system for less densely populated suburbs and the means for generating the initial transmission current for the central city D.C. systems. This basic compromise lasted for the next 30 years. As far as the battery was concerned, therefore, the new battle of the systems of the 1890s centered on the question of whether or not to add a battery to a D.C. generating station. For those D.C. systems in which the primary generation was A.C., the question was whether to add batteries to the rotary converter substations. Such debates centered on careful cost accounting using the data mentioned above. A good example of one such debate took place at the National Electric Light Convention in Cleveland in 1895. The debate occurred in the discussion period at the end of a paper delivered by Nelson Perry; an Edison-system manager.[4] Perry defended the traditional Edison method of meeting peak loads with auxiliary generators, and presented elaborate capital and operating cost data to show that batteries gave only small savings for relatively high expense. He admitted the utility of batteries for stations with unusually unfavorable load-factors, but his choice of a 25 percent load-factor for the theoretical station upon which he based his calculations was unrealistically high for the time and a bit unfair to the advocates of battery use.

One of those present at the discussion was Herbert Lloyd, president of the Electric Storage Battery Company of Philadelphia. Lloyd attacked Perry's anti-battery position, and presented equally detailed cost data to prove the benefit of batteries.[5] Moreover, as Lloyd pointed out, Perry's figures would have appeared more favorable for batteries if they had been calculated on the basis of peak-load use, in which the discharge would have only lasted 2-3 hours, instead of the unrealistically long 10-hour discharge Perry used.

The important thing about this debate is not that Lloyd and Perry reached a consensus about battery use, which they did not, but that the debate was a vehicle for revealing to the contemporary American electrical engineer the proper as well as the improper use of batteries in central stations. Perry used

---

[4] Ibid., 2 March 1895, pp. 274-276.

[5] For example, Perry put the cost of a power station battery at $35 per horse-power hour, and gave maintenance costs at 10 percent per annum. Lloyd claimed that these figures should have been $24 and 6 percent per annum. In fact, both sets of figures would have been accurate for different companies and different central stations at that time.

his data to evaluate and condemn the older conception of battery D.C. systems, that is, in which the battery was used to store all or most of the current generated by the dynamos. Perry's negative conclusions about such systems were accurate, although, for the time when he was speaking, he was flogging a dead horse. Lloyd, defending the peak-load use of batteries, was reflecting the best contemporary European practice and the practice which was then becoming standard in the United States as well.

The statistics presented by Perry and Lloyd, although divergent, were detailed and verifiable. Their value for the American storage battery industry can be gauged from a comment made by one of the participants in the Perry paper discussion:[6]

> It is well known that Europe has far excelled America in the development and applications of the storage battery. This has not been entirely due to apathy manifested in this country towards the storage battery, but rather to positive prejudice. This prejudice has been so marked in the past that those interested in the development of the storage battery have been accorded derision instead of encouragement. In such bad repute have they been held that no scientist dared risk his reputation in thorough investigation of this class of electric apparatus. But a change seems to be taking place; the storage battery is rapidly becoming an important factor in the economies of electrical engineering.

The prejudice which Professor Stine mentioned was being broken down in the mid 1890s in the United States by such theoretical discussions as those between Lloyd and Perry. Of even greater importance, however, was the fact that by the mid 1890s several American central stations were already operating with batteries, and quantitative reports of their performance were appearing in the technical press. For example, the operators of the Germantown, Pennsylvania three-wire system installed a battery in December 1892, and one report of its operation was presented during the above-mentioned Perry-Lloyd discussion.[7]

Germantown was run on a 24-hour basis. It had such a low day and late night load that the low load-factor made the station a financial failure.[8] The battery was installed to cure this

---

[6] Comment by Prof. Stine, *National Electric Light Association Convention #18* (Proceedings) pp. 80-81.

[7] *Electrical Engineer* (London) 6 April 1894: pp. 404-405; *Electrical World*, 2 March 1895: 274-276.

[8] Average day-load was about 50 amps, or the equivalent of only 100 16 candle-power lamps.

and was run on what was called the "minimum load" system: from dusk until about 11 P.M. the generators ran all the lamp circuits direct and the battery sat idle. From 11 P.M. until dawn the generators used an increasing percentage of their output to charge the battery. At dawn the generators were turned off entirely and the batteries carried the full day-load until dusk, when the cycle began again. This method of using batteries is not the same as the previously discussed "peak-load" technique, but the principle of improving load-factor was the same.[9]

The 1892 battery set-up at Germantown was a small one since the owners considered it only experimental and were uncertain of its results. Therefore, they kept meticulous records of the first year's operation. The results of this experiment are illustrative of both the practicality of batteries for load-leveling and the ignorance of American engineers of the proper working of this technology in 1892. The records revealed a saving of $2,500 because of the battery, and an overall profitable year for the company, as contrasted with red ink in the past years. The records also showed, however, a depreciation rate for the battery of 25 percent per annum; much in excess of current European practice.

The reason for the rapid deterioration of the battery was that its total capacity had been calculated on the basis of the total day-load. That is, the day-load which the battery was supposed to handle was X ampere-hours, and therefore the designers of the system had installed a battery of X ampere-hours capacity. Consequently, the battery had to be completely discharged each day, producing maximum sulphation, maximum swelling of the active materials, and all the other ills which such excess leads to. In February 1894 the Germantown

-----

[9] As a rule, smaller stations tended to use the Germantown "minimum load" systems, while the larger, city stations used the "peak load" method. The chief reason for this was the lack of "motor load" in small towns. The use of electric motors in large cities gave central stations a greater day-load relative to night-load than was the case in suburbs such as Germantown. The use of batteries to shave the peak-load thus gave city stations greater cost leverage in reducing load-factor losses than if they had used the battery to carry the minimum day-load. It will be apparent that the day-load versus peak-load problem was yet another skin on the onion of possible alternative designs facing the early electrical engineer. He first had to choose between A.C. and D.C. If he selected D.C., he had to decide whether or not to use a battery, and if he designed the station for the battery, he then had to decide which load to use the battery for. As the 1890s progressed, these technical "onion skins" were peeled away as a standardized body of storage battery application techniques was evolved and codified in textbooks. The detailed reports of central station operation, appearing in the technical journals, formed the groundwork for such codification. See: e.g. *Electrical World*, 25 January 1896: 101.

company, realizing the source of its durability problems, increased the size of the battery 2.5 times.[10] From the detailed discussion which took place in the American technical press in 1894 about the Germantown experiment, it is clear that the company's engineers recognized the reasons for the initial low life of the battery. More importantly, it would have been clear to a contemporary reader of these articles that American engineers were coming to understand and solve the detailed problems of storage battery operation; consequently, they gained confidence in the profitable use of batteries and ceased thinking of them, as Thomas Edison had done, as useful only to stock swindlers.

A second American central station to install batteries during the early 1890s was the three-wire D.C. plant of the Boston Illuminating Company. Detailed statistics of this plant's operation were also revealed at the 1895 Cleveland convention of NELA. The battery and installation were the work of AFA, who agreed to maintain it for ten years at 6 percent per annum. The leading German storage battery manufacturer was responsible for a number of early American installations. The Boston system went into operation early in 1894 and was used to carry the peak load for 1½ hours early each evening. At the end of a year of careful bookkeeping, the company found that the load-leveler saved 10 percent on their coal bill. They were pleased with this performance and ordered an additional 200 percent battery capacity.[11]

The expense of such an installation as that at the Boston station reveals why engineers were reluctant to experiment with batteries before reliable design data were available. The Boston battery, including the auxiliary apparatus cost $130,000 for a 9,900 ampere-hour battery. The management calculated, however, that the cost of extra boilers, engines, and dynamos for handling the 90-minute peak load would have been double that of the battery for the same amp-hours.

Although storage batteries were used at a few American central stations before 1895, this year marked a watershed in the thinking of the American electrical engineers towards battery auxiliaries. This watershed was caused by the publication of central station statistics, replacing the grandiose, but usually valueless pro-battery advertisements of the 1880s and early '90s.

---

[10] *Electrical Engineer* (London) 6 April 1894, pp. 404-405.
[11] *Electrical World*, 11 January 1896: 54-55.

6. Trolley Car Economics and the Battery

During the final years of the last century the American storage battery industry began to match its European elders in size and importance. In fact, by the early years of the new century, the leading British manufacturer became a wholly-owned subsidiary of the leading American producer, thus reversing the situation which had existed in the 1880s between EAC and EPS. Yet we cannot explain this situation solely on the basis of the material presented in the last section, since the use of batteries in lighting central stations never became as widespread in the United States as in Europe. The answer to this problem lies in the development of the electric streetcar, whose early evolution was almost totally American.

The development of the electric streetcar had two important but very distinct influences on the evolution of the storage battery. One of these influences was the use of batteries as on-board power sources for self-propelled streetcars. The other influence was the use of batteries in trolley car power stations as load-levelers; essentially, this was a spin-off technology from battery load-leveler use in lighting power houses. Because the design features of streetcar power stations were very similar to those of lighting stations, the transfer of technology from the one to the other was extensive and not limited to the use of batteries alone. Among other things, the problem of load-factor was common to both, and the difference between peak and average load was as serious an economic problem in trolley stations as in the older lighting power stations.

In the case of streetcars, however, the problem of load-leveling was different than in lighting stations. In lighting, there is generally one peak period every 24 hours. This lasts for a significant period of time—one to three hours—and its timing can be accurately predicted. Consequently, extra boilers, steam engines, and dynamos could be used instead of a battery; these being started up and shut down in rhythm with the 24-hour cycle of consumer demand. Edison favored this approach. The design engineer's job was to calculate whether this system or the battery system was more economical for the particular 24-hour cycles his station would have. In practice, this was usually a question of balancing the initial capital costs of the two methods (the batteries were usually cheaper here) against the efficiency, maintenance, and depreciation costs of the two methods (batteries always more expensive here).

For the streetcar stations, on the other hand, peaks and troughs of demand occurred continually, unpredictably and for very brief periods. These variations are caused because the current consumed on startup by car motors is much greater than that for cruising. Startup current for a fully-loaded car on a steep upgrade could be several times that of the same car cruising level or downhill. A number of examples demonstrate this. The average daily load on the electric railway of the Baltimore Belt Line Tunnel in 1902 was 175 kilowatts. When all the cars were moving at cruising speed, the station needed 525 kw, but the peak load when all motors started up at once was 1300 kw.[1]

At about the same date, on the Plymouth, England trolley system, the average demand was 80 amperes, with the peaks and troughs being 200 and 0 amps respectively. It is obvious that the generators at the power station would be alternately uneconomically underloaded and damagingly overloaded by this kind of demand. It is also obvious that the classical Edison solution of using auxiliary steam plant and dynamos to meet the peaks would have been unsuitable here. Consequently, the use of batteries as load-levelers became more important for trolley service than for lighting stations.

Streetcar load-leveling batteries worked analogously to those in lighting stations. The battery "floated" across the mains of the powerhouse between the dynamos and the streetcar trolley wires. In this way, the output of the generators could be kept stable. Although battery use for streetcar service was more valuable than for lighting, it also required more sophisticated development of the technology of battery control. For example, the simple, manually-operated boosters of the lighting stations were clearly too slow acting to handle the rapid fluctuations of streetcar service. Consequently, a series of automatic booster designs were developed during the '90s. Most of this work was American, carried out by engineers employed by the larger battery firms. For example, Justus Entz of the Electric Storage Battery Company of Philadelphia took out many of the fundamental automatic booster patents.[2] The complexity of this design work illustrates another reason why the small battery firms of the 1880s, with perhaps only a single engineer running the company, could not have formed the basis for the large-scale industry of the 1890s.

A number of forms of automatic boosters were manufactured, each designed for varying kinds of load fluctuations.

---

[1] *Electrical World* 39 (1902): 297.

[2] U.S. Patents: #625,099—16 May 1899; #625,098—16 May 1899; #625,100—16 May 1899.

The most commonly used type at trolley stations was the self-excited differential booster. The differential booster's function was to speed up the normal charge or discharge rates of the battery so that it could meet the very rapid and violent voltage fluctuations encountered in trolley service, that is, since the rates of discharge were often greatly in excess of those which the battery's own EMF could produce, the differential booster provided the additional EMF. The figure below shows a simplified schematic of a differential booster circuit:[3]

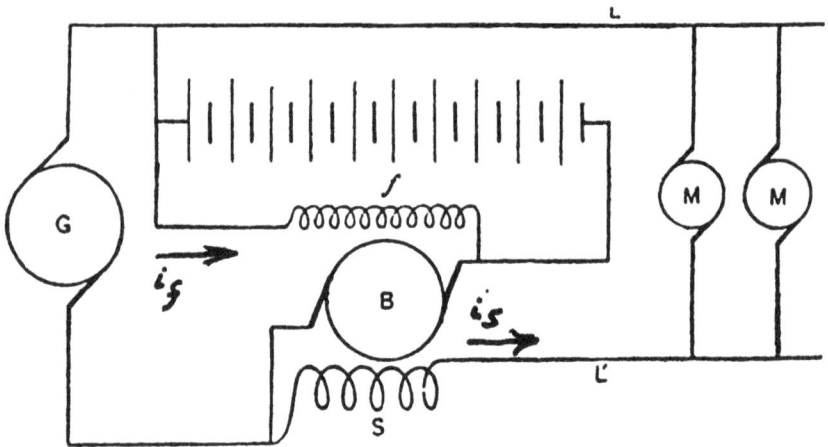

The booster B has two independent sets of field coils; one in parallel (f), and one in series (s), with the load M. The load is a group of trolley car motors. The field coils are wound to oppose each other. When the load exactly matches the design output of the main power generator G, The currents i_f and i_s energize the coils f and s equally, causing zero torque in the booster motor and no voltage output by the booster generator, and the battery will neither charge nor discharge. When the load increases, however, i_s rises, i_f falls, and the effect of coil s overpowers that of f. Since s is wound to cause the battery to discharge, the battery is aided in carrying the difference between the normal generator output and the present high demand. The reverse procedure occurs when the load is less than the generators', and the battery is recharged.[4]

---

[3] Lyndon, *Battery Engineering*, p. 339.

[4] This simplified description suffices to demonstrate the principle of the differential booster, although it does not illustrate the actual operation of the physical machines used in power stations. The most commonly used differential booster, for example, had two separate series coils for more precise maintenance of constant speed on the generator. See: ibid, pp. 322-443 for one of the clearest treatments of this technology.

This new use for the battery provides us with an object
lesson about the unpredictability of technological evolution.
As we have seen, the idea of using batteries as load-levelers in
central lighting stations was one of the chief motivating factors
behind the battery manufacturing enthusiasm of the late
1880s. During this period, however, fears were expressed that
the growing use of electric motors, particularly for streetcars,
would increase the day-load of lighting stations and thus reduce the utility of batteries as load-levelers. In fact, the opposite turned out to be true. By the turn of the century, the sale of
storage batteries for trolley stations in the United States was of
considerably greater economic importance for the battery
manufacturers than for lighting stations.

As in the case of electric lighting, the storage batteries could
be installed either in central stations themselves, or in rotary
transformers or rotary converter sub-stations. The generator-
load (dotted line) and battery-load at an American trolley substation around 1904 is shown below: [5]

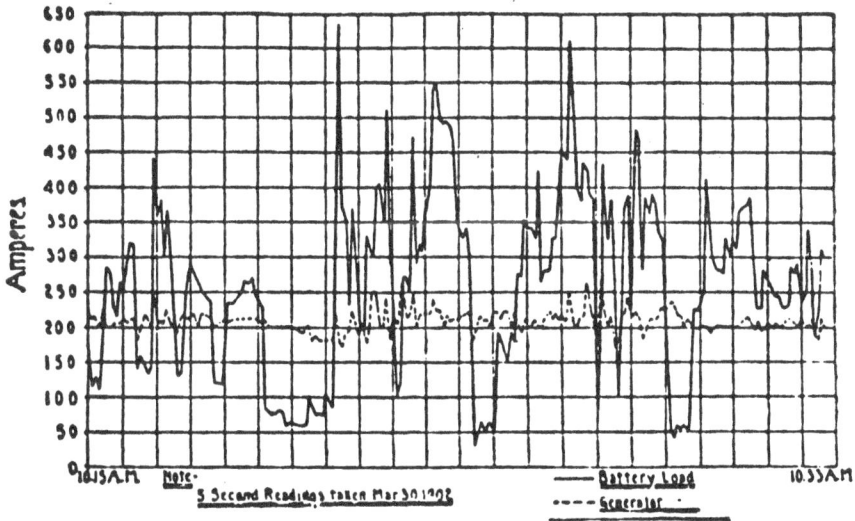

The shape of the battery load curves on this chart indicates
something else about a fundamental change which began in
storage battery manufacturing in the 1890s. Most battery design until the mid '90s stressed the total energy storage capacity of the cells. The shape of the curve on p. 152 reveals why
this was so—each day the battery was called upon to store a
significant percentage of its overall capacity for several hours

[5] W.C. Gotshall, *Notes on Electric Railway Economics and Preliminary Engineering*, p.
141.

and then give back all this energy during peak or minimum load. This kind of operation is called deep-cycling. The curve above however, reveals a very different kind of operation. Here the discharges, although often violent, are quite brief; none is longer than a minute.

Batteries designed for deep-cycling must be constructed differently from those intended for short, sharp discharges.[6] In order to serve the new motor-load market, battery companies began designing such batteries in the mid 1890s. The *Street Railway Journal* of February 1899, reported that the reason for the recent large-scale use of batteries in trolley stations was the development of rapid-discharge-rate batteries during the prior few years.[7]

The commercialization of high discharge batteries was a factor in the introduction of electric motors into other kinds of industrial processes. For example, in steel mills the use of electric motors simplified the back-and-forth rolling of ingots, since the reversal of current replaced the mechanical shifting of gears. In 1908 the Edgar Thomson works of Carnegie Steel installed such an electric rolling mill. During the start, stop, and reverse operation, the load on the motors varied from 0 to 15,000 amperes in a matter of seconds, creating the same electrical supply problems faced by the streetcar stations. A 10,000 amp storage battery-booster combination was therefore installed to smooth out the fluctuations.[8] Early electric motor elevator systems also used such auxiliaries if the power was supplied by a private generator on the premises.[9]

The increasingly economic significance of streetcar power stations for the United States storage battery industry can be gauged from the following statistics:[10]

Year	Number of Trolley Systems	Miles of Track
1890	200	1400
1891	280	2500
1894	610	8000
1895	880	10600
1904	750	22000
1907	900	34000

---

[6] Deep cycling batteries, for example, need a large reservoir of electrolyte and fairly heavy plates. Cells for sharp discharges, however, need less, but more concentrated electrolyte, and more, but thinner plates.

[7] Vol. 15: 110.

[8] *Electrical World* 51 (1908): 403.

[9] Alexander Gray, *Principles and Practice of Electrical Engineering* (New York, McGraw-Hill, 1917) p. 180.

[10] C. Francis Harding, *Electric Railway Engineering*, p. 11.

The reasons for this phenomenally rapid growth in electric streetcars after 1889 were two-fold: first, most of the early installations involved the conversion of existing horsecar lines to trolleys, which required little capital, and, second, the extraordinary differential between the profitability of trolleys versus horsecars and cablecars was being revealed in the publication of the records of the early systems installed in 1888 and 1889. The timing of this mushrooming growth was also critical; it occurred contemporaneously with the first publication of European data on the cost effectiveness of electric lighting load-leveling batteries and the initial attempts by Americans to use the technology.

During the 1890s the United States became Europe's equal in the production and design of storage batteries. Much of this catching up can be credited to the simple copying of European technology, but much of American progress, stemmed from the need to develop storage battery technology along paths not yet trod in Europe. Batteries designed for trolley station use were introduced more slowly in Europe because of the much slower rise of the trolley industry in Europe. For example, in 1895 there were 36,121 electric trolley cars running on American streets. As early as 1890 there already had been 5,592 cars.[11] Yet, for all of Europe, there were only 1,236 cars in 1895, of which 625 were German, 152 French, 129 Austrian and 125 English.[12] In 1896, 17 electric roads were in operation in Britain, with 29 more in progress.[13] The data for the United States vary considerably on this point, but the minimum number of roads then operating in this country was 900.[14] Massachusetts alone had as many trolley lines operating in 1891 as Britain had in 1896; in August 1891 Boston had 800 cars running—not many fewer than ran in all of Europe four years later.[15]

It therefore remained for American engineers to develop the kinds of batteries and auxiliary control devices needed for the streetcar industry. The importance of the development of this technology for the emergence of a native American industry can be appreciated from the following statistics: in 1904, one American company—the Electric Storage Battery Company of Philadelphia—accounted for 95 percent of *all* lead-acid battery

[11] Philip Dawson, "Electric Railways and Tramways," p. 585.
[12] *The Electrician* (London) 29 March 1895: 666.
[13] Ibid., 3 January 1896: 321.
[14] Harding, *Engineering*, p. 11.
[15] *Electrical Engineer* (London) 28 August 1891: 193; *The Electrician* (London) 24 April 1891: 751.

production in this country.[16] Two years earlier, when this percentage was probably the same, the company did the following business: [17]

Nature of Business	Estimates Given	Contracts Closed	Value of Contracts	Kilowatt Hours	Percent of Business ($)
Streetcar Station Battery	115	59	$1,579,700	31,884	45%
Lighting Central Station Battery	26	10	$367,328	8,440	16.6%
Private Lighting Plant Battery	113	25	$118,371	1,919	6.4%
Miscell.[18]	——	—	$1,445,045	——	32.0%

Even as these 1902 statistics were being compiled, however, the dominant position of streetcar batteries in the American storage battery industry was about to come to an end. The streetcar load-leveler held its position of economic importance for about a decade—from the mid 1890s to about 1905. That decade was just long enough, however, to give the American industry a good start on life, since after this time the use of batteries on automobiles more than made up for the rapid decline in the trolley market.

There were a number of reasons for the decline of the use of batteries in trolley powerhouses. One reason was the growth in the number of cars on individual trolley lines. As the lines grew larger and installed more cars, the need for fluctuation-controlling batteries became less important, since the peak and trough loads of the individual cars tended to average out.[19] Moreover, as trolley lines grew longer and more complicated, the same pattern of mixed A.C./D.C. operation was introduced here as in the big city lighting systems. Due to the difficulty of developing variable speed A.C. motors, the trolley lines themselves ran on 500 volt D.C., but after the mid 1890s the practice became common of transmitting the current from a central,

---

[16] *Philadelphia Electrical Handbook* (Phila., Amer. Inst. of Elec. Eng., 1904).

[17] Report of Proceedings, Second Convention, Electric Storage Battery Company, Colonnade Hotel, 13-16 October 1902 (typescript of proceedings in possession of author).

[18] For electric automobiles and trucks, telegraph and telephone common battery systems, R.R. car lighting, laboratories, dentist's drills and electric cautery, etc.

[19] A good deal of consolidation of smaller trolley roads into larger systems took place beginning in 1895. C. Francis Harding, *Electric Railway Engineering*, p. 11.

high voltage A.C. generating station to rotary converter substations spread out along the line.[20]

These substations were also equipped with boosters, which drew additional current from the powerhouse during surges in their sector of the line, and which returned current during trough periods. If a line had a significant number of such substations, the sum variation in total load on the powerhouse tended to balance out. The use of batteries would still have provided smoother operation, of course, but they were twice as expensive to install as an equivalent amount of booster-rotary converter equipment, were 10 percent less efficient, needed more maintenance, and depreciated at a faster rate. Around the turn of the century, therefore, electric transportation authorities began to advise the use of purely elec-tromechanical apparatus at substations.[21] The use of batteries at substations involved different cost-comparison considera-tions than at the generating stations themselves. As indicated above, the cost per equivalent kilowatt capacity of batteries was twice that of rotary converters at substations, but the comparative costs at the generating stations were roughly equivalent.[22] The reason for this difference was the extra ex-pense of installing boilers and steam engines at the power-house, which would not be so at the substation.

Louis Bell, writing in 1900, described a trolley power plant which would have required 100 kilowatts of generating ca-pacity if no battery was used, and only 50 kw with battery. No capital cost savings were realized here, however, since the 50 kw battery cost just as much as the extra 50 kw of generating capacity would have cost. The savings from the use of the battery came from the fact that the 100 kw generating capacity plant would have operated with a 50 percent load-factor, whereas the 50 kw battery plant operated at 90 percent load-factor. Thus, current in the two plants cost 6¢ and 4¢ respec-tively.[23] It is clear, however, that with less fluctuation in the load, the 50 percent factor in the 100 kw station would have been raised, with consequently less need for the load-leveling battery. As systems added more cars and substations, load-factor improvement increased.

[20] Ibid.

[21] Louis Bell, *Power Distribution for Electric Railroads*, p. 132.

[22] Ibid.

[23] Ibid., pp. 133-135.

The use of batteries as load-levelers was not their only application in streetcar electrical systems. Another popular use was as voltage controllers at the end of long feeder cables from the power station. The same principle was used in early lighting systems. On early streetcar systems, it was often difficult to prevent a serious fall in voltage as the conductors were run to a considerable distance from the generators. The overhead conductors were often made of thin wire and there was usually only a single wire carrying current from the station to the farthest point on the line; early streetcar companies did this to reduce capital costs by keeping the amount of copper in the transmission network to a minimum. With single, thin conductors, therefore, it was often difficult for cars far from the powerhouse to move if a large number of cars nearer the generating station were drawing current, or if unusually severe conditions existed near the end of the line.[24]

These were the conditions on the Milford & South Framingham Street Railway line when two battery substations were installed in 1901.[25] This was a pure D.C. line[26] without parallel feeders. The twelve-mile length alone would have caused voltage-drop problems, and these problems were aggravated because the power station was located at the extreme Milford end of the line, while the South Framingham end had a steep, three mile long incline. These problems were so severe that the design EMF of 500 volts could be reduced to 220 or even 150 volts during difficult operating conditions at the Framingham end.

Originally, the line had made good the voltage drop by purchasing power from a local Framingham trolley line. This, however, proved uneconomical. In 1901, therefore, the Electric Storage Battery Company of Philadelphia was commissioned to install a battery substation at the Framingham end.[27] The battery consisted of 240 refrigerator-sized batteries; enough to make up for a total voltage loss if necessary. The installation

---

[24] During the twentieth century these problems were reduced as the single wire conductor was replaced on many lines by the parallel feeder method. Here, a number of wires led directly from the powerhouse to various points on the trolley wire itself. These points were at varying distances from the powerhouse; most of them at the extreme end of the line where voltages could be expected to fall. In addition to the parallel feeder technique, thicker trolley wires were also used as trolley technology evolved.

[25] *Street Railway Journal*, 25 July 1903: 120-122.

[26] That is, using D.C. for both transmission and motor current.

[27] *Street Railway Journal*, 25 July 1903: 121.

saved the company $1,200 a year over buying power from out-
side sources. A similar installation was used on the Milford
and Uxbridge branch of the Milford system. This line was only
6.5 miles long, but the end farthest from the powerhouse was
also hilly and ended at a lake-side amusement park. On summer days "seven cars frequently start from one of the parks at
the same time, and without the assistance of the battery it
would be impossible to move them, owing to the low voltage
that would ensue."[28]

From a purely technical point of view this use of storage
batteries was perfectly respectable engineering; it did what it
was supposed to do. Yet from the perspective of economics it
was not so ideal; the batteries were an expensive solution, and
their use was not encouraged. It was reasoned that since automatic boosters had to be used in streetcar substations anyway, these devices might as well be used to do the whole job of
voltage control, and so dispense with the trouble, size, and
expense of batteries. Nevertheless, during the early years of
streetcar practice, many such battery voltage controllers were
used in the United States.

This large-scale use of batteries in streetcar engineering reflects an almost complete change in the attitude of American
electrical engineers towards storage batteries. Although the
1880s had seen a deep prejudice against the use of batteries,
the American engineer of the late 1890s and first years of the
new century was almost too ready to adopt them. Their use
reflects the feeling of security with which engineers regarded
the battery's stored power; storage ability which the booster
did not have, regardless of its efficiency, reliability and small
size. In case of equipment breakdown, the batteries could always carry the load for a time. After the turn of the century,
such security against breakdowns was not really necessary,
but the readiness with which engineers adopted batteries, if
not the best design attitude, helped to maintain the financial
health of their manufacturers.

---

[28] Ibid.

7. New Companies and New Technology

As we have seen, American electrical engineers started to
accept batteries into powerhouses in 1887-1888. These in-

stallations, however, were of small size and therefore failed to make a significant impression upon the minds of the engineering community. The engineering journals reported on such installations, but only in the form of tiny news notes of a sentence or two on their back pages. These notes, of which there were dozens in the 1887-1888 period, were merely press releases by the battery companies. Neither the journals nor the engineering societies at their conventions saw fit to study or report upon the operation of these installations.

This situation changed by 1894-1895. After the economic downturn of the early 1890s, the American battery companies were again obtaining power station contracts. The difference was that the technical journals and societies were now following up these stories with avid interest. An American electrical engineer of the late 1880s who read his journals but avoided the fine print on the back pages might have been totally unaware of the scope of American interest in battery lighting installations; the same engineer six or seven years later could hardly open an American electrical engineering journal without seeing enthusiastic editorials on batteries or extensive, detailed articles on the latest Edison station battery, complete with photographs, graphs, and cost-saving calculations. Passer's "significance of scale" was the crucial factor here; once the Edison system dropped its opposition and adopted batteries on a significant scale, the AIEE and the technical press stopped treating storage batteries as esoteric curiosities of impractical European origin, and started to give them serious attention.

The *Electrical World* indicated this in a review of progress in 1895:[1]

Electric lighting presented no marked developments during the year, if we except the progress made in the education of central station men with respect to the utility of storage battery auxiliaries. The excellent results achieved during the year with such plants in the New York, Brooklyn, Boston and Lawrence Edison stations (note that only Edison stations are mentioned as being important, although many non-Edison stations also had batteries) and the publicity which the subject received from the discussion on storage batteries at the November meeting of the American Institute of Electrical Engineers, will probably fix 1895 as the missionary year to what promises to be a wholesale conversion of central station managers to the use of the storage battery.

---

[1] *Electrical World*, 4 January 1896: 3-4.

Although the prediction about "wholesale conversion" was an exaggeration, the rest of this editorial was substantially correct. The middle 1890s was the takeoff period for the American storage battery industry. During these years, for example, the companies which still dominate the field today were created. The mid 1890s created a feedback effect for the further growth of the industry—the large-scale use of batteries by Edison stations induced the journals and societies to report about them, and the publicity generated convinced increasing numbers of engineers to try the batteries, thus producing more articles and symposia.

If the "takeoff" period for the American storage battery industry was the mid 1890s, however, the importance of the two earlier periods of commercialization should also be noted—namely, 1882-1883 and 1887-1888. These might be called periods of "failed takeoff." After the mid 1890s, the storage battery industry in the United States expanded continually.[2] The opposition of the Edison interests during the 1880s made it difficult for the industry to get the chance to establish a reputation, and this, in turn, made it hard for the industry to get any business at all during times of economic slowdown.

Nevertheless, the two early periods of failed takeoff were important since they established the technological base upon which the American battery firms of the 1890s grew. The 1882-1883 period saw the creation of the EAC, the Brush storage battery and the Viaduct Company's battery. In the period from late 1886 to 1888 such firms as the American Julien company and the Bradbury-Stone storage battery company were formed. Brush himself did not actively advance the technology of his battery after the mid 1880s, but the Julien company did. After lengthy patent litigation, the Julien Electric Company bought out the Brush patents, and, in 1890 changed its name to the Consolidated Storage Battery Company.[3] This amalgamation was not unlike that which had occurred in England almost a decade earlier when the Faure and Swan-Sellon-Volckmar patents had been combined with the agreement between EPS and FEA. As in that case, one company held priority over the basic patents, but wasn't working them aggressively (Brush), whereas the other firm (Julien) was in precisely the reverse position.

---

[2] Between 1894 and 1897, the production of storage batteries in the United States increased by more than an order of magnitude; ibid., 20 August 1898, p. 178.

[3] Testimony of Bracken; Brush & Consolidated Electric Companies versus Milford & Hopedale, etc., op. cit., p. 39.

Although it is difficult to be precise about the operation of these early American storage battery companies, it appears that Consolidated, along with EAC, was the most technologically sophisticated of the early companies. For example, as early as 1892 Consolidated's engineers had developed a mechanical pasting machine, in a day when all other companies used hand pasting with wooden trowels.[4] Neither Consolidated nor EAC survived past 1894, but this was because both companies were bought up in that year by the Electrical Storage Battery Company of Philadelphia (ESB), and the technology and patents of these companies passed into control of ESB, along with a number of the engineers of these firms.

Bradbury-Stone was a smaller version of the Julien-Consolidated company; it appeared during the second "failed takeoff" period and concentrated on batteries for self-propelled streetcars. Bradbury-Stone was also absorbed by ESB in 1894. Although a number of similar small battery companies were also absorbed into ESB, there were a substantial number of American battery firms which did not survive long enough to be absorbed. One of these was the Electric Storage Company of Baltimore (Viaduct). Nevertheless, the expertise developed by this firm was not lost. In the 1890s, the chief engineer of Electric Storage, A.H. Bauer, turned up again as the chief electrician of the Pullman Company, where he was responsible for installing his system of storage battery train lighting on the famous Pullman cars.[5]

During the late 1890s and the early years of the twentieth centruy, ESB dominated over 90 percent of the American storage battery market. For all practical purposes, ESB *was* the American storage battery industry, and, in the days before anti-monopoly legislation, the company frankly admitted seeking a total monopoly of the market.[6] There were uniquely American aspects to the creation of ESB, but, in a broader sense, the creation of this company was mirrored by similar developments occurring almost simultaneously in Britain, Germany, and France.

In each country, one firm rose to dominance during the 1890s. During the years and decades which followed, this

---

[4] Ibid., p. 40.

[5] *Electrical World*, 23 January 1892: 65.

[6] ESB Convention typescript, op. cit., paper by Charles Blizard, p. 1.

premier firm steadily absorbed the assets and technology of its rivals until it controlled the vast bulk of the battery business in that country. Although this process may appear to imply a kind of survival-of-the-fittest competition, the actual process of amalgamation was on the whole a peaceful and rational adaptation to market conditions. A limited market, the convergence of technology and the inability of small companies to handle the complex engineering of power station battery installations all contributed towards amalgamation of storage battery companies.

The origins of ESB lay in the gas lighting industry. In 1882 a group of Philadelphia capitalists financed the creation of the United Gas Improvement Company (UGI), which manufactured gas-producing equipment and operated as a holding company for gas utilities in a number of eastern states.[7] The chief moving force behind UGI was a man named W.W. Gibbs. Gibbs was a former dry goods dealer who moved into the gas-lighting business in the late 1870s. His initial work was analogous to that of the L.C. Kinsey company in electric lighting; Gibbs installed small gas plants for factories and small towns. In 1881 he teamed up with an inventor named Thaddeus Lowe to create the National Petroleum and Water Gas Company. Lowe was the holder of patents for the enrichment of lighting gas by means of petroleum. UGI was an expanded form of this earlier company.

During the early 1880s UGI concentrated its activities on the gas business, buying up companies and patents in that field. After the mid '80s, however, the firm expanded into the field of electric lighting. This was not an unusual practice; a number of gas companies, both in the United States and Europe, diversified into electric lighting in the 1880s and 1890s, thus assuring themselves of control of the competition.[8] UGI was especially skillful in hedging its bets against the technological future; at the same time that it entered the electrical field UGI also purchased the American rights to the Welsbach mantle. In 1887 UGI created the Welsbach Incandescent Light Company in Gloucester, New Jersey; this was the company which supplied most of the gas mantles for the American market in the next decades.[9]

---

[7] Andrew C. Irvine, "The Promotion and First Twenty-Two Year's History of a Corporation in the Electrical Manufacturing Industry," Master's Thesis, Temple University, 1954 (unpublished) p. 28.

[8] Passer, *Electrical Manufacturers*, p. 37.

[9] Irvine, "Promotion," p. 31.

During the next year, UGI created another subsidiary, the
United Electric Improvement Company (UEI), whose role was
to buy up promising electric utilities and manufacturing companies. In 1889 UEI bought the Germantown Electric Company.[10] The connection between the growing American
awareness of the value of storage batteries as load-levelers and
the publication in 1894-1895 of the technical operating data of
this station has been discussed previously. It will now be apparent, however, why this particular central station installed
one of the first significant sized central station batteries in the
United States—that is, that the UEI owned both the station
and the company (ESB) which supplied the batteries and control devices. The installation was created in 1892; only two
years before the ESB was to make its successful bid for nationwide control of the battery industry. The Germantown project
possessed the *scale* to attract the attention of a nationwide
audience of engineers. It was the first American battery installation to be written up at length in the technical journals.
With reference to the Passer "significance of scale" argument,
the Germantown station did for the American storage battery
what Niagara Falls did for the American A.C. equipment industry. There is little doubt that the UEI planned the station for
such publicity value.

In 1889 UEI purchased the Heisler Electric Light Company,
which gave it the patents for producing complete incandescent
lighting systems.[11] At the same time, a young engineer named
Stanley Currie was employed to direct the technical activities
of the company; Currie had previously been involved in incandescent lighting. The story of ESB was part of this process
of acquisition. In 1887 a French inventor named Clement
Payen came to the United States to sell the patents to a new
form of electrode plate he had invented. W.W. Gibbs met
Payen, and, in the same spirit in which he had seen the potential in the Welsbach mantle, Gibbs bought the Payen rights
and hired the inventor to commercialize the process. UEI was
impressed by the performance of the batteries and on 7 June
1888 created the ESB to manufacture the batteries.[12]

The process which Payen sold to Gibbs was a variant on one
of the "brute force" techniques developed in the days before

---

[10] Ibid.

[11] Ibid., p. 32.

[12] S. Wyman Rolph, "Exide—The Development of an Engineering Idea," *Newcomen
Society of the United States* (1951).

manufacturers learned to use the "setting" properties of lead oxides. A simple way to make a coherent, durable plate is to cast it from molten material. The Crompton-Howell Planté-type plates were an application of this idea. Numerous inventors in the 1880s played variations on this theme to get plates of large surface area and yet preserve the durability of a plate cast from molten lead. The difficulty with this approach, however, is that molten lead possesses no inherent porosity; therefore, the porosity must be created by some mechanical means, as in the laborious and tricky Crompton-Howell ladling technique.

One way of getting around this problem was by casting the plates from some fuzible lead salt. Once cast, this material could be reduced to metallic form, producing porosity. This was the variant Payen chose. Lead chloride is a material which melts at a low temperature—501°C. A number of French and English inventors had proposed its use for batteries ever since 1882. The idea was to cast lead chloride into pellets and imbed these in lead grids.[13] Once cast, the chloride pellets could be held firmly in place by having a lead plate cast around them from molten lead. Zinc chloride, which was added to the lead chloride to increase porosity, could then be dissolved out of the pellets. Finally, the lead chloride was reduced to spongy lead by temporarily connecting the plates up like the cathodes on a primary battery with plates of metallic zinc.

The "chloride process" was the most popular of the so-called "molded pellet" storage battery plate-making processes. It figures prominently in the patent literature of the 1880s and 1890s, and major new storage battery firms were created in France, England, and the United States to use this process. This popularity may appear puzzling, if we consider the complexity of the process compared with the ordinary pasting of grids with litharge or red lead. The popularity arose, however, from the appeal that the molten casting of the active material had for engineers seeking a simple and sure way of making coherent electrodes.

In 1885, Clement Payen and E. Leclerc in France developed a slight modification of the basic chloride process which simplified manufacture somewhat. Payen and his partner had a falling out, and Payen left for the United States to sell the patent rights in this country. Payen's connection with ESB was brief. He was apparently not an easy man to work with, and when Gibbs borrowed Herbert Lloyd, a chemist from the

---

[13] F. Maxwell Lyte, 1882; see: Wade, *Secondary Batteries*, p. 17.

Welsbach factory, to oversee Payen, the French inventor departed. Although Payen remained in the United States and took out a number of further battery patents in the 1890s, he appears to have had no significant influence on the course of battery technology after leaving ESB.

The early history of ESB was bleak. From its founding in 1888 until 1894 the company sold very few batteries. The manufacturing was carried out in a corner of the Welsbach plant, and the technical personnel were also "on loan" from Welsbach. Workmen were hired when an order came in and fired when it was completed. Basically, the company was just being maintained at survival level by Gibbs in anticipation of the return of economic prosperity.

Technologically, the development of ESB was similar to the evolution of Viaduct and EAC. For a year, Herbert Lloyd tried unsuccessfully to turn the Payen patent into a practical battery. Finally, in 1889, he made the same pilgrimage to European battery plants that A.H. Bauer had made years earlier. In France he visited the factory of the Société Anonyme pour le Travail Eléctrique des Metaux (TEM). TEM was organized by Rothschild interests at the same time that Gibbs was creating ESB in the United States. TEM also used the chloride process, but developed under the patents of an inventor working independently of Payen.[14]

Lloyd also went to England, where he visited the major battery manufacturing plants and convinced the battery engineer William Taylor to join ESB.[15] Taylor had been one of the designers of the Colchester system.

By the early part of 1890 ESB was ready to sell a few batteries. It did not sell many, however, and until 1893 its sales were inconsequential. Nevertheless, the economic position of the American storage-battery market was improving during the 1890s, after the downturn which had begun during 1889-1890. Statistics are unavailable, but a number of battery companies began reporting good business in 1891. The Gibson Storage Battery Company reported good prospects in January of that year, and Pacific Electrical Storage Company reported good business a month later.[16] Perhaps more impressive than these reports was the sudden upsurge in the appearance of brand

---

[14] Those of A.M.F. Laurent-Cély.

[15] M. Barak, "History of the Chloride Group" (unpublished typescript) p. R-3.

[16] *Electrical World*, 17 January 1891: 51; 21 February 1891: 141.

new battery companies in early 1891 after an almost complete absence of such startups in the past two years. The following is a non-exhaustive list: [17]

Date of Formation	Company
2/91	Ford-Washburn Storelectro Co., Cleveland
3/91	Cambridge Accumulator Co., Boston
early/91	New Era Electric Storage Co., Chicago
11/91	Western Storage Battery Co., Kansas
12/91	Acme Storage Battery Co., New York
1/92	National Storage Battery Co., Troy
4/92	Syracuse Storage Battery Co., Syracuse
5/92	American Battery Co., Chicago
7/92	San Francisco Electrical Storage Co., San Francisco
4/93	Globe Storage Battery Co., Chicago
4/93	Electric Power & Storage Co., Ashland, Wis.
4/93	Coleman Battery Co., Chicago
6/93	Pumpelly-Sorley Storage Battery Co., Chicago
7/93	International Storage Battery Co., Detroit
2/94	Chicago Accumulator Co., Chicago

As in the 1886-1888 period of storage battery enthusiasm, the development of a self-propelled streetcar battery played an equal role with the development of stationary batteries in motivating the entrepreneurs and inventors of the post 1891 period. For example, in the reports of the formation of new battery companies appearing in the technical press starting in 1891, those firms which specified their chief commercial goal usually indicated self-propelled streetcars, for example, Ford-Washburn, Bradbury-Stone, River & Rail, Cambridge, Syracuse, American and Chicago.[18] This enthusiasm arose from the post-1887 American boom in electric streetcar investment. Another stimulus was the fact that transportation battery engineering was more exciting and (apparently) simpler than the older, more mundane business of load-leveling stationary batteries. Load-leveling technology, for example,

---

[17] Ibid., 14 February 1891: 123; 17 March 1891: 201; 18 April 1891: 297; 28 November 1891: 407; 12 December 1891: 439; 16 January 1892: 47; 9 April 1892: 251; 14 May 1892: 341; 16 July 1892: 47; 1 April 1893: 253; 29 April 1893: 328; 24 June 1893: 473; 15 July 1893: 54.

[18] *Electrical Review* (New York) 18 May 1889: 23.

did not attract the attention of the dozens of inventors who set up the small, ephemeral battery companies. Most of these companies, including the majority of those listed above, built a few batteries, ran a few experimental battery streetcars, and disappeared.

The more substantial companies, such as ESB, took a more sophisticated and diversified approach to the streetcar market. Although ESB experimented with self-propelled streetcars, most of its early effort was concentrated on the development of stationary batteries, automatic boosters, and end-cell switches designed for the trolley power station load-leveling market. The self-propelled streetcar was a gamble and the overhead conductor trolley a sure thing.[19] In addition to ESB, other more substantial American battery companies also diversified their approach towards the battery market and did not concentrate solely on traction cells. Both the Julien-Consolidated and the Bradbury-Stone companies started out by concentrating on streetcar propulsion, but later shifted towards stationary batteries. In 1893, for example, Bradbury-Stone announced it was beginning to manufacture larger, European-style central station batteries.[20] Indeed, the economic outlook for all applications of storage batteries looked rosy in the United States in the early to mid '90s. One market which was finally beginning to reap rewards for some manufacturers was the domestic and medical battery field. Ever since the time of Gaston Planté, some inventors had proposed using miniature lead-acid cells in the home for such purposes as driving fans, or in the doctor's office for cauterization or in polyscopes. Few such sales were made during the 1880s, however, due to the availability of good, cheap primary batteries for these same purposes.

By the early 1890s, however, a number of companies reported good sales of so-called "cabinet" batteries. For example, the commercialization of miniature electric motors allowed Bradbury-Stone to do a brisk business in lead-acid cells for dentist's drills, electric fans, sewing machines and phonographs by mid 1892.[21] In February 1892 the Storage Battery Supply Company of New York was created for the sole purpose of manufacturing and selling such small electrical appliances, together with sales of the batteries of the Consolidated Storage

---

[19] *Electrical World*, 7/1900, advertisement section.

[20] Ibid., 28 October 1893: 353.

[21] Ibid., 2 July 1892: 16.

Battery Company, of whom Supply Company was a franchised dealer.[22]

The market for railroad car lighting batteries was also doing well, as was that for private lighting plants.[23] And, of course, there was the newly favorable attitude of American central station managers—particularly the Edison people—towards batteries. The economic results of these increases in demand can be seen not only in the lists of new incorporations mentioned above, but also in the expansion of established companies. In 1891, for example, Bradbury-Stone announced that its factory in Lowell, Massachusetts was too small so that they were erecting a larger one. In November, the company announced the creation of a branch plant in Sioux City; then, in September 1892, another in East St. Louis; and, in December, another in Portland, Maine.[24] At this time, Bradbury-Stone was one of the most aggressive American storage battery firms—it was involved in all phases of the battery market, including the installation of central station lighting plants in collaboration with the Thomson-Houston Company.[25]

None of this economic activity was lost on W.W. Gibbs or the other financial backers of UGI and UEI. During 1893 and 1894, these men matured a plan whereby ESB was to become the dominant force in the American storage battery industry. This was an ambitious idea, since at the start of 1894 there were essentially three major American companies in the field—Consolidated and EAC, and Bradbury-Stone making up a smaller, but ambitious third. ESB was a barely surviving member of the large pack of tiny, ephemeral firms such as Pumpelly-Sorley, American, and Chicago. Then, overnight, the entire shape of the American industry changed.

Gibbs raised $4 million for the recapitalization of ESB. In December 1894 this money was used to purchase Consolidated, EAC and Bradbury-Stone, as well as the assets and patents of a number of smaller companies. In all, over 500 American storage battery and battery-related patents were purchased, including those held by Brush, Morris, Salom, Julien

---

[22] Ibid., 27 February 1892: 130.

[23] The railroad car storage battery was very similar in its use to that of the private lighting plant battery—it provided constant current for the bulbs from a generator which ran from the train's axles, which obviously would produce a varying current depending on the train's speed.

[24] Electrical World, 23 May 1891: 383; 28 November 1891: 407; 24 September 1892: 202; 31 December 1892: 431.

[25] Ibid., 8 August 1891: 101.

and Mailloux (that is all the Consolidated's patents), General Electric, the Edison Electric Light Company, Thomson-Houston, Tudor, Starr, Van Depoele, Philippart, Sorley, Swan, Griscom, Laurent-Cély, Howell, Edison, Sellon, Faure, Meserole, and many others.[26]

Other purchases of patents were to come later. Essentially, the ESB's policy was to buy up every significant American battery and battery-related patent, thus controlling the entire technology in this country. The original 1894 purchases, however, were by far the most important, since they eliminated all serious competition. After 1894, therefore, the ESB essentially had a vacuum into which to expand, and it expanded very rapidly indeed: [27]

Year	ESB Gross Sales (Millions of Dollars)
1890	neg.
1891	neg.
1892	.01
1893	.02
1894	.08
1895	.35
1896	.65
1897	.85
1898	1.10
1899	2.15
1900	3.30
1901	2.65
1902	3.00
1903	4.04

W.W. Gibbs made an astute move when he purchased this large collection of patents in 1894. To understand what Gibbs did, it is necessary once again to draw a careful distinction between patents and technology. Patents control the application of technology, but they are not the technology itself—ESB could only obtain a workable knowledge of storage battery technology by hiring experienced engineers trained in the field. Moreover, patents need not be bought to obtain the ideas contained in them, since anyone can read the patent literature.

---

[26] *Electrical Engineer* (London) 28 December 1894, p. 738: *Electrical Engineer* (New York) 19 December 1894, p. 508; *The Electrician* (London) 21 December 1894, p. 224; Unpublished list of patents, files of ESB Company, Phila., Pa.

[27] Irvine, "Promotion," p. 88.

The motivation for Gibbs's purchase originated in the continual patent interference activity which had bedeviled the application of storage battery technology in the United States since its inception. The Brush patents were the chief means by which suits were brought and manufacturers enjoined from making batteries. The Brush patents described an unworkable technology; yet, they were very effective in controlling other storage battery technology. Nevertheless, various manufacturers over the years believed that they could evade the Brush patents by developing new patents sufficiently different from the Brush method so as to avoid infringement. Consequently, patent fights were continual from the early 1880s on. The tiny, ephemeral battery companies were generally not bothered by patent suits, but they had little influence over the evolution of the technology anyway. It was only the larger companies, such as Julien or EAC, which were sued. These fights absorbed a considerable amount of the capital of the American battery companies, none of whom were very well capitalized to begin with. Moreover, the constant litigation tended to scare off customers, who feared buying the products of a company which could be ordered to stop manufacture; the continual supply of replacement plates is crucial for proper maintenance.

It is true, however, that the influence of the Brush patents in stifling economic activity in the American battery industry before the early 1890s was less important than the prejudice of American engineers against storage batteries. In the mid 1890s, this prejudice rapidly disappeared. Accordingly, the Brush patents from that time on became a roadblock to further economic advance, rather than simply one more nuisance for an already marginal industry. Gibbs bought the Brush patents, therefore, not to use them, but to remove their controlling effect on the industry. Moreover, he bought all the other American patents for a similar reason; it was much simpler and cheaper to buy up these patents than to fight them endlessly in the courts. The clearing up of this patent logjam cost the ESB promoters a good deal of money, but for the first time in the history of the American storage battery industry the prospects for a reasonable return on the investment looked good.

The unimportance to the ESB of the technology contained in most of these patents can easily be demonstrated. During its early years of manufacture, ESB concentrated on the chloride process of plate manufacture; no other American manufacturer ever did this. Any technology transferred to ESB from the older

American companies came from engineers hired by ESB or associated with ESB in some other way. The early battery-making technology used by ESB came almost exclusively from technology-sharing agreements with European manufacturers.

The influence of European manufacturers on the American industry was not new, but the agreements worked out by ESB in 1894 represented a new level of sophistication in this process. As was indicated earlier, a single storage battery firm rose to dominance in each of the major manufacturing countries in the 1890s. The agreement in which ESB participated was a cartel between these firms designed to divide up the world market and to share technology. The participants were: ESB (U.S.); TEM (France); AFA (Germany); and the Chloride Electrical Storage Syndicate (hereinafter: Chloride) (U.K.). As far as the American firm was concerned, this cartel ended the threat of German competition, which had worried American producers since the late '80s. A few of the first American central station batteries were AFA projects.

Before the creation of the cartel, Adolph Müller of AFA had come to the United States in 1893 to investigate the market. He visited a number of central stations, pointing out the value of load-leveling. A large contract was made with the Boston Edison station and this installation began operation in 1895.[28] Nevertheless, Müller apparently became aware of the ESB plans at this time and realized the difficulty of competing against an efficient, well-capitalized American company, particularly in the face of high tariffs and the transportation costs involved in sending a solid lead cargo across the Atlantic. Consequently, AFA and ESB agreed at this time to establish "freundschaftliche Beziehungen."[29]

The second important consequence of the friendly relationships between the four firms was the unrestricted transfer of technology. The surviving records are too scanty to tell much about the precise way in which processes were shared between the firms. It is clear that one way in which the technology was transferred was by engineers from the companies paying visits to each other's plants, some of these visits lasting months. Since all of these companies, except AFA, used the chloride process during their early years, the transfer of lessons learned on the factory floor was of importance, particularly in view of the fact that all of these firms were the first ones

---

[28] Nadolny & Treue, *VARTA*, p. 49.
[29] Ibid.

to use the chloride process in their respective countries. The four firms also had an automatic patent-sharing agreement.

The 1894 agreement was evolved over a series of years, beginning in 1889 when Lloyd paid his first visit to TEM. The Chloride Company was formed in 1891 and immediately established close relations with both TEM and ESB. The creation and early development of this British company closely parallels that of ESB in the United States.

As with ESB, Chloride was created as part of a plan of industrial diversification. The original idea came from William Mather, head of the engineering firm of Mather & Platt.[30] Mather & Platt specialized in making machinery for the Manchester textile industry. During the early '80s, Mather became interested in the possibility of electrolytic manufacture of bleach to replace the traditional sulphuric acid-based methods. Being an admirer of Thomas Edison, Mather went to America to consult Edison about the use of his dynamos for chlorine manufacture.

As a result of his initial interest in bleach manufacture, Mather became deeply interested in electrical technology. He purchased the rights to the Edison dynamo in England. Mather's chief engineer, Edward Hopkinson, induced his brother John to work on the problem of dynamo design; the result being the famous Edison-Hopkinson dynamo.[31]

During a second visit to the United States, Mather visited the ESB works and was impressed by the chloride process. He purchased the British rights to the Payen patents.[32] At the same time, he also acquired rights to the similar Laurent-Cély patents from TEM. The Chloride Electrical Storage Syndicate was incorporated in 1891 on the basis of these patents.[33] Actual manufacture started about two or three years later.

Mather's growing interest in storage batteries had not dampened his original plan to make electrolytic chlorine. Sometime after 1891 he made contact with Hamilton Castner, who was then developing his famous process for electrolytic production of sodium and chlorine from fuzed sodium chloride. Mather purchased an abandoned aniline dye works

---

[30] James Grieg, *John Hopkinson, Electrical Engineer* (London, The Science Museum, 1970) p. 18.

[31] Ibid., p. 19.

[32] The chronology is uncertain.

[33] Barak, "History," p. C-1.

at Clifton Junction and set up an experimental installation of Castner-Kellner cells here. The Chloride batteries were also manufactured here. Although there is no direct evidence, Mather's attraction to the chloride process of battery manufacture must have been influenced by his prior interest in chlorine manufacture, since the output of one could be used in the manufacture of the other.

Although there are no records to indicate how technology was transferred between the three chloride companies, an examination of the designs of their batteries demonstrates the close interaction which must have taken place between their engineering staffs. Initially, all three companies used the chloride process to make both the positive and negative plates. The manufacturing process was as follows: litharge powder was dissolved in acetic acid, converting it from the oxide to the acetate of lead. Hydrochloric acid was then added, precipitating the lead as chloride, and regenerating the acetic acid to be recycled to the first step.[34] The powdered precipitate was then dried, mixed with zinc powder, melted, and cast into pellets. Although there were initially some differences between the techniques used by the three companies, by the later 1890s all companies were using the identical method.[35]

Having been formed, the flat pellets were arranged in a mold and a molten alloy of lead and antimony was cast around them under pressure, forming a frame to hold the pellets rigidly. The geometrical design of the chloride pellets used by the three companies differed somewhat, but all other details were identical.[36]

Although the three chloride companies initially used the chloride process for making both plates, they soon abandoned this procedure. As was discussed in the last chapter, the durability of storage batteries is chiefly a function of the positive plate; under the stress of expansion, the lead peroxide of the positive tends either to crumble or to warp the grids. The crystallized pellets of the chloride process were probably held too tightly by the cast grids and could therefore not expand properly. Moreover, it is likely that the active material did not

---

[34] Wade, *Secondary Batteries*, p. 407.

[35] The original ESB PbCl process consisted of blowing molten lead through an orifice to produce fine powder. This was then dissolved in HNO_3 and precipitated by adding HCl. The rest of the process was as described above. *The Electrician* (London), 8 February 1895: 425.

[36] Wade, *Secondary Batteries*, p. 408, 463; Treadwell, *Storage Battery*, p. 40.

possess the durable, fibrous structure of cured, pasted litharge plates, and, being purely crystalline, was even more brittle when converted to PbO_2 than the ordinary pasted positive plates.

Therefore, during the early '90s it was found that chloride positive plates lacked durability. In order to avoid this problem, the Chloride Company introduced a Planté-type positive plate in 1895, and used this in combination with the chloride negative, which proved satisfactory. The new positive consisted of strips of pure lead, crimped and rolled into spirals. These were forced into pre-cast holes in a thick sheet of lead-antimony alloy. The design of this plate is derivative of the Elieson and Walter grids discussed in the last chapter: [37]

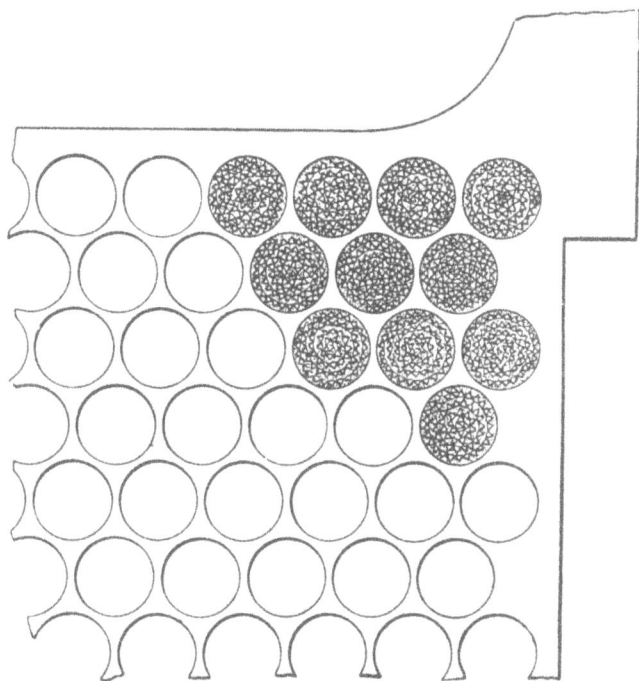

Shortly after Chloride's introduction of this "rossette plate," it was adopted by ESB and renamed the Manchester positive. TEM also copied Chloride's idea of a Planté positive soon after 1895. The French company, however, altered the form of the crimped strips slightly, leaving them straight instead of rolled.

---

[37] Wade, *Secondary Batteries*, p. 409.

Again, as with the design of the chloride plates, the three companies copied each other's basic technology, but altered some details of plate geometry.[38]

From the mid 1890s until the turn of the century, ESB, TEM and Chloride used the mix of Planté positives and chloride negatives in all their batteries. Then, around the turn of the century, all three companies abandoned the chloride process. There were a number of reasons for this. First, the rather high weight of the Planté-chloride combination was only a minor problem during the '90s, when the only applications of economic significance were for stationary batteries. Then, however, near the end of the decade, the electric automobile created a major new market for lightweight, transportation batteries.

When men like Biggs and Mather had first considered the chloride process in the late 1880s and early 1890s, they contrasted its ability to produce coherent active material masses with the poorly developed technology of producing well cured, coherent, pasted litharge plates. By the end of the decade, however, the advantages of the chloride process over the traditional pasted techniques were seen as being more apparent than real. The cast pellets possessed inherent coherence, but so did properly pasted and cured litharge or red lead grids. Moreover, the pasting technique was much simpler and cheaper than the chloride. Finally, the chloride process depended upon the addition of two materials to the active mix—chlorine and zinc—which were difficult to completely remove after manufacture, and which caused serious local action problems in the finished battery if they were not completely removed.

In 1899-1900 ESB brought out two new, non-chloride storage batteries. One was for stationary use and the other for the electric automobile. The stationary form retained the Manchester positive, but substituted a new form of pasted litharge grid for the chloride negative. This was called a "box negative." The box negative, or Kastenplatte, was a recent development by AFA which ESB copied in 1900. As was discussed in the last chapter, AFA engineers had specialized in pure Planté plates in the early 1890s, but after 1895 they had adopted a Planté positive-pasted negative mix. This was in the same year that ESB and Chloride adopted a similar mix of plates, although using a chloride negative instead of the positive.

---

[38] Ibid., p. 465.

It was also noted in the last chapter that the German battery firms pioneered in the development of complex casting techniques for the manufacture of the so-called double and triple grids. The evolution of the box negative was an outcome of this pattern of evolution. This is a typical early box negative of German manufacture: [39]

Like the open grids, the box negative had the virtue of light weight, but not to a great degree. Unlike them, however, the durability of the plate depended not so much upon the internal coherence of the active material mass as on the fact that the mass was held inside miniature sieve-like cells. This design may appear to be a technical retrogression towards the "brute force" techniques of the 1880s and away from the development of more delicate, weight-saving grids.

---

[39] H. Beckman, *Zur Geschichte*, p. 254.

In fact, however, the evolution of the Kastenplatte was a symptom of the increasing sophistication of storage battery engineering as evidenced by the development of special purpose grids for different applications. The box negative was invented to counter the tendency of the negative active material—sponge lead—to cake together. So far in this study we have concentrated on the bad physical properties of the lead peroxide, but this does not mean that sponge lead is without fault. The physical behavior of sponge lead is in many ways the exact reverse of lead peroxide—the peroxide is hard and brittle; the sponge soft and ductile. Neither substance, however, is elastic.

With the sponge lead plate, the problem is that the sponge tends to pull in upon itself, or shrink, in the course of many cycles. This produces a lumping of the active mass, with undesirable loss of porosity, so the active material supporting members in a negative plate should ideally be designed to resist the centripedal tendency of the active mass. It is obvious that Planté-type grids would not do this, nor would the usual type of open-work grids, since the active material tends to bunch-up around the delicate support bars. In the box negative, on the other hand, the supports surround the active material, rather than being surrounded by it, and the caking action of the sponge is resisted.

Battery engineers came to recognize the problem of negative plate caking considerably later than the much less subtle problems of the lack of coherence of active material pastes and the resultant collapse of the pasted positive plate. Consequently, most of the attention of these engineers during the 1880s and early 1890s was on the development of positive plates which would not rapidly fall apart. The various "brute force" techniques were one sympton of this. Since the early negative plates had a much longer lifespan than the early positives, engineers ignored its problems. Indeed, it was the conviction of many of them that negative plates experienced no changes in use at all.[40]

During the 1890s, however, increasing attention was paid to the negative plate. The most obvious reason for this was the development of more durable positives, which allowed battery engineers to pay more attention to other design questions. Of more subtle importance, however, was the introduction dur-

---

[40] Ibid., p. 253.

ing the early 1890s of the battery maintenance contract. Battery manufacturers began to offer such contracts to their customers as soon as they felt that they possessed reliable, quantitative data on the expected lifespans of their products. In return for a yearly fee—usually 6 to 10 percent of the initial capital cost— the battery company agreed to replace all defective parts during that year. In some cases the contracts also specified that the manufacture would perform routine maintenance. This, however, was uncommon, and in the more usual contract the customer agreed to abide by a strict regimen of maintenance procedures if the contract were to be honored by the manufacturer.

The importance of these contracts was that the staff of each power company became, in effect, surrogate laboratory assistants of the battery companies. The battery firms specified the rules of maintenance that the customer was to follow—the batteries were continually tested and examined, and their performance recorded on forms specified by the contract. These forms were collected and examined by the manufacturer.[41]

To the modern reader, who is acquainted with the structure of our present day high technology companies, this method of data collection may not seem very important, Today, almost all large manufacturing companies maintain research and development centers where such data can be obtained with far greater precision, detail and flexibility than if it were collected as outlined above. Yet this situation was far different in the 1880s and '90s. At that time the concept of the industrial research center was only being born. A number of manufacturing companies operated laboratories, but these were rarely more than chemical quality-control operations, where the raw materials coming in were tested for the presence of the more obvious impurities. Indeed, most of the early battery companies operated on such a marginal economic level that even simple safeguards like the analysis of lead ingots for copper, iron, and zinc impurities were looked upon as advanced techniques. Some early batteries failed simply because the manufacturers did not check for high metallic impurities in the supposedly pure lead ingots he bought.[42]

A few early companies, such as EPS, did have facilities for conducting test on samples of their products, or for developing

---

[41] ESB Convention typescript, op. cit.

[42] *The Electrical Accumulator*, 8/1906, pp. 150-151.

improvements in those products. This we saw in the work of
Drake and Reckenzaun. Most of the companies, however,
lived a hand-to-mouth existence. ESB and Chloride occupied
corners of larger factories and manufactured cells only for immediate sale. The testing which was carried on must have been
of a most elementary and inexpensive kind. The lack of technical data on storage batteries in the technical press bears this
out; the only early report of storage battery performance
quoted in the journals came from university laboratories.

The battery manufacturers paid careful attention to the
maintenance records which they received from the contract
holders. The most important reason for this was that they were
acutely aware of the direct relationship between proper
maintenance and battery life; a fact which was well known at
least as early as the publication of Drake's famous study.
Moreover, the manufacturers found that the publication of
these data in the technical press was excellent advertising. The
grandiose but unsubstantiated claims of battery promoters in
the '80s were replaced in the '90s by the publication of maintenance report results.[43]

It is not surprising that battery engineers began in the '90s to
take cognizance of the more subtle problems of storage battery
design. For example, the masses of data collected from the
maintenance forms revealed such unexpected phenomena as
the slow but steady loss of a battery installation's overall storage capacity, *despite* the fact that the worn *positive* electrodes
were replaced. This directed the attention of the engineers
towards the gradual deterioration of the negatives as a result of
the shrinkage of the sponge lead mass. The collapse of the
positive grids was an "eyeball obvious" problem, whereas the
deterioration of the negatives could only be seen microscopically, or through the medium of careful analysis of quantitative data. It is no coincidence, therefore, that it was the
German storage battery industry which first recognized the
negative electrode problems and developed the box grid to
counter them, since it was they who pioneered in the development of the maintenance contract.[44] Both practices
quickly found their way to the other major manufacturing
countries.

[43] For example, *Electrical Engineer* (London) 17 July 1891, pp. 58-59; 10 November
1893, pp. 438-440.

[44] For example, see paper by John R. Williams of ESB; "Maintenance with Various
Forms Offered by Our Company," ESB Convention typescript, op. cit, p. 52: "Our
German friends are responsible for introducing the maintenance contract into the
battery business. . . ."

ESB and Chloride soon adopted the box negative, although Chloride did so by a more round-about process. Until 1901 Chloride was not the dominant giant of the British storage battery industry that it subsequently became. It was simply one of several British manufacturers. In that year, however, ESB bought control of Chloride and began turning it into a British version of itself. A number of American engineers and managers were sent to the Clifton Junction factory, and, over the next several decades, Chloride obtained control of most of the other British manufacturers.[45] In 1903, Chloride adopted the AFA box negative from ESB.

The ESB-Chloride adoption of the box negative allowed both companies to totally abandon the chloride process and to take full advantage, for the first time, of the more practical and mature technology of litharge and red lead pasted batteries. At the same time, both companies also began manufacturing batteries with plates of the more traditional, open, double-grid design. These were also pasted. This technology also was a transfer from AFA. The box negative-Manchester positive batteries were designed for stationary use, while the pasted double-grid cells were intended for the rapidly expanding market of the electric automobile.

Chloride remained a subsidiary of ESB until 1954, and until that time the batteries made by both companies were almost carbon copies. Subsequent to 1954, some divergence of technologies has taken place, but this has not been extensive. TEM was never financially linked to either Chloride or ESB, but its technological evolution was nonetheless similar. It also abandoned the chloride process around the turn of the century and opted for a pasted, double-grid type plate.[46] It is not known whether it adopted the box negative.

---

[45] Barak, "History," pages marked T.

[46] E.J. Wade, Secondary Batteries, p. 462.

8. The Battery Versus the Trolley: An Almost Technology

The early 1890s proved that three methods of electric streetcar propulsion were practical: the trolley (single wire overhead system); the conduit (single conductor underground); and the on-board battery system. The '90s were also to prove that although all three systems were workable, the trolley was far and away the most economical. As the century ended, interest in battery streetcars stopped and the new century saw very few such schemes put forward. Nevertheless, up until the mid 1890s the proponents of battery traction were still fairly optimistic, and, although they admitted the lead taken by the trolley, they felt that the battery might yet carry the day if its durability could be improved and its weight reduced.

Thus, for example, Pedro Salom, late of the Julien and Consolidated Companies, delivered the following peroration in a paper to the Franklin Institute in mid 1892: [1]

> The trolley may be introduced at the present time, pending solution of the difficulties under which storage batteries have come into disrepute, but that it (i.e. the trolley) can obtain in the long run against the many and obvious advantages of the storage system is incredible. Let the public once understand that there are no insurmountable difficulties connected with the storage system, beyond the fact that it costs a few cents more per car mile than the trolley, and the demand for its introduction will be irresistible.
>
> The public are not interested in the cheapest and most objectionable method of transit, especially when they derive no benefit from the economics effected, but they are interested and entitled to a safe, reliable and absolutely unobjectionable method of transit, which is cheaper than horses at the present time, and which may in a few years, from the further knowledge and experience gained from actual use, almost, if not quite, complete in cost with the trolley.

This is certainly a brave defense of the battery car, and Salom may actually have believed it himself, though only two years later, Salom abandoned battery streetcars to concentrate on the new electric automobile. [2] Nevertheless, a large number of battery streetcar projects were attempted through the '90s, many reaching the commercial stage. This activity led to increased effort on battery design.

The large-scale activity in battery streetcar projects during the 1890s requires an explanation, since the experiments conducted during the 1880s were hardly encouraging to further effort. This explanation requires a two-part answer: first, why did electric traction itself arouse such great interest in the '90s; and second, why did so many inventors cling tenaciously to the battery idea despite persistent failure to match the success of the trolley.

---

[1] *J. of the Franklin Inst.* 134 (1892): 150-152.

[2] Ibid., 141 (1896): 280. Specifically, in June 1894.

To the first question there is an easy answer: money. As soon as the stockholders reports of the first trolley roads became available, it was obvious that a gold mine had been struck. The earnings-to-cost ratios for trolleys doomed the horse and cable roads overnight. These statistics were all the more impressive since the comparisons between trolley and horse roads were easy to make and devoid of unsuspected variables. This was because the data were usually obtained from the same streetcar companies, before and after switching from horses to trolleys.

One such comparison came from the Rochester Railroad Company, which switched to trolleys in 1891.[3] With horses, costs per car mile had been 10¢ while receipts were 12¢. After conversion, costs rose to 11.4¢, but receipts soared to 23.2¢. This gives a gross receipts/expenses ratio of 1.2 to 2.0 before and after conversion. Although this degree of improvement was greater than that experienced by most roads, it was not unique.

A more common range of improvement was reported by the West End Company of Boston in 1891, at a time when it still used both horses and trolleys on different routes. Although the trolleys were used on those routes which had been traditionally the least profitable, the receipts/expenses ratio of the trolley and horse line were 1.71 and 1.43 respectively.[4]

Boston was the leading center of early trolley construction, so it will be well to focus our attention on the record of this city. In 1893 the railroad commissioners for the State of Massachusetts reported that the "electric system is on the whole preferable to the horse system." At this time the state had a total of 9,372 miles of street railway, of which 8,611 miles were horse and 761 electric. Almost all the electric lines were trolley, except for a short battery line.[5]

The commissioners based their recommendations on the following data:

	Horse System	Electric System (including battery)	Percentage Increase
Net Earnings per Passenger Carried ------	1.02¢	1.62¢	62.5%
Net Earnings per Car Mile Run ------------	5.79¢	10.04¢	73.6%
Net Earnings per Mile of Track ------------------	$2,521.00	$3,968.70	57.4%

---

[3] *Electrical Engineer* (London) 2 October 1891: 330-333.

[4] Ibid.

[5] Killingworth Hedges, *American Electric Street Railways*, p. 165.

Moreover, as a result of the gradual changeover of the Boston system from horses to trolleys, the gross receipts/operating expenses ratio for *all* streetcar lines rose from 1.23 in 1888 to 1.44 in 1893. The annual reports of the West End Street Railway Company of Boston reveal the following patterns:[6]

As of Sept. 30 of ---	1888	1889	1890	1891	1892	1893	1894
Miles of Track		233	235	244	260	268	273
Miles Electrified	0	0	65.5	81.2	148	183	212.5
Percent Electrified	0	0	28	33	57	68	78
Horse Cars	1584	1794	1694	1662	1226	826	606
Electric Cars	0	47	337	469	1028	1346	1509
% Electric Cars	0	2.5	16.6	22	45.6	62	71
Gross Receipts/ Expense	1.22	1.22	1.30	1.34	1.41	1.47	1.51
Total Cars	1584	1841	2031	2131	2254	2172	2115

The next to last two rows of this table reveal an almost perfect direct relationship between the percentage of electric cars and the receipts/expense ratio. Indeed, if we extrapolate to the 100 percent electric cars point, the gross receipts/expense ratio is 1.65, which agrees fairly well with the 1.71 figure for the company's pure trolley lines reported in 1891, as mentioned previously.

Although such receipts/expense ratios tended to vary widely between individual lines, averages of a number of lines tend to agree with the figures presented below. For example, the receipts/expense ratio for a series of seven American trolley lines reported in 1897 was 1.68, while the ratio for two horse lines operated by two of these seven trolley companies was 1.14.[7]

In 1899 William Clark presented a detailed analysis of over a dozen American streetcar lines which possessed varying operating conditions.[8] Excluding those which Clark identified as unusual, we are left with six more or less average systems which had all used horses in 1887 and had all been converted to trolleys by 1896. The average gross receipts/expense ratios for 1887 and 1896 respectively were 1.3 and 1.74. For two other systems which used horses in 1887 but still retained them in

---

[6] Dawson, "Electric Railways," p. 14.

[7] Ibid., p. 9. Identities of lines not given, except to indicate that all are large.

[8] *Cassier's Mag.* 16 (8/1899): 518-526.

1896, the respective ratios were 1.33 and 1.33. Using these two systems as a control, it is clear that the improved economics of the former six systems came from introduction of electricity.

An interesting picture emerges if we try to break down the receipts/cost ratios to see if the increase in this ratio with trolleys came from a decrease in the denominator, or an increase in the numerator, or both. In other words, did the switch from horses to trolleys lower the costs of carrying passengers or make it possible to carry more passengers at the same cost, or a combination of these? An analysis of individual cases shows that different cases give widely differing answers to this question, even though the ratios themselves remain surprisingly constant for different cases.

For example, in the Rochester system, car mile costs rose, but receipts rose much more dramatically. This indicates that the new technology was used to carry more people per car. For the Boston West End Company in 1891, however, the situation was more complicated. Here, costs per car mile fell from 24¢ to 21.5¢, while receipts per car mile rose from 34.3¢ to 36.7¢. On one of the systems analyzed by Clark above, receipts per car mile decreased over the period 1887-1896 from 30.62¢ to only 22.10¢, yet operating costs fell even more significantly from 23.70¢ to 13.64¢. This was the opposite of the change which occurred in the Rochester system.

These different patterns demonstrate a crucial factor which led to the rapid adoption of trolley technology during the '90s—its greater flexibility over horses. Notice that the costs per car mile for the West End Line in 1891 were 21.5¢; these are among the highest costs ever reported for a trolley line. In 1894, for example, an English authority quoted 12¢ per car mile as the average cost for American trolley roads. Of 22 trolley roads analyzed by Carl Hering in 1892, the highest cost per car mile was 23¢, the lowest 7.8¢, and the average 11.2¢.[9] Despite its high operating cost, the West End Line had a very respectable earnings/costs ratio of 1.71. By comparison, the line quoted in the preceding paragraph which had operating costs of only 13.64¢ per car mile had a lesser earnings/cost ratio of 1.62, but and most importantly, both of these lines improved their earnings/costs ratios by about the same amount as a result of changing from horse to trolley technology—20 percent for West End and 25 percent for the second line.

---

[9] Carl Hering, *Recent Progress in Electric Railways*, p. 110.

The difference between these two lines came from their method of operation; the one intensive operation, and the other extensive. In 1889, when it was totally horse-operated, the West End Company had 233 miles of track. By 1893, with almost 75 percent electric cars, the line had increased its trackage to only 268 miles and its total number of cars had risen from 1,584 to only 2,172.[10] The rate of increase of passengers carried per annum was about the same and equaled about 155 million in 1895.[11] By comparison, the second line, which served a much smaller population of about 60,000, carried only about 10,000 passengers in 1896. Data for the two lines in 1895 and 1896 are shown below:

	West End (1895)	Suburban Line (1896)
Miles of Track	274	60.2
Car Miles Run	26 million	2.3 million
Passengers/Mile of Track	578,000	166,000
Gross Receipts/Car Mile	33¢	22¢

Between 1888 and 1896 the suburban line increased its trackage from 18.4 to 60.2 miles and the number of passengers carried was increased from 2.7 to 10 million. The passengers carried *per car mile*, however, decreased during this period from 5.7 in 1888 to 4.2 in 1896. This accounts for the loss in receipts per car mile, as follows: 30.62¢ to 22.10¢: 4.2/5.7 × 30.62 = 22.6¢ (off by ½ cent).

By contrast, the West End Line increased its receipts per car mile from 1888 to 1895, although there was a fall towards the end of the period:[12]

Gross Receipts per Car Mile–West End Line

1888 — 28.92¢	1892 — 32.52¢
1889 — 30.52¢	1893 — 30.92¢
1890 — 31.02¢	1894 — 29.82¢
1891 — 31.56¢	1895 — 29.74¢

The virtue of trolleys over horsecars can be seen in these two divergent sets of data. When the West End Line adopted the new technology, it used the greater power of the electric cars to carry more passengers, thus increasing gross receipts, and at

---

[10] Hedges, *Street Railways*, p. 162.

[11] Dawson, "Electric Railways," p. 578.

[12] Ibid.

the same time it reduced costs by benefiting from the greater economy of the new cars. Yet for the West End Line, the costs were not reduced substantially over horsecars, since the heavy, urban, stop-and-go traffic put almost as severe a strain on dynamos and motors as on horses. Indeed, this was the chief reason for the great popularity of load-leveling batteries in large, urban trolley power houses, the data presented above show more clearly than previously the economic appeal of batteries.

The suburban line, on the other hand, took full advantage of the greater speed potential of the trolleys to extend its lines, thus attracting more customers and simultaneously reducing car mile costs by a spectacular 43 percent. The non-urban location of the line meant lower receipts per car mile than the West End system, and as the lines were extended from 18.4 to 60.2 miles the passenger density fell quickly. Yet this was more than compensated for by the operating cost savings, as lightly-loaded cars moved long distances without stopping, thus giving the power house a good load-factor and not straining dynamos or motors.

The preceding discussion of the economics of the trolley is not the product of modern historical hindsight. The arguments which have been presented, although cast in new form, are essentially the same as those presented at electrical engineering society meetings and in the pages of the technical press beginning about 1890. The erection of new trolley lines and the conversion of horsecar lines after 1890 was stimulated by these mathematically-calculated predictions of the economic benefits.

The proven economic success of the trolley system stimulated not only its own further development, but also the development of the battery system. After all, both were forms of electric traction, and it was reasoned by many inventors that the economic advantages of battery-traction should not be much different from that of the trolley, if only the battery could be made more efficient. Battery enthusiasts realized that horsecars and battery cars both suffered from the fact that their motive power sources were extremely susceptible to strain. Thus, in order to keep operating expenses low, both forms of cars had to be driven slowly, with light loads and avoiding much stop-and-go traffic. But this kind of operation also kept receipts low. Therefore, the relative inflexibility of both these methods of transport, as contrasted with trolley, usually meant a low receipts/costs ratio.

Serious as this objection was, it was not so depressing for
the battery promoters as for the horsecars. There was not much
to be done about changing the inflexibility of horsecar operation, since horses are not easily improved by human effort,
and the horsecar body itself had long since reached the limits
of strength and lightness that contemporary materials technology could give it. Storage batteries, on the other hand, were
undergoing rapid improvement. Moreover, the timing of storage battery and electric streetcar innovation was crucial for
maintaining a positive attitude towards the potential of battery
traction. Investment in all forms of electric streetcar systems
was low until 1887, when Frank Sprague built his famous
Richmond line, and opened the floodgates of investment.
Until this time, investor attitudes towards all forms of electric
traction were of a wait-and-see nature. But attitudes towards
the storage battery followed a similar pattern. Improvements
in the technology of the battery had been made since 1882, but
it was only around 1886-1887 that the battery became an
economically-practical device. At the same time that Frank
Sprague's trolley technology was making its favorable impression
upon the horsecar companies, the increasing durability of such
batteries as those of EPS and EAC was making a less dramatic,
albeit important impression upon the minds of such men as
W.W. Gibbs, William Mather, the Rothschilds and the Berlin
financiers of AFA. The storage battery and the electric streetcar
began to mature at the same time, and therefore the concept of
using the one to drive the other caught the imagination of
dozens of inventors in this country and in Europe.

This is not to deny the fact that the success of the Sprague
overhead trolley caught the imagination of many more inventors and investors than did the battery system. Moreover, it is
doubtful if many of even the most die-hard battery streetcar
supporters believed after 1890 that their system would ever be
able to compete against the trolley on the basis of its appeal to
streetcar company directors. Trolley economics were simply
too favorable. Yet despite the clear economic superiority of the
one technology over the other, it should be noted that the
evolution of technologies, like that of living things, tends to
produce diversity if the environmental niches are similarly diverse. Battery streetcar technology survived during the 1890s
since its promoters felt that there were a significant number of
such niches in the streets of American and European cities
where the trolley would not be able to adapt itself.

The quote by Pedro Salom with which we began this section refers to this question. During the early days of the trolley, many people considered it an objectionable, dangerous eyesore which should not be allowed to reduce the quality of city life. Ever since the middle of the century, the space above city streets had become clogged with an ever-expanding mass of wires—first from the telegraph, then the telephone, then arc lights, then incandescents, and finally the trolley. In larger cities, utility poles sometimes had a dozen or more crossarms and over a hundred individual wires.

During the early decades of electrical technology most European and American cities tolerated overhead wires. By the 1880s, however, resistance to the increasing unsightliness of vast numbers of communication wires, many of them long since unused, and the danger of power circuits, led to the passage of laws against overhead construction. This movement began earlier in Europe than in the United States, but by the late 1880s even New Yorkers were beginning to feel that things had gone too far. The trolley appeared at just this time. The telegraph, telephone, and power lines darkened the space above the sidewalks, and now the trolley threatened to do the same for the middle of the street. A well-designed system of overhead trolley wires was not as offensive a sight as the masses of telegraphic and telephone wires on the utility poles. Nevertheless, before many people had a chance to see what a trolley system looked like, a good number of municipalities voiced concern about the dangerous mass of wires needed for its operation.

One of the chief reasons for the persistence of battery streetcar developers was their belief that the trolley could never be used in many cities because of the growing movement for total abolition of overhead wires. Despite the rapidity with which the trolley replaced the horse on American streets, a number of engineers and streetcar executives during the early years of the trolley refused to consider its adoption, since they felt it would be impossible to overcome municipal opposition to overhead conductors. Perhaps the most interesting example of this occurred in Boston.

Boston was to become the leading center of trolley development in the world by the 1890s. Yet when Henry M. Whitney organized the West End Street Railway Company in 1886 his thoughts were far from the trolley system. Although Whitney initially equipped the line with horses, he commissioned the

electrical expert Franklin L. Pope in early 1886 to investigate
the economic potential of electric traction. Whitney instructed
Pope,[13] "that I must confine my researches to conduit (i.e.
underground third rail) and storage (battery) systems; that
anything in the shape of an overhead line was out of the question, and would never be tolerated in Boston for a moment."

Pope studied the dozen or so electrical lines then running in
the United States and presented Whitney with the following
estimates of probable car mile costs for the four systems considered. All estimates are low since they do not include repair
costs: Horse—11.07¢; Battery—7.34¢; Conduit—6.02¢;
Overhead—3.28¢.

This report was submitted to Whitney in April 1886. He
instructed Pope to build an experimental storage battery car.
This was put in operation on the Boston-Cambridge line in
October 1887, and was soon found to be about as expensive to
operate as a horsecar of equal capacity. In 1888 Whitney au-
thorized an experimental conduit road, and, finally, an overhead line. The results were so favorable to the trolley that the
objections of the citizens of Boston were overcome and Boston
was on its way to becoming the chief center of trolley innovation.[14] By 1890 the London *Electrical Engineer* reported that:
"There are probably more electric cars in Boston, U.S. than in
any other town. The wires are overhead, and no one is understood to complain about them."[15]

Although Whitney's fear of the prohibition of trolleys in
Boston turned out to be exaggerated, it was by no means
groundless. Many cities, particularly in Europe, either banned
the trolley outright, or at least strongly encouraged the installation of the alternative battery or conduit systems. In this country, the leading center for the battery car was Manhattan. The
following news note from the 24 May 1889 London *Electrical
Engineer* shows why this was so,[16] "It is curious to find that
while New York (i.e. Manhattan) is insisting upon all overhead wires being taken down, at Boston, U.S., authority has
been granted for the introduction into the heart of the city of
the overhead wire system. . . . "

---

[13] *Cassier's Mag.* 16 (8/1899): 518-526. Although these calculations are derived from
Clark, they have been altered for purposes of argument.

[14] *Electrical Engineer* (London) 13 March 1891: 258-260.

[15] Ibid., 6 June 1890: 439.

[16] P. 405.

Municipal opposition to overhead wires was intense in Manhattan. Although the trolley system was allowed in the other boroughs, it was never allowed on Manhattan itself. The depth of early sentiment can be gauged from these comments in the technical press:[17]

> 10/11/1889—Overhead wires will be out of the field (on Broadway)
>
> 8/1/1890—The overhead system, of course, never can find its way into New York City, and indeed provokes a good deal of unfavorable comment wherever it has been tried in very populous districts. . . .

In August 1892 the New York Board of Fire Underwriters passed a resolution calling for the prohibition of overhead trolleys in New York.[18] It is not surprising to hear of the opposition of so many New Yorkers to any sort of overhead wires in the late 1880s and early 1890s; New York's overhead wire situation was probably the most offensive of any city in the world at this time. As the most populous American city, and the financial heart of the nation, it had more circuits than any place else. Moreover, American companies were notorious for the cheapest, most unsightly and dangerous utility pole construction. Visiting British engineers have left us some picturesque descriptions of this rough-and-ready workmanship and it is not hard to imagine the angry reaction of New Yorkers against the business ethics which produced such engineering. For example, the British electrical engineer G.L. Addenbrooke gave the following impressions of American overhead work after his visit in 1889:[19]

> When posts are used, they are rough pine, never painted, often out of the straight, and warped or bent. . . . During my whole journey from San Francisco to New York, I can hardly recollect seeing one ship-shape, neat and smart-looking post, whether for arc lamps, or for any other electrical purpose. Everything has a temporary and expedient look about it, which is very offensive to English eyes. . . . Owing to the length of the (cross) arms, and their not being very securely fixed to the poles, the weight of the circuits often pulls them out of the horizontal. To see a warped and bent pole, set crookedly in the ground, and with the arms at

---

[17] *Electrical Engineer* (London): 282; *The Electrician* (London): 359.

[18] *Electrical World*, 27 August 1892: 139.

[19] *Electrical Engineer* (London) 6 December 1889: 453-457.

various angles to one another, and the wires on it all hanging in different curves, seems about as dismal and woe-begone a piece of engineering as can be imagined.

Washington D.C. also banned the trolley; Congress passed a law forbidding the erection of overhead conductors in 1891, so a streetcar company which had installed one of the earliest Thomson-Houston overhead trolley systems in that city ordered the equipment for six storage battery cars from the EAC.[20] Despite early fears of widespread abolition of trolleys, few American cities actually carried out such bans. It was in Europe that opposition to the trolley led to bans in many cities. Consequently, the creation of storage battery roads on a commercial, rather than merely an experimental basis, was more common in Europe than in the United States. For example, between 1890 and 1902, the percentage of all electrical streetcars in the United States using battery propulsion never rose above 2 percent. In Britain, however, the percentage in 1890 and 91 was 33 and 20 percent respectively, and in Europe (including Britain) the percentage in 1894, 1897 and 1899 was 6, 8 and 8 percent respectively.[21]

Even in Europe the trolley system increasingly became the dominant traction technology, representing 84 percent of all electric roads in 1899. Nevertheless, sufficient economic incentive was provided here for continuous battery car experimentation throughout the 1890s. In 1891 the London *Electrical Engineer* editorialized:[22]

American citizens are not in most cases living in overcrowded cities, their aesthetic tastes are not allowed to stand in the way of business requirements, hence overhead wires are permitted to a greater extent than in England. In our highly civilized land overhead wires are not allowed promiscuously. . . .

At a meeting of the Tramways Institute of Great Britain in 1891, one speaker confidentially stated: "It may be taken as a foregone conclusion that neither in Great Britain, nor in most of the principal Continental towns, will the overhead system ever be tolerated."[23] In June 1889 the *Electrical Engineer* de-

---

[20] Ibid., 27 February 1891: 204.

[21] *J. of the Franklin Inst.*, 7/1897: 156; *Street Railway J.* 15 (5/1899): 284; *Electrical World*, 14 February 1891: 11; Hering, *Recent Progress*, p. 34; Oscar Crosby & Louis Bell, *The Electric Railway*, p. 235.

[22] *Electrical Engineer* (London) 12 June 1891: 578.

[23] Ibid., 10 July 1891: 38-40.

scribed the overhead system as: "a system practically impossible of introduction in English towns."[24] These stout denials of the impossibility of the introduction of trolleys into European towns continued to appear regularly in the European technical press until that system had just about totally replaced all its competitors—horse, cable, and other forms of electric traction.

The following is a list of storage battery car use from 1889 to 1900, exclusive of those discussed in the last chapter. The list is not complete, but does include all significant experiments or installations. Those marked with an asterisk were experimental lines; the others were regular commercial service.

YEAR	CITY AND NUMBER OF CARS, IF KNOWN
1889	London (5)*; Beverly & Danvers, Mass. (6); Birmingham, Eng. (1)*
1890	Phila. (1)*; Paris*; Liverpool (1)*; Phila. (4)*; Dresden*; London (2)*; Germany (1)*; Paris (1)*; Paris (1)*
1891	Hague (6); London (5); Washington (1)*; Lyons (1)*
1892	Rome; New York (1)*; New York; Washington (same as 1891 above)
1893	Croydon (2)*; Cleveland (1)*; Paris; Paris; Paris; Milan
1894	Oakland, Ca. (1)*
1895	Nice, Vienna (2)*; Hagen, Ger. (4); Hanover; Berlin (3)*; New York; New York
1897	Chicago
1898	New York; Wurtemburg
1899	Washington
1902	New York*
1911	New York (35)

It is clear from this list that there was a rapid decline in interest in battery streetcars after 1895. Indeed, if we consider comparative cost data reports of trolleys and battery cars which appeared regularly in the technical press from 1889 on, an even more rapid decline in battery experiments might have been expected. Such comparisons almost always favored the trolley by a wide margin. The comparisons listed below are from various reports which appeared in the technical press from 1891 to 1895:[25]

---

[24] Ibid., 7 June 1889: 448.

[25] Ibid., 27 February 1891: 210-211; 13 March 1891: 258-260; 10 July 1891: 36-37; 6 July 1894: 18-22; 21 December 1894: 695-698; 21 September 1895: 326-327.

DATE	SYSTEM	CAR MILE COST
2/27/91	Trolley	5.1¢
	Battery	10.6¢
3/13/91	Trolley	3.28¢
	Battery	7.34¢
	Conduit	6.02¢
7/10/91	Trolley	12.7¢
	Battery	29.0¢
1893	Trolley	9.6¢
	Battery	17.3¢
7/6/94	Trolley	12¢
	Battery #1	16¢
	Battery #2	33¢
	Conduit	10¢
12/21/94	Trolley	9¢
	Battery	18¢
9/21/95	Trolley	15¢
	Battery	15¢

By the end of the 1890s, as the details of trolley technology became standardized and textbooks were prepared which taught engineers how to design roads for specific, predetermined car mile costs, the leading authorities recommended designing roads for costs of about 10¢ per car mile.[26] This figure seems to be a fair average of the expenses of trolley roads built since 1890. The average for battery cars, on the other hand, was about 16¢. For example, an estimate prepared in 1897 from the data for the Paris systems, which had used both trolley and battery cars for years, gave the expenses per car mile for these two systems as 9.24¢ and 16.66¢ respectively.[27]

Yet despite the clear implications of these data, many municipalities ignored them due to the opposition to the trolley. The data which they were interested in were those of horsecars and conduit roads versus battery cars, and here statistics often showed an economic advantage for batteries, at least in comparison with horses. For example, for the Paris systems mentioned in the preceding paragraph, the costs for horses and batteries were 17.1¢ and 16.7¢ per car mile respectively. The promoters of battery cars during the 1880s had usu-

---

[26] Hering, *Recent Progress*, pp. 100-136.
[27] Dawson, "Electric Railways," p. 502.

ally claimed enormous cost savings for their system over horses, sometimes up to 100 percent. By the '90s, however, these claims had been toned down to a more reasonable 1 to 3¢ per car mile differential. Actual operating cost figures for conduit roads were often lower than those for battery lines. On the other hand, the capital cost of installing an underground conduit system was almost as high as that for a cable car. Moreover, the technical difficulties connected with the proper insulation of a third rail exposed to the elements in an underground trough were never totally solved. Because of these various factors, those cities which gave up horses but refused to adopt trolleys were about equally likely to adopt battery cars as conduit cars. In 1899, for example, of 204 street railway systems operating in Europe, 16 were battery, 8 were conduit, 8 were surface-type third rail, and 174 trolley.[28]

A final reason for the relative popularity of battery traction schemes until the mid 1890s was that the promoters greatly underestimated costs. Typically, battery traction would be sold to a streetcar company by claiming costs equal to or only slightly in excess of those of the trolley system. In the early 1890s, for example, the Jarman company convinced a number of English lines to try their system on the basis of a maximum expense per car mile of 9¢.[29] This was the rate at which the General Electric Traction Company agreed to operate battery cars in Canning Town in 1890 for the North Metropolitan Streetcar Company. The next year, however, they had to ask for an additional 2¢ per car mile.[30] Then, within another year, the Traction Company told the Metropolitan Company that it would either have to raise the expenses, or else Metropolitan would have to buy the equipment and run it themselves. Traction wanted expenses raised to 15¢. The companies failed to agree on a contract and the Canning Town service terminated in September 1892.[31]

This pattern was typical of most storage battery lines—an initial low estimate of costs, usually under 10¢, and then steadily increasing corrections as service was begun. The Waddell-Entz battery cars, which ran in a number of cities during the mid 1890s, fit this pattern. At first, they were re-

---

[28] *Street Railway J.* 15 (5/1899): 284.
[29] *Electrical Engineer* (London) 27 June 1890: 500.
[30] Ibid., 24 April 1891: 402.
[31] Ibid., 19 August 1892: 188-190.

ported to cost a mere 9.3¢ per car mile. The Electric Storage
Battery Company cars running in Chicago in 1897 were reported as needing only 8¢ per car mile during their first year of
service.[32] The same rate was claimed for the Acme storage battery car which ran on Amsterdam Avenue in Manhattan during 1892-1893.[33] The Eckington & Soldiers Home Railway in
Washington D.C., which began regular service in the spring of
1892, claimed 9¢ per car mile for the first five months of operation.[34] All of these estimates were at or below the actual cost of
operating a successful trolley line, and yet all of these battery
systems failed after a year or two.

One of the best documented of full-scale battery traction
installations was begun in Birmingham, England in 1889. It
illustrates the typical pattern of initial low cost estimates with
rising costs and discouragement to follow. Birmingham began
in 1889 to experiment with new systems to replace its horsecars. Cable, conduit, and battery cars were tried, along with
miniature steam locomotives. These systems were compared
with the still-running horsecars. The ECC built the battery
cars, with Thomas Parker in charge of the project. A series of
different batteries were tried—EPS, Julien and Epstein.[35] The
first experimental runs were made on the rails of a twelve-
year-old Birmingham horse road in November 1889.[36] These
runs were successful, so that during early 1890 a decision was
made to convert the line to regular operation with batteries. A
special car barn was designed for efficient switching and
charging of batteries, with elevators to raise and lower them
from the cars.[37] The line was to have a full complement of
twelve cars. These were delivered in June and the line opened
for service in July 1890.[38]

Originally, it was hoped that the line would operate at about
the same cost per car mile as the London Canning Town
Line—9¢. A year later, the expenses were found to be 19.8¢,
versus 22¢ for steam cars, 19.6¢ for horses, and only 12.7¢ for

[32] Ibid., 17 November 1893: 459-460; *Street Railway J.*, 1/1898: 43.

[33] *Electrical World*, 3 January 1894, p. 158-159.

[34] *Electrical Engineer* (London) 19 August 1892: 188-190.

[35] *Electrical World* 15 February 1896: 177-178; *Electrical Engineer* 20 June 1890: 487; 11
July 1890; 26-27; 15 August 1890: 137-142; 14 July 1893: 37; 21 July 1893: 61-62; 28 July
1893: 85-86.

[36] *Electrical Engineer* (London) 1 November 1889, p. 342.

[37] Ibid., 15 August 1890, pp. 137-142.

[38] Ibid., 25 April 1890, p. 323; 15 August 1890, pp. 137-142.

the cable.[39] A year after that, the expense figures were: batteries—30.8¢; steam—24.1¢; horses—19.9¢; and the cable cars—12.4¢.[40] The battery line ran at a deficit of $8,500 that year, while all the other lines ran fairly profitably.[41]

During 1893 the battery line managed to get its costs down to 23.2¢. A breakdown of these costs for that year dramatically demonstrates the source of the cost problem: Wages—6.74¢; Fuel—3.52¢; Repairs & Maintenance—10.98¢; Misc.—1.94¢; Total—23.18¢. If the costs for repairs and maintenance, which were mainly for the batteries, are removed, and the cost of wages reduced by the extra amount needed to recharge and switch batteries, the costs would have been reduced to around 13-14¢. This was at the upper limit for a trolley line.

The same pattern of costs appears for the London Canning Town line in 1890 and for the Paris TEM cars in 1893:[42]

EXPENSES	CANNING TOWN	PARIS CARS
Wages	6.2¢	2.42¢
General Expenses		.41¢
Fuel and Oil	2.4¢	
Power		5.66¢
Battery Depreciation and Labor	5.7¢	
Motor Depreciation and Labor	.9¢	2.82¢
Other Depreciation and Labor	1.2	
Maintenance & Handling Batteries		5.10¢
Miscellaneous		.27
	16.4-5.7=10.7¢	16.68-5.1=11.58¢

The Paris car mile cost minus battery costs gives a figure which is about 2.5¢ above that of contemporary Paris expenses for trolleys—9.24¢. The extra cost probably resulted from the additional power needed to move the great weight of lead around in the cars.

The difference between the favorable cost estimates of the battery car promoters and the dismal record in actual operation was largely due to honest underestimates on the part of the promoters of the great maintenance expenses of the batteries. A comparison with the operation of horsecars is instructive here. For example, in 1895, Geneva, Switzerland operated both a horse and a trolley system. The car mile expenses of these

---

[39] *Electrical World*, 8/1891: 168.

[40] Ibid., 17 September 1892: 186.

[41] *Electrical Engineer* (London) 16 August 1892: 205.

[42] Hering, *Recent Progress*, p. 131; Dawson, *Electric Railways*, pp. 501-502.

lines were: 7.9¢ (trolley) and 18.7¢ (horse). Almost 80 percent of the difference between these costs was due to renewal costs for the horses themselves—8.3¢. This sum represented 44 percent of the total cost of operating the horsecar line. By way of contrast, the expense of battery renewals on the 1890 Canning Town cars was a fairly similar 35 percent of total costs. Depreciation of trolley equipment, on the other hand, cost much less—only .46¢ per car mile for cars, motors, and running gear. This works out at only 5.8 percent of total expenses for the Geneva line.

Therefore, the chief reason that the battery car turned out to cost almost as much to run as a horsecar is that the battery suffered from the same technological limitation as the horse. The horse had to be replaced on the streets of late nineteenth-century cities because of the inverse ratio between its life and the amount of service it could be called upon to perform. The cars needed to go faster and carry more passengers as the cities expanded. The horses could not do this without shortening their lives to an unprofitable extent.

The storage battery is precisely like the horse in this respect. Forcing it to bear short, intense discharges is one of the best ways to reduce its life. This is what happened to the batteries in stop-and-go traffic with fully-loaded cars. The positive plates of traction batteries rarely lasted six months. On the aforementioned Eckington & Soldiers Home line in 1892 they lasted two months.

The huge size of the stationary batteries used in trolley power stations was needed not for their total energy storage capacity, which was rarely if ever drawn upon, but for the ability of the cells to handle a short, intense drain without suffering physical shock. Their size assured them of long life; by 1897 the rate of depreciation of load-leveling batteries was calculated at 9-11 percent per annum, which was only slightly greater than the amortization of the dynamos and steam engines of such plants (5-10 percent), or that for the trolley motors (5-8 percent).[43] Moreover, such batteries were used not only to save power at trolley central stations, but also to extend the life of dynamos by relieving them of strain. In a well-designed trolley plant, therefore, the carrying capacity of the system did not inherently lower the life of the equipment.

For battery streetcars, this was impossible. The weight of the cells of the original cars of the 1880s had been reduced, since

---

[43] Dawson, "Electric Railways," p. 566.

these usually represented half the weight of a loaded car. By the early 1890s the weights of the cells in some cars was only 20 percent of the total weight, and 25 percent was common. Such advances could only be bought, however, with thinner grids, which were more easily subject to warping under heavy discharge. Attempts to use lighter metals than lead were tried, but these proved unsatisfactory. The use of smaller cells was counterproductive, since it put greater strain on the remaining batteries and meant increased labor costs because of the need for more frequent recharging.

The great advantage of the trolley was that it could run slow or fast, fully loaded or empty, cruising or stop-and-go, on level or on fairly steep grades, and do any combination of these without seriously shortening its life. Thus it could maximize receipts while keeping the increase in operating costs to a reasonable rate. It was a flexible technology; entrepreneurs could use it in a combination of ways to make their profits. The battery car, like the horsecar, did not possess this flexibility. Given the need to operate sometimes empty and cruising, and sometimes over-loaded and stop-and-go, the battery responded like the horse and lived a short life.

Nonetheless, the low cost predictions of 7-10¢ per car mile which battery car promoters gave at first were honest estimates. The initial, experimental runs usually did not place severe strains on the cars. Often, the tests were carried out for such short periods before commercial service was begun that the short life of the cells could not be determined. This was the case with the Birmingham cars. Nevertheless, it would be an oversimplification to suggest that the collapse of the battery plates was the only cause of the failure of storage battery streetcars. A number of roads reported that the batteries held up well, but that the cars' power costs were excessive, because of the heavy dead weight of the batteries, and the tendency of a battery to discharge too rapidly thus curtailing its energy efficiency sharply, that is, making it operate considerably below the 80-85 percent obtainable from large, stationary batteries.[44]

Whatever the specific cause of their failure, most of the battery streetcar lines disappeared quickly, especially those in cities which did not oppose the erection of trolley systems. Cities which forbade the trolley had only the conduit or cable

---

[44] Hering, *Recent Progress*, p. 201.

to adopt as alternatives, and both of these required very heavy capital investment. Accordingly, every effort was made in these cities to give the battery lines a chance of improvement. In Dubuque, Iowa, where trolleys were not forbidden, a much-heralded nine-car battery line was begun in May 1891. As in Birmingham, maximum effort was made to minimize labor costs—a battery-changing house was designed in which it took only 30 seconds to replace the discharged cells.[45] Nevertheless, by March 1892 the streetcar company was already changing over to the trolley system.[46]

In Europe, on the other hand, where opposition to trolleys was stronger, battery cars were retained longer. In Birmingham, despite their poor performance, the last of the battery cars ran until 1901, when the service was replaced with the conduit system. The Paris system of battery cars, which was perhaps the most extensive ever built anywhere, also was long-lived. The Philippart brothers took up battery streetcar work again in the late 1880s and TEM took it over from them about 1890.[47] TEM ran a series of experimental cars between 1890 and 1892. In 1892 TEM used the results of these tests to construct two commercial lines, and by 1900 there were six Paris battery lines, run by three separate streetcar companies and all powered by TEM traction cells.[48]

In addition, by 1900 Paris also had six more lines operating on the so-called mixed system. This was a system pioneered by the Germans in 1895. It had a European vogue during the late 1890s—15 continental roads used it by 1897.[49] The mixed system was a compromise between the speed and efficiency of the trolley and the aesthetic appeal of the battery car. It was used in cities such as Hannover, Berlin, and Paris for main lines which ran from the city center out into the newly-built suburbs. In the suburbs, where municipal control was not strong, the cars ran from overhead conductors, but inside the

---

[45] *Electrical World*, 25 July 1891: 61.

[46] Ibid., 19 March 1892: 207. The life of the batteries varied from 40 to 90 days, and the original complement of 50 cells per car had to be increased to 80 for successful operation. T. Illingworth, *Battery Traction on Tramways and Railways*, p. 10.

[47] *Electrical Engineer* (London) 3 October 1890, p. 279; 14 November 1890, p. 421; 20 November 1891, p. 482-483; 20 November 1891, p. 485.

[48] Ibid., 30 September 1892: 334-335; 4 August 1893: 100; 8 December 1893: 535-538; 11 May 1894: 558-559; *Street Railway J.* 14 (4/1898): p. 210; 14 (6/1898): 342: *The Electrician* (London) 6 April 1900: 84.

[49] *Electrical World*, 6 February 1897, pp. 204-205.

walls of the traditional city the wires stopped and the cars drew power from their batteries. Usually the overhead conductor sections of these mixed lines made up far over half of the total length of the line. Since the downtown run was short, batteries were recharged from the overhead run. They could thus be made lighter than the 30-35 mile ones used on pure battery-drive cars.

Berlin used the mixed battery cars from 1893 to 1901. Paris began switching to conduit cars after the turn of the century, but it did not get rid of the last of its battery cars until 1910.[50]

In New York also the history of battery cars was a long one. Trolleys were used in all the boroughs except Manhattan, where almost all the horse and battery roads had been converted to the conduit system by the early years of the new century. Nevertheless, until the time of the First World War there were sporadic attempts to use battery cars on various New York lines. For example, in 1902 the Brooklyn Rapid Transit Company was experimenting with battery cars because "popular agitation against overhead wires has reached a point such that municipal government is expected shortly to order the removal of the trolley wires from the streets of that borough."[51] The company estimated that it would have cost $25 million to install conduits and hoped to save this by the use of battery cars. As in the case of Boston, however, the feared prohibition never materialized and nothing more was heard of the Brooklyn battery cars.

During the early years of this century, battery streetcars survived for a while in Manhattan on crosstown shuttle routes. This service proved to be marginally acceptable for battery cars, since the routes were short and the passengers carried were not as numerous as on the avenue traffic. Prior to the twentieth century, when streetcar managements were more optimistic about the battery car's prospects, Manhattan avenue traffic was often converted from horses to batteries, while the horsecars were retained for the less arduous cross-town traffic. Thus, for example, all of the early battery experiments took place on the avenues: the Julien cars on Madison Avenue in the 1880s; the Acme car on Amsterdam Avenue in 1892; the Waddell-Entz cars on Second Avenue in 1892; cars with ESB cells on Madison and Fourth Avenue in 1895.

[50] Illingworth, *Battery Traction*, p. 7.
[51] *Electrical Review* (New York) 12 April 1902: 459.

By the turn of the century, however, all the avenue traffic was conduit, and the crosstown shuttle routes, which were too marginal to justify the expense of a conduit installation, became the new target for battery car experiments. In 1902 a battery line was operating on 34th Street.[52] In 1910 the 28th and 29th Street lines were experimenting with cars equipped with the new Edison storage battery. A year later, the 3rd Avenue Line purchased 25 new lead-acid battery cars for its crosstown shuttle, which brought its total complement of shuttle cars to 65.[53] Although such shuttle cars continued to run on a number of lines throughout the world for many years, their history is of no importance for the technological development of the storage battery, since by the turn of the century the market for electric automobile batteries had become the dominant factor in traction battery design.

In itself, the evolution of the storage battery streetcar was of little importance. It had no influence over the development of trolley technology and the number of commercially-run battery cars was insignificant compared with the number of trolley lines. If the battery streetcar had never existed, the rate of introduction of electric traction into European and American cities would not have been much changed, since the conduit system was always the alternative where trolleys were banned.

Yet the importance of the evolution of any species extends beyond the continued existence of that species and includes the future developments made possible by that evolution. Today, the trolley is also a largely extinct technology, and yet its influence is still strongly felt in the impetus it gave to the design and commercial manufacture of electric motors. Likewise, the importance of the storage battery streetcar can still be seen today in the construction of the modern automobile starting battery.

One important influence that the battery streetcar had on storage battery design was on the development of the containing box. The boxes of the stationary batteries were designed for maximum durability and simplicity. The usual materials of construction were a heavy, acid-resistant wood such as teak combined with an inner lining of lead. Sometimes glass was used. The boxes were open at the top to facilitate inspection,

---

[52] Ibid.

[53] Ibid., 19 February 1910: 373; 6 January 1912: 10.

cleaning, and maintenance. Essentially, the design of stationary battery boxes involved nothing more than simple yet
rather massive cabinet-making and sheet-lead soldering skills.

The battery streetcar, on the other hand, stimulated the development of a sealed battery. A serious problem of the battery
streetcars was the escape of sulphuric acid fumes and spray.
Since the batteries on most cars were in recesses directly below
the passenger's seats, riding in such cars could be an unpleasant experience, particularly in hot weather. In addition, the
traditional wood and lead boxes were excessively heavy and
tended to develop cracks and leaks. These problems led to the
introduction of the first molded hard rubber boxes, and to the
use of the first covers for the tops of the cells. Such boxes were
used on the Paris cars in 1892, and there may have been earlier
applications.[54] These molded rubber boxes were very expensive when first introduced, but in the twentieth century, with
the start of a mass market for starting batteries, they became
the universal container for most storage cells, and remained so
until the introduction of plastic boxes after World War II.

In the last chapter we saw that early transport battery electrodes collapsed readily. To correct this problem, a new kind of
electrode was designed during the 1890s. This electrode was
intended originally for self-propelled streetcars, but after the
mid 1890s, it was applied to the design of automobile batteries. This new electrode was called a *tubular* electrode.

The history of the tubular electrode is a complicated one. It is
impossible to identify any one man or company as *the* inventor
of the design; as with many inventions which arise from established industries where a number of companies are all at about
the same level of development, this invention appeared independently in a number of forms from a number of inventors at
the same time. By the first decade of the present century, these
various but similar ideas blended into an industry-wide
synthesis.

The earliest tubular-plate batteries appeared in 1890. Donato
Tommasi in France and H. Woodward in the United States
patented designs which embodied the basic tubular electrode
concept—a thin, central conducting rod of lead alloy was surrounded by a layer of active material paste, which in turn was
held in place by an outer shell or tube of electrically and chemically inert, porous material. In the Woodward and Tommasi

---

[54] *Electrical Engineer* (London) 26 July 1889: 62.

patents the tubes were one piece sleaves of hard rubber or
celluloid. The figure below shows a tube of the Tommasi
plate: [55]

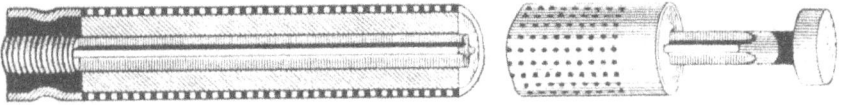

In 1891, Stanley Currie of ESB took out a series of American
patents based on the idea of giving the tubes a microporous
structure by making them of a woven material. Central rods of
lead alloy were placed in the center of woven tubes of asbestos.
The molten lead-zinc mixture used in the Payen process was
then poured into the space between asbestos and lead: [56]

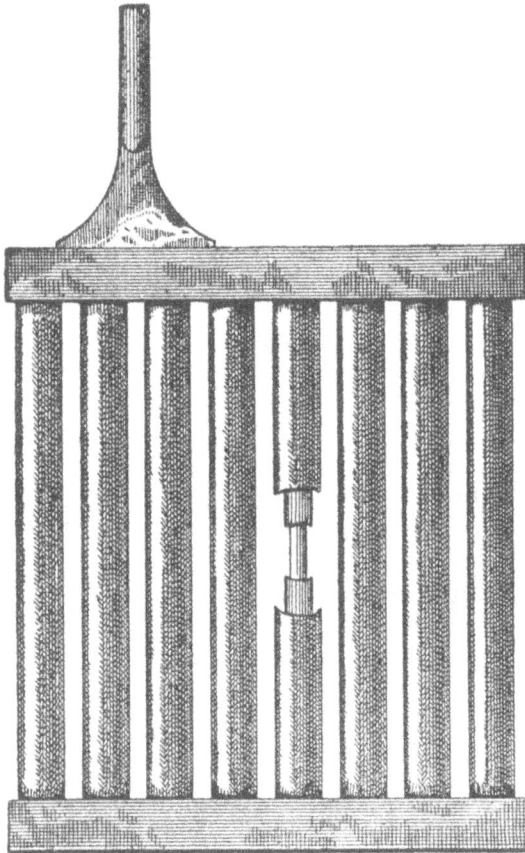

[55] Wade, *Secondary Batteries*, p. 60, fig. 53.
[56] Ibid., p. 61, fig. 54.

Tubular electrodes were a particular favorite with several French battery transport promoters. The Philippart brothers, for example, were interested in the idea. Despite their failure in the early 1880s, Simon and Gustav had not lost confidence in the future of battery streetcars. In 1889 they formed the Société Française d'Accumulateurs Electriques, which, like the earlier venture, was chiefly involved with traction schemes. During the next decade, Gustav developed a series of light, pasted grid batteries for this company. These batteries were first intended for electric streetcars, but during the later 1890s, the Philipparts turned their attention to the electric auto. Gustav developed his first tubular electrode battery in 1897, and commercial production began the next year.[57]

The plates of this battery were constructed by stringing pin-like rods of antimony-lead alloy with a series of thin, hard-rubber washers. The annular space between washers and rod was then packed with the traditional red lead or litharge paste, and the ends of the assembly sealed with pierced lead discs. A group of these tubes was then inserted inside a leakproof hard-rubber box to form the complete battery. These were sold in France under the name *Phenix*:[58]

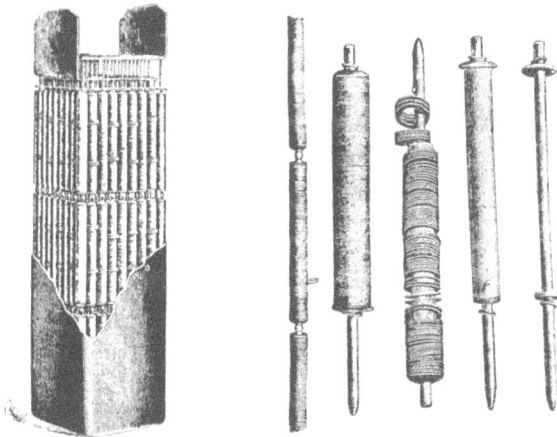

Most of the early tubular electrode designers used the tubes for both the positive and negative electrode, as in the examples shown above. For the positive, the tubular construction offered real advantages. First, the active material is given the

[57] French patent #271,392—16 October 1897; British patent #1194—15 January 1898; U.S. patent #620,172—28 February 1899.
[58] Wade, *Secondary Batteries*, pp. 476-477.

opportunity of expanding outward during discharge, thus relieving the grids supports of strain. The strains are borne by
the hard rubber, which is more elastic than lead. Second, these
central rods are protected from excessive oxidation. In a storage battery the strongest oxidizing agent is the lead peroxide.
When the metallic grids and the PbO_2 are in contact, the grids
tend to be oxidized, resulting in a plate which will eventually
collapse since it is all flesh and no skeleton. In the tubular
design, the central grid rod is protected from immediate contact with the outer coating of PbO_2.

Nevertheless, from our discussion of the "densification"
tendency of sponge lead, it will be apparent that the early
practice of using tubular electrodes for the negative as well as
the positive plate was not a good idea. The negative active
material does not need to be held tightly against the central
grid, and the idea of providing the negative with a central rod
against which the lead sponge crystals can densify is clearly
poor design. The failure of early tubular plate designers to
recognize this fact is another reflection of the neglect of negative plate design and the single-minded concentration on the
positive plate's problems.

Towards the end of the '90s, a number of tubular plate designs appeared which indicate that the designers of transport
batteries were learning to adapt their designs to the particular
characteristics of each electrode—this was the same lesson that
stationary battery designers were learning at the same time.

The most popular solution to the problem of finding a substitute for the tubular plate for the negative electrode was the
so-called "protected electrode." Around the turn of the century, the idea of using a mix of tubular positive and protected
negative began to take hold, and it has remained the most
popular design for electric vehicle batteries ever since.

It is perhaps an exaggeration to describe the protected electrode as a new idea, since the earliest Faure batteries were
essentially "protected." A protected electrode is simply a
pasted, flat-plate electrode which has some sort of inert, porous material attached to its faces to hold the active material in
place. Yet although the first Faure cells were equipped with felt
sheets on their faces, this practice was abandoned in most of
the stationary batteries of the '80s. The faces of most stationary
batteries were left open; they could be better examined and
maintained this way, and the large space between each electrode left plenty of room for pieces of loose active material to
fall to the bottom of the cell.

With the revival of interest in battery streetcars in the late 1880s, the idea of using inert retainers on the faces of the electrode plates was also revived. For example, in 1890, Gustav Philippart patented a fairly standard flat pasted grid upon which was placed a perforated celluloid sheet for each side. Such electrodes, bound securely together in a horizontal stack, were used for both the positive and negative.[59]

Celluloid was one of the only "plastic" materials available to early battery designers. Unfortunately, it was subject to oxidation if placed next to the peroxide plate. Later designs often used a compound layer of protection. Thus Herbert Lloyd of ESB developed a series of designs for protected covers for the company's chloride pellet plates, in which there was an inner layer of woven asbestos and outer sheets of perforated hard rubber held on by rubber bands. These designs were patented in 1892-1893:[60]

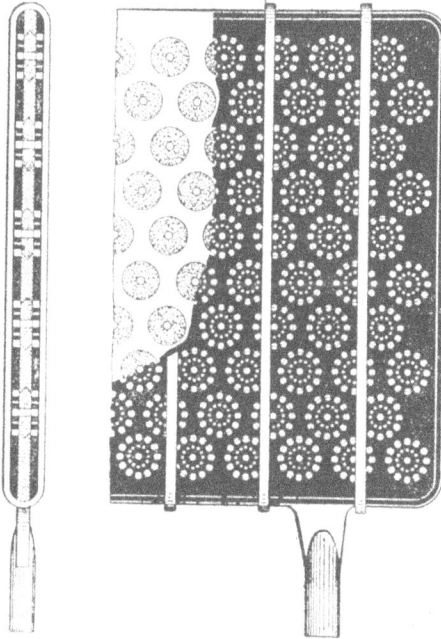

The Hess Electric Storage Company developed a similar plate in 1891, but with glass wool employed for the inside layer—this is the material usually used today for this pur-

---

[59] U.S. patent #425,957—15 April 1890, filed 6 January 1890.

[60] Wade, Secondary Batteries, p. 51; U.S. Patent #477,182—14 June 1892, filed 27 February 1892; #490,254—17 January 1893, filed 30 August 1892.

pose.[61] In 1894 Hess patented a curious variant on this in which the outer envelope was made of metal and the inner layer was a mixture of quartz sand with a binder.[62] Donato Tommasi, who was a well-known French battery inventor, invented a flat-plate electrode at this same time which closely resembled that of Philippart:[63]

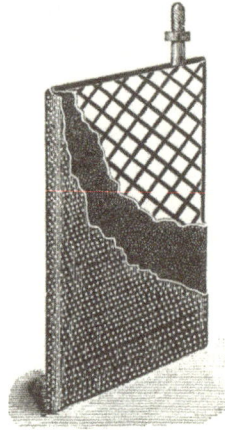

In 1895 Frank King and Camille Faure developed the electrodes for a traction battery for the EPS company. Again, the plates consisted of an inner pasted grid covered with asbestos, and the whole surrounded by perforated celluloid.

During the late 1880s and '90s the design of such protected electrodes was a popular activity with battery inventors in both Europe and the United States.[64] Initially, these protected plate designs were used for both electrodes, just as the tubular plates were. However, towards the end of the decade a number of combinations of these designs were produced. For example, the "Pope" or "Sherrin" battery was an electric automobile battery which enjoyed a brief popularity in France and England just before the turn of the century. This cell was based on the patents taken out by J. Vaughan-Sherrin in 1896. The figure below shows one form of the tubular positive used in these

---

[61] Wade, *Secondary Batteries*, p. 50.

[62] Ibid., p. 55.

[63] Ibid.

[64] For example, U.S. patents: #399,535—12 March 1889 (J.L. Huber); #400,226—26 March 1889 (I. Kitsee); #438.827—21 October 1890 (E.N. Reynier); #552,322—31 December 1895—(A. Krotz & W. Spencer); #550,480—1894 (W.A.B. Buckland); #540,185—28 May 1895 (C. Payen).

batteries, in which hard-rubber tapes were wound spirally around a central spiral, H-cross section lead-antimony core. The interstices were filled with the traditional oxide pastes. For the negative plates, flat, protected grids, like Figure #5, were used. In the next chapter we will see how this tradition of combining tubular positives and protected negatives was developed through the first seven decades of the twentieth century.[65]

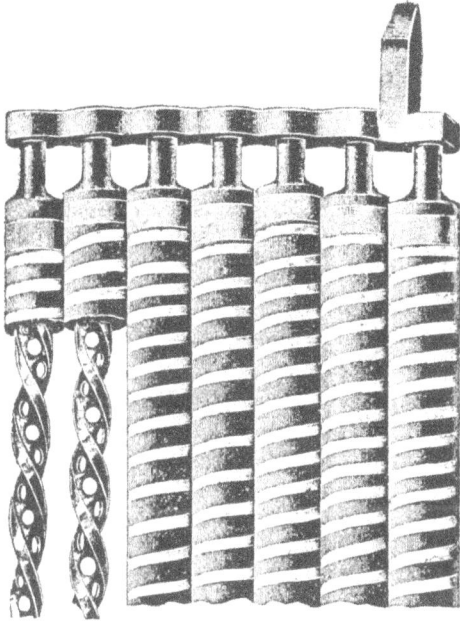

---

[65] Wade, *Secondary Batteries*, p. 437.

IV. The Automobile and the
Battery: 1900-1970

1. The Electric Car: The Beginnings

The storage battery industry entered the new century on a note of high optimism. Most markets for its products were doing well—sales of load-leveling installations for lighting stations and trolley powerhouses were high; the market for private plant batteries was doing well, and the use of storage batteries in telegraph and telephone offices was increasing rapidly. While it was true that perceptive battery people could see that the advance of A.C. generation would soon decrease the market for lighting station batteries, the market for trolley station batteries was even bigger than that for lighting stations at the turn of the century. Few electrical engineers were perceptive enough to see that the demise of the trolley would not long follow the death of D.C. generation itself.

The source of the greatest optimism for the storage battery industry at the turn of the century, however, was the automobile. The growth of United States motor vehicle registrations was explosive:[1]

YEAR	No. Car & Truck Regis.	YEAR	No. Car & Truck Regis.
1895	4	1908	194,400
1896	16	1909	305,550
1897	90	1910	458,377
1898	800	1911	618,727
1899	3,200	1912	901,596
1900	8,000	1913	1,190,393
1901	14,800	1914	1,664,003
1902	23,000	1915	2,332,426
1903	32,920	1916	3,367,889
1904	54,590	1917	4,727,468
1905	77,400	1918	5,554,952
1906	105,900	1919	6,679,133
1907	140,300	1920	8,131,522

---

[1] *The Storage Battery Manufacturing Industry* (1970 Yearbook) (New Jersey, Assoc. American Battery Manufacturers, 1970).

Fortunately for the traction storage battery, enthusiasm for the automobile began to eclipse the need for battery streetcars. American periodicals from 1895 on began to predict confidently that the car would soon replace the horse.[2] The cheap automobile which everybody could afford was a reality in the American press beginning in the mid 1890s—years before Henry Ford made it a reality on the road. Many inventors believed that the electric automobile would capture the lions' share of that reality.

Of course, this did not happen. In fact, the electric car never even captured a significant minority of the market, and, after the first decades of the new century, it had no part of the market at all. Moreover, it would seem that this future pattern of development was perceptible even to the very earliest automobile inventors of the early 1890s, since only a small minority of them ever concentrated on the design of electric cars. Finally, we may wonder if the failure of the battery-powered streetcar did not sound a cautionary note in the ears of those mid-1890s inventors who were tempted to develop a commercial electric automobile, especially since many of these had been involved in streetcar projects.

Nevertheless, a good number of inventors and entrepreneurs did invest their talents and money in the interests of the battery-driven car, and if we examine some of the variables of the electric-versus-internal combustion equation around the turn of the century we can understand something of the sanguine outlook of the battery car's promoters.

These men were aware, that the rapid failure of the streetcar batteries was due largely to the stringent demands of streetcar service—continual stop-and-go traffic combined with a mixture of damaging overloading during rush hours and energy-wasting underloading during slack hours. On shuttle routes where traffic was not as heavy, the runs were shorter, and the over- or underloading of cars easier to prevent, battery streetcars were successful. Indeed, the use of battery streetcars and buses for shuttle service has been used throughout this century in Europe, albeit on a small scale.

Electric automobiles were designed to operate similarly to shuttle battery streetcars. Their load was always roughly the same; they did not operate stop-and-go fashion; and most of them were deliberately designed for short cruising ranges—

---

[2] James Flink, *America Adopts the Automobile* (Cambridge, MIT Press, 1970) p. 36.

25-35 miles was average.[3] Some cars were designed for longer
cruising ranges of 60, 100, or even more miles per charge.[4] This
was found to be undesirable, however, because of the more
than proportional increase in weight of battery, weight of carriage, and the very substantial increase in cost which both
these expensive additions caused. That from the first, electric
car designers chose short cruising ranges indicates that they
had learned the lessons of the battery streetcar.

Unfortunately for the electric car, its inherent short range
was the chief reason for its defeat by the gasoline car. We today
can appreciate the seriousness of this limitation, but this was
not so apparent to the electric car's early promoters. They conceived the electric car as a means of urban transportation, and
in those days, urban dwellers who used some method of
transportation usually confined such traveling to the city limits
where roads were usually good, in contrast to the usually un-
paved ones of the countryside.

The electric car promoters understood the technical reasons
for the short range of their vehicles versus the long range of the
gasoline cars, but they failed to anticipate the enormous implications inherent in the long range of the latter. This failure is
not surprising, since the implications inherent in the gasoline
car went far beyond the merely technical and economic and
were of a social and psychological nature. Long cruising range
implies freedom, and freedom is an exhilarating concept.

The technical characteristics of the electric car, on the other
hand, implied not freedom, but constraints and stability. It
was ideally adapted to the needs of the urban, upper, and
middle classes of the later nineteenth century. In evaluating
their technology, therefore, electric car inventors tended to
measure it against the existing means of transportation used
by those particular social classes; that is, to measure it against
the horse and carriage. Thus, for example, C.E. Woods, in a
1900 treatise on the electric car, claimed that the 25-40 miles
range of electrics was not a limitation on their acceptability
since no *horse-drawn* vehicle could be made to go more than
twenty miles without overtaxing the horse.[5]

Yet the horse and carriage was used only by a small percentage of the urban population, and it was largely to this group
that the electric car had appeal. The electric car promoters did

---

[3] C.E. Woods, *The Electric Automobile* (New York, Herbert Stone, 1900) p. 14.

[4] *Proceedings Inst. Civil Engineers* (London) 1903: 35.

[5] Woods, *Electric Automobile*, p. 100. Woods was one of the first commercial makers
of electrics, having begun in Chicago in 1895.

not anticipate how fast the rest of the population's appetite for pleasure driving and "touring" would be whetted by the availability of long-range gasoline cars. As a result, many of them were optimistic about the battery car's chances of defeating its gasoline rival—they tended to pay attention to the more immediately obvious advantages of the electric over gasoline, but to miss the more subtle psychological disadvantages. Thus, for example, inventors such as Thomas Edison stated flatly that people would not choose a smelly, noisy, unreliable and complicated gasoline car to the quiet, clean, reliable and simple electric one. What Edison missed, of course, was that the gasoline car gave its owner freedom as well as odor.

A second major problem with the electric car was its cost. Electric pleasure vehicles were alway more costly than comparable gasoline models, although many gasoline cars were as expensive as the costliest electric ones. The high cost of the electric was partly due to the battery, which made up a substantial part of the total price of the car. Electric cars priced below a thousand dollars were rare; of 18 models listed in October 1914, the average price was $2,950 and the lowest $2,250.[6] At this time a Ford roadster sold for $440; a town car for $640; a sedan for $975; and touring car for $1,400. The electric was always described as a rich man's car and the cost differentials between electrics and gasoline cars have been used as explanation for the demise of the electric.

Nevertheless, a caveat must be added. It is true that part of the cost differential was due to the high price of the storage battery, but much of the differential came from factors not immediately related to the electric system. For example, the body and furnishings of electrics usually were more luxurious than those on gasoline cars. A more careful study should be made to determine what percentage of the gasoline-electric car differential was due to electric-related and unrelated costs. It is clear that one of the reasons why the electric was a rich man's car was that few attempts were ever made to produce lower-priced models; the electric never had its Henry Ford. The early electrics, like the early internal combustion cars, were built as pieces of fine coachwork and unlike the gasoline model which evolved into metal-bodied models, this costly tradition continued.

---

[6] *Horseless Age*, 14 October 1914: 568-569.

Nevertheless, not all electric car manufacturers were so conservative. By 1910, several manufacturers were imitating the design of stylish gasoline cars, thus lowering the excessively high prices.[7] The price of this gasoline car style Columbia Electric Vehicle Company runabout in 1914 was $785:[8]

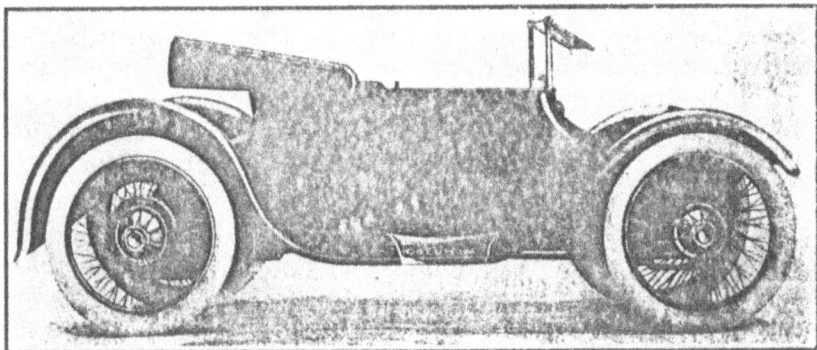

Since these lower priced models do not seem to have sold noticeably better than the higher priced electrics, it was probably the short range and low speed of the electric car which hurt its development more than its price.

Against these limitations, however, the promoters of the electric saw several advantages. First of all, although the inventors of the gasoline car were working in technologically unexplored territory, the technological ingredients of the electric automobile were well developed *before* its invention—the design of electric transportation motors had a decade and a half of evolution behind it in the trolley industry, whereas the gasoline engine was both inherently more complicated and far less developed by 1900. In 1902, for example, the journal *Horseless Age* attributed the American lead in electric car design to the fact that trolley technology was far more advanced in the United States than in Europe.[9]

The electric automobile enticed much early investment since it seemed near perfection; for example, commercially-obtainable components were soon available. An examination of the circuit diagrams of the electrics shows them to have used

[7] *Electrical Review* (New York), 12 February 1910: 350-354.

[8] The Storage Battery Mfg. Industry, 1970, New Jersey Assoc.

[9] 1 January 1902: 3.

the same motors, controllers, switches, and, of course, batteries, as the streetcars, albeit in smaller size.[10] The situation
with regard to gasoline car transmissions, spark plugs, piston
rings, carburetors, and so forth was one of starting from
scratch. The electric car promoters made the mistake of assuming that their gasoline rival would remain unreliable, unsafe,
unclean, and unquiet much longer than it in fact did. To again
quote the 1902 *Horseless Age*:[11]

> This deceptive appearance of perfection in our early storage bat-
> tery vehicles has been the cause of much harm to the industry.
> How this vehicle, in its first stages of development, was seized
> upon by promoters and capitalized at several hundred millions
> of dollars is a matter of history. . . .
>
> Very little remains today of these gigantic paper corporations.
> Especially during the last year has there been a thorough weed-
> ing out of the unfit.

A large number of inventors produced electric cars in the
1890s and the early years of the new century. The first American electric car experiments began in 1891 and the first commercial manufacture started in 1894.[12] It was in 1895, however,
that the automobile passion began and a flood of electric vehicles soon poured from garages and workshops. For our purposes, however, most of this work can be ignored, since it did
not directly influence the evolution of storage battery design.
    There was one car which did have a major effect on the
history of the battery, however, and this was the Electrobat.
This was a creation of the team of Morris and Salom, who first
became acquainted with electric vehicles by designing streetcars and batteries for the Julien and Consolidated companies.
Morris, a mechanical engineer, designed the motors, gearing,
frames and other mechanical parts of the vehicles, while
Salom, the chemist, concentrated on the batteries. The two
began work on an electric automobile in Philadelphia in June
1894.[13] By August the first Electrobat was ready. This was a
rather heavy affair of 4,250 pounds, of which 38 percent was

---

[10] See, for example, *Horseless Age*, 9/1898: 16.

[11] 1 January 1902: 3.

[12] *Development in Electrically Powered Vehicles* (Wash., Bureau of Power, Federal
Power Commission, 1967) pp. 2-3.

[13] *Journal of the Franklin Institute* 141 (4/1896) p. 280.

battery. The motor came from General Electric, and was a standard type designed for use on storage battery-powered boats (during the 1890s a sizable market developed for electric pleasure boats and commercial components for these craft were available in both the United States and Europe). The Electrobat went between 50 and 100 miles on a charge and could go up to 15 miles per hour.

Morris and Salom had no intention of commercializing this clumsy vehicle; its purpose was to provide a basis upon which to design a practical automobile. The two men drove it around Philadelphia in the winter of 1894/95 and "have never been compelled to attach a horse to bring it home, for any reason whatsoever"—a remark they probably couldn't have made if they had worked on the internal combustion car.[14] Of course, the years of experience which Morris and Salom had with battery streetcars must be credited for their quick success with the Electrobat.

The battery for the Electrobat came from the ESB company. ESB began patenting transportation battery designs in 1890-1891, at which time its cells were used in the Fiske Warren car, the earliest American electric car.[15] No records of the batteries used by Morris and Salom survive, but they probably were some form of the company's standard chloride type. The two inventors had used Julien or Consolidated batteries for their electric streetcar work, but in 1894 Consolidated was bought out by ESB. Moreover, their decision in 1894 to shift from battery streetcar to battery automobile inventing was at least partly prompted by the ESB itself. Along with its other major activities in 1894, ESB created the Electric Vehicle Company (EVC), while Morris and Salom set up their own company— the Electric Carriage & Wagon Company.[16] At first, EVC and Electric Carriage technically were separate companies, but the money for both came from ESB sources and the leaders of ESB sat on the board of Electric Carriage.[17] Morris and Salom were not directly employed by ESB, as they had been employed earlier by Julien and Consolidated, but the real relationship was probably similar.

---

[14] Ibid., p. 281.

[15] *Journal of the Franklin Institute*, 141 (4/1896): 283.

[16] *Horseless Age*, 5/1896: 22.

[17] Ibid., 24 January 1900: 11.

With the lessons learned from Electrobat #1, Morris and Salom began building Electrobat #2 in the spring 1895. This machine was quite different from the first—it weighed only 1,650 pounds and the body was built by a firm of carriage builders. Every precaution was taken that "no machinery of any kind is in sight. . . . and it is exempt, therefore, from the criticism generally passed on motor vehicles that they look more like a piece of machinery than a pleasure carriage."[18] Electrobat #2, in other words, was designed to impress the public. On Thanksgiving Day, 1895, the Chicago *Times-Herald* sponsored the first American automobile race, and Morris and Salom built Electrobat #2 specifically for this media event.

This car held 160 pounds of batteries—quite a difference from the 1600 pounds of Electrobat #1.[19] The ratio of battery weight to total weight was less than 1/10, which was extremely low; commercially-produced electrics in the period 1895-1910 usually had battery weights in the ratio of 30 to 45 percent of total weight, with the percentage tending towards the upper limit at the end of this period, as manufacturers tried to compete with gasoline by extending the range of their cars.[20] The 9.7 percent battery weight of Electrobat #2 reveals the aforementioned lack of concern of the early electric car makers with range; the car had a maximum reported range of 25-30 miles.

In 1896 Pedro Salom told the Franklin Institute why electrics should be designed for short ranges:[21]

> We believe that the most practical application to which these vehicles can be put, at the present time, is for service in parks and as delivery vehicles. We do not consider it practicable, at the present time, to send out such vehicles broadcast over the country before any proper arrangements have been made for their intelligent care and maintenance. We think that the proper plan for their introduction is to construct a sufficient number of vehicles of one kind to warrant the building of a charging station, where the batteries can be charged and the vehicles kept when not in use. This will enable a systematic and intelligent inspection of the various parts of the machines, such as all mechanical contrivances require, and will prevent all the troubles that

---

[18] *Journal of the Franklin Institute,* 141 (4/1896): 282.
[19] 12 cells of ESB batteries @ 13 lbs, plus a 4-lb tray to hold them.
[20] *Journal of the Franklin Institute,* 141 (4/1896): 284.
[21] *Journal of the Franklin Institute* 141 (4/1896): 284-285.

would naturally arise from inexperienced persons handling them.

This proposal puts forward a quite different approach to the marketing of electric vehicles than that used by most of Salom's contemporaries. Most of them—Baker, Woods, Riker, the Studebaker brothers, et al—sold electrics directly to consumers the way Henry Ford was to sell gasoline cars. Morris and Salom conceived a somewhat more grandiose plan. Their company was to secure agreements with carriage builders and motor, battery, and component manufacturers for the supply of the various parts. The company would then assemble the cars and ship them to a series of large, urban stations, also owned by the company, where the cars would be cared for on the model of the livery stable. The company would pay the attendants, operate the charging equipment and retain ownership of the cars. Customers would lease or rent a car from the station. All maintenance, charging, and repairs would be covered in the rental charge. Once the customers had leased their cars for a while and had been instructed in the proper care of batteries, they could "then be justified in making their own personal arrangements."[22]

The wisdom of this scheme probably arose from the years of experience which these two engineers had running the Julien battery streetcars in Manhattan. It is significant to note that the manufacturers who simply sold their cars outright to consumers, such as A.L. Riker and C.E. Woods, had not been connected with the earlier battery streetcar developments. It is also clear, however, that the capital base required to launch such a scheme was prodigious—Salom indicated that he planned on starting with a fleet of over 100 cars. The connection of Electric Carriage with ESB was to be crucial here.

Electrobat #2 did not win the *Times-Herald* Thanksgiving Day race. That prize was taken by the gasoline-powered Duryea. Salom's concentration on design paid off, however, when the car was awarded a gold medal for appearance. It was an attractively designed car, even for modern tastes.

The black cylinder by the left-hand seat is the speed controller; a standard piece of equipment on contemporary trolley cars. The inventors predicted that such cars could be manufactured in quantity at between $1,200 and $1,500 apiece, while some-

---

[22] *Horseless Age*, 1/1896: 24.

what less elegant delivery vehicles, weighing 1,000 pounds, would cost $600 to $800 each:[23]

Electrobats #3 and #4, constructed apparently at the same time as #2, were again experimental. Morris was dissatisfied with the idea of using horse-carriage technology for the frame and worked on adapting the emerging technology of the bicycle to the electric. This car weighed only 1,180 pounds with battery.[24]

In May 1896 the Electric Carriage and Wagon Company published its plans for putting Morris and Salom's ideas to work. The first central station and car barn was to be built in Manhattan.[25] By January 1897 the station was opened at 140 W. 39th Street with a complement of thirteen hansom cabs.[26] It is not clear if Electric Carriage ever tried to rent its vehicles according to Salom's original idea. In any case, the actual operation of the company was always as a pure cab-running firm. The reason for this may be that from 1897 on Electric Carriage ceased to exist as a separate company, Morris and Salom began to fade into the background of its leadership, and it was purchased outright by Isaac L. Rice.[27]

---

[23] *Journal of the Franklin Institute* 141 (4/1896): 282.

[24] *Horseless Age,* 1/1896: 24.

[25] Ibid., 5/1896: 22.

[26] Ibid., 1/1897: 3.

[27] *Electric Vehicle News* (8/1972): 26-27.

Issac Rice was a friend of W.W. Gibbs and a cofounder of
ESB with him. Rice was most interested in storage battery
transportation and was the moving force behind the 1894 creation of EVC. Both Rice and Gibbs sat on the board of Electric
Carriage. With its purchase in 1897, Electric Carriage became
part of EVC.

This move, however, was only the beginning of a much
larger story. Both ESB and EVC were soon themselves to become the target of acquisition attempts which would propel
them into the heady and complex world of late nineteenth
century high finance. The man chiefly responsible for this was
William C. Whitney.

Whitney was one of the larger than life characters of the
Gilded Age like J.P. Morgan or Cornelius Vanderbilt.[28] During
the late 1880s and early '90s Whitney built a streetcar empire in
New York not unlike that of Yerkes in Chicago. By the late
1890s his syndicate's Metropolitan Traction Company (MTC)
owned lines on Broadway, Lexington, Columbus, St.
Nicholas, Amsterdam, Madison, Lenox, 2nd, 4th and 6th
through 9th Avenues, in addition to seventeen crosstown
lines.[29] It was Whitney's MTC which supported most of the
New York battery streetcar projects discussed in the last
chapter.

Beginning in 1896, the MTC began buying up large blocks of
ESB stock.[30] In 1897 Rice took over the presidency of ESB from
Gibbs and fought the Whitney syndicate's attempts to gain
control. The takeover struggle was fierce, with the stock soaring from $20 to $120. By late 1898, however, Whitney succeeded in getting control of the battery company, and by June
1899 Whitney's people had control of the ESB board and Rice
was forced out as president.[31]

The entry of Whitney into the orbit of EVC considerably
expanded its horizons. Morris, Salom, Rice, and Gibbs had
been operating a slowly expanding fleet of cabs from one station in one city. Whitney had much bigger plans. The EVC was
to become the headquarters for a worldwide network of branch
EVC's. These local companies would be franchised in their

[28] Mark D. Hirsch, *William C. Whitney, Modern Warwick, passim*.

[29] Ibid., pp 439-440.

[30] Ibid.

[31] *Horseless Age*, 24 January 1900: 11; Irvine, "Promotion," p. 61. For an account of
the complex financial details behind the ESB takeover, see: *Horseless Age*, 27 January
1900: 11; 27 September 1899: 7-8; *New York Times*, 13 May 1899: 4; Hirsch, *Whitney,
passim*.

localities to operate cabs and/or sell cars and trucks built by EVC. For this right, the local firms were to pay an inital lump sum to EVC, and were to pay an additional 2½ percent of gross earnings after that.[32] The initial payments were in the stock of the franchised firms.

As the major stockholder in EVC, ESB was to be the capstone of this pyramid. Like the Philippart pyramid of a generation earlier, this complex exercise in incorporation was built upon a minimal technological base. The Whitney interests were less concerned with the methodical commercial exploitation of a new technology than with the legal and commercial dominance of the field. This is not to say that the engineers connected with the firms, such as Morris, Salom, or Stanley Currie, were not seriously concerned with developing electric car or transportable storage battery technology. The Whitney interests, however, were chiefly concerned with using their financial leverage and stock manipulation to create a monopoly over the emerging automotive industry.

Nevertheless, the enthusiasm of these promoters for their prospective empire is understandable. News reports about the advance of the electric automobile were frequent during 1897-1898. For example, in these two years, all of the following famous inventors turned their attention, at least for a time, to this field: Elihu Thomson, Ferdinard Porsche, Elmer Sperry, Thomas Edison, Ransom Olds and the Studebaker Brothers.[33] To quote John B. Rae: "The year 1897 marks the effective start of the automobile industry in the United States."[34] In addition to the electric cab company established in New York, 1897 also saw the creation of the London Electric Cab Company, which had 50 battery cabs in service the next year.[35] In 1898 electric taxi experiments were also going on in Paris.[36] The star of the automobile, electric and otherwise, was rising rapidly in these years.

Nevertheless, with its control of ESB and EVC, Whitney still lacked an important part of his empire, namely, control of a

---

[32] *Horseless Age*, 27 September 1899: 8.

[33] John B. Rae, *American Automobile Manufacturers-The First Fifty Years* (Phila., Chilton Co., 1959); Walter H. Nelson, *Small Wonder, the Amazing Story of the Volkswagen* (Boston, Little, Brown & Co., 1967); Thomas P. Hughes, *Elmer Sperry, Inventor & Engineer* (Balt., Johns Hopkins Press, 1971); Harold Abrahams & Marion Savin, *Scientific Correspondence of Elihu Thomson* (Cambridge, MIT Press, 1971); Flink, *America Adopts*.

[34] Rae, *American Automobile*, p. 21; Flink, *American Adopts*, pp. 25-31.

[35] *Horseless Age*, 5/1898: 14.

[36] Ibid., 6/1898: 11.

means to manufacture electric cars. Morris and Salom had built their taxis from components purchased from other manufacturers and assembled these in their New York garage. Accordingly, in early 1899 the EVC was reincorporated, this time in New Jersey and with a capitalization of $25 million. This was a considerable increase over the original capitalization of $300,000 two years previously in New York.[37] Part of the money was used to purchase the Siemens & Halske Electric Company of America, located in Chicago; the car's electrical equipment minus batteries was made there.[38]

The cars themselves were made by the Pope Manufacturing Company of Hartford, Connecticut. Colonel Albert A. Pope was one of the pioneer manufacturers of bicycles in the United States, having set up a factory in Hartford in 1878.[39] This became the largest bicycle factory in the country by the end of the century, manufacturing the famous Columbia bicycles. In 1897 Hiram P. Maxim,[40] at that time Pope's chief engineer, developed an electric car. The Columbia Electric weighed a bit more than the Electrobats—1,900 pounds—and carried a higher percentage of ESB batteries—850 pounds or 45 percent of the total. Yet the Columbia's frame, which was almost identical to the later Electrobats, also revealed the influence of bicycle technology.[41] These cars went about thirty miles on a charge.

Maxim began work on the design of the Columbia Electric just shortly after Morris and Salom began their work—January 1895.[42] There were several similarities between the work of Morris & Salom and Maxim, but with this important difference: by 1897 the Columbia works had established a large factory for the manufacture of its cars.[43]

Pope manufactured the cars for direct sale until early 1899, having nothing to do with the New York cab project. At that time Pope became a partner of the Whitney syndicate. The Pope Manufacturing Company and the ESB agreed to pool their electric car and car battery patents and to place these under the ownership of a new company—the Columbia Automobile Company (CAC). Stock in this firm was owned 50-50 by its parents. The manufacturing patents of CAC, excluding

---

[37] Ibid., 3/1899: 7.

[38] Ibid., 27 September 1899: 8.

[39] Frederick Anderson, *Bicycling, A History* (New York, Praeger, 1972) pp. 41-42.

[40] Son of Sir Hiram Maxim.

[41] *Electrical World*, 15 May 1897: 614-615.

[42] *Horseless Age*, 4/1897: 4.

[43] Ibid., 4/1897: 5.

those for batteries, and the patents of EVC were then united and placed under the care of yet another new corporation—the Columbia & Electric Vehicle Co. (C&EVC). C&EVC stock, in turn, was held 50-50 by its parents, CAC and EVC, both of which continued to exist independently.[44]

C&EVC became the manufacturing branch of the operation. It therefore purchased the Columbia factory from CAC, as well as the plant of the New Haven Carriage Company and the already mentioned Chicago plant to Siemens & Halske. EVC became the central distributing company for these cars and was given the sole right to buy the output of C&EVC. The actual cab-operating function of EVC was removed and placed under control of another new corporation—the New York Electric Vehicle Transportation Company (NYEVT).[45] NYEVT in turn paid for its license to do so to EVC in the form of its stock, as already indicated. At the same time, franchised local cab companies were created in other major cities: Boston (New England EVT); Philadelphia (Pennsylvania EVT); and Chicago (Illinois EVT). Despite the capitalization of these non-New York companies at a total of $56 million, their actual activities never appear to have passed beyond the paper stage.[46]

The complexity of the inter-company relations outlined above was more apparent than real. Its chief function was to provide a legal framework for the monopoly Whitney hoped to erect and it also served to float new stock issues for the vast amount of capital needed—in May 1899 one newspaper article indicated that the syndicate was planning a capitalization of $200 million. The apparent complexity disappears, however, if we examine the men behind the companies. EVC, ESB, C&EVC, and so forth all had the same boards of directors, with slight differences. Thus, for example, George H. Day, the secretary and treasurer of CAC, was president of ESB.[47] It was in this way that the fortunes of the leader of the leading American storage battery company were tied up with the success of the Whitney electric car empire. This linkage was to have a major influence on the evolution of the storage battery.

EVC wasted no time in beginning its new operations. In September 1898 the New York cab operation was running only fourteen taxis and still had the one original garage.[48] Already by February 1899, however, 200 more cabs were on order and a

---

[44] Ibid., 27 September 1897: 8.

[45] *New York Times*, 25 February 1899: 1.

[46] *Horseless Age*, 10 May 1899: 6.

[47] Ibid., 3 May 1899: 7.

[48] Ibid., 9/1898: 9.

second New York garage had been opened.[49] The next month EVC announced plans to open two more New York garages, and in yet another month the New York company was planning to introduce 1,500 more cabs, while the EVC was telling the press that it soon would have 15,000 battery cabs running nationwide.[50] By April EVC was placing some very large contracts with other manufacturers to supplement CAC output.[51]

By May the EVC had incorporated franchised companies in each of sixteen states and the District of Columbia, and it announced that it would soon cover every state and United States territory.[52] It is interesting to speculate what the future of the American automobile industry might have been like if Whitney had sought to dominate gasoline car manufacture as he did electric. The EVC bought the Selden patent in 1899, thus acquiring control of the basic American internal combustion car patent. However, unlike his attempt to dominate the electric field by his control of the manufacturing of electrics, Whitney never manufactured internal combustion cars himself, but only licensed others under the Selden patent.

This, of course, was a major mistake; Whitney was backing the wrong horse—a mistake that began to make itself felt during 1899 when it was discovered that the electric cabs were operating poorly. The stock of the syndicate's companies rose until mid-year, when rumors of the poor operation of the cabs began to be heard. The batteries were failing speedily, causing the New York company to operate deep in the red. Moreover, it is clear that these failures could not have been recent phenomena since the company had been operating with the same type batteries for almost two years prior to the Whitney syndicate takeover. The records of these two years cannot have been any better than those of 1899. It may be that Isaac Rice suppressed this data, but in the absence of detailed information on this matter, we can only wonder at the careless abandon with which the syndicate's creators poured millions of dollars into a technology which they had taken no steps to evaluate. In addition the failure of so many battery streetcar ventures should have made them proceed with caution, instead of jumping in with both feet.

---

[49] Ibid., 2/1899: 13.

[50] Ibid., 3/1899: 7; 12 April 1899: 11.

[51] Ibid., 19 April 1899: 12.

[52] Ibid., 24 May 1899: 11.

Rumors about the bad performance of the electric cabs continued and found their way to Wall Street. Towards the end of 1899 the stocks of all the syndicate's companies were plunging. ESB lost thirty-six points during November and December.[53] To add to the difficulties, the EVC was in the unenviable position of having been born at the beginning of massive public reaction against monopolies. The press was soon calling it the Lead Cab Trust and its every wart and blemish was seized upon and published.

The journal *Horseless Age* was particularly unrelenting, with its editor frankly conducting a campaign to undermine the EVC. A few examples will show the temper of his comments:[54]

> The Lead Cab Trust is going into the funeral business in Mexico. The Mexicans, it seems, use the street cars as hearses. The Lead Cab Trust will do away with these and substitute up-to-date lead cabs with storage battery coffins. Inasmuch as the trust's principal business here (i.e. New York) has been carrying "dead weight," they are well prepared for the funeral business in Mexico. The promoters themselves might act as chief mourners. . . ."

> Designing men, emboldened by the prevailing speculative boom and conjuring with the magic word "electric," had concocted a gigantic scheme of plunder to loot the motor vehicle industry. . . .

> The integrity of the (automotive) industry was threatened. Men high in financial and political circles (Whitney was involved in New York politics) had fastened like vampires upon it and were draining its blood. They had organized for this purpose the most gigantic stock-jobbing scheme in the history of Wall Street. . . . The storage battery, on the exploitation of which the vast scheme was based, is of limited use in motor vehicles at best, and . . . is admitted by all competent authorities to be unfitted for this service. On such a rotten foundation was this Babel Tower of speculation built. The rottenness has finally been exposed and the whole structure is toppling over upon the builders.

> (Character studies of Hiram P. Maxim) This puny champion of that "unmitigated nuisance," the storage battery. . . .
> . . . . this blind and infatuated partisan of the lead wagon. . . .
> . . . . this flabby storage battery puff. . . .

---

[53] Ibid., 20 December 1899: 8.
[54] 25 October 1899: 7; 3 January 1900: 7-8; 24 January 1900: 10; 18 October 1899: 6.

Horseless Age took particular delight in describing the falling stock values in late 1899 and early 1900: [55]

> The Lead Cab Trust is apparently in desperate straits. From the chief conspirator, the Electric Storage Battery Company, of Philadelphia, down through the long list of dummy manufacturing and operating companies, its stocks are on the decline. . . .The Lead Cab is a predetermined failure. . . . The next slump will be still more sensational.

> Lead Cab is on the decline in Wall Street. . . . When it finally does fall it will splash (the promoters) with mud.

The vehemence of these attacks appears to have had nothing to do with the more famous and bitter struggle over the Selden patent monopoly which Whitney was to create, since the EVC did not buy the patent until many of these editorials had been written, and legal proceedings did not begin until all of them had been published. Moreover, no mention is made in these editorials to Selden.

As if these attacks were not enough, in June 1899 the Automobile Club of France began a series of well-publicized endurance tests of electric-car batteries. Among those batteries entered were TEM cells, roughly equivalent to the ESB's. The tests were designed to duplicate the kind of service that batteries would be called upon to perform for an average electric car owner. Within six months, all but four brands had been disqualified. TEM had failed, but of those which succeeded, one was a pure tubular plate battery (Phenix); one was a mixed cell with tubular positives and the standard pasted double-grid negative (Pope); and two others used a mixture of Planté positives and pasted double-grid negatives (Blot-Fulmen and French Tudor).[56] By contrast, the cells used in the EVC cabs were simply smaller versions of the ESB's load-leveling batteries—chloride pellet negatives and Planté positives.

In the final results of the French tests, which appeared at the start of 1900, none of the batteries received a good rating, and none lasted longer than six months, even when the manufacturers took pains to give them special care during the tests.[57] *Horseless Age* drew its conclusions:[58] "After this and many

---

[55] 20 December 1899: 8; 25 October 1899: 7.

[56] *Horseless Age*, 8 November 1899: 21.

[57] Ibid., 7 February 1900: 14; 14 February 1900: 10.

[58] Ibid. At least one company similar to the EVC took this advice; in July 1900 the London Electric Cab Co., went bankrupt. *The Times*, 13 July 1900: 3.

indubitable proofs of the shortcomings of the storage battery for vehicle work, the Lead Cab promoters should revise their figures and clothe themselves in sackcloth and ashes."

The result of all these discouraging developments was to force the ESB into a long-term research effort to design a truly durable, practical automobile battery. While ESB was interested in battery transportation since the start of the 1890s, little actual work was carried out towards developing a special battery for this purpose. ESB engineers drafted a number of patents during the 1890s designed for automotive use, but the actual batteries used in the EVC cabs differed in no way from small load-leveling cells. Then, however, faced with the imminent collapse of the EVC because of the collapse of its batteries, ESB began seriously to design a specialty transportation cell.

The first fruits of this work appeared in early 1900 when ESB introduced a new battery called the Exide. The development of the Exide was concurrent with ESB's abandonment of the chloride-process plates for its stationary batteries and the Exide also represented an abandonment of chloride technology. The Exide used pasted plates for both electrodes. The figure below shows a drawing of the Exides plates, with positive on left and negative on right:[59]

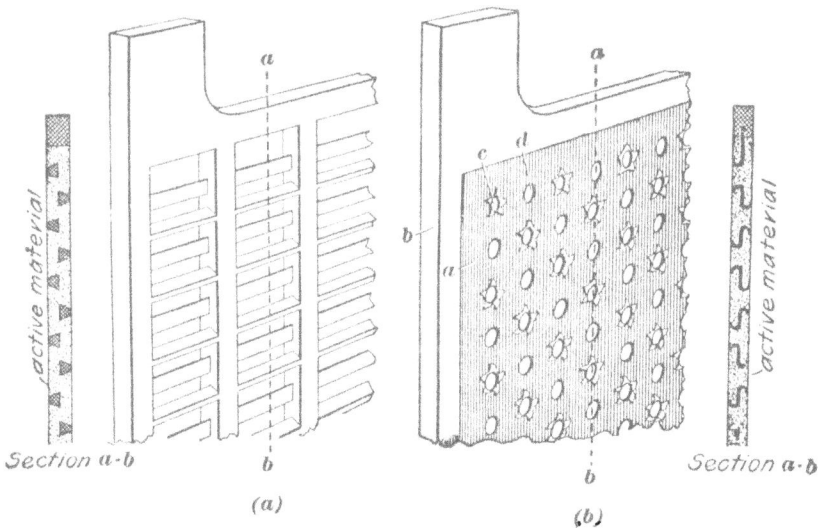

Section a-b   *b*   *b*   *Section a-b*

(a)   *(b)*

[59] *International Correspondence School Reference*, p. 24.

The adoption of this design was an extraordinarily radical departure from the company's past practice. The older ESB batteries, both stationary and traction, had been made with massive lead plates. Even the adoption of the box negative in 1900 for stationary batteries was not so radical a change as the above, since the box plate was still a moderately heavy design. The Exide, on the other hand, used two very delicate electrodes—the positive was of the traditional pasted double-grid pattern, while the negative was an even more delicate design formed from a single, thin sheet of lead. The sheet was perforated with a multiple punch which raised burrs. These burrs were then turned over to form hooks, which held the negative active material in place. To complete the battery, perforated hard rubber plates were placed around the positive to hold the active material more securely while thin wood veneer separators were inserted between the plates.[60]

In a manner similar to its adoption of the box negative, the ESB obtained most of the details of this design from other companies or inventors. As with the box plate, very little in this design resembles anything the ESB had done before. The only exception to this was the use of the hard rubber plate on the positive, which Herbert Lloyd had patented during the early 1890s.[61] On the other hand, the various components of the design of these first Exides had been previously developed by other companies. The double-grids, of course, were a common feature of European storage batteries by 1900, and it is probable that ESB obtained detailed knowledge of this technology from AFA. The thin, punched-sheet negative was a less common design than the double-grid, and it did not survive long as a design concept. Nonetheless, a number of American and European inventors experimented with this idea during the later 1890s, all of them stressing the idea that the light-weight punched lead support sheet was good for electric automobile use.

One such battery was designed by Elmer Sperry, who was an enthusiast for the electric automobile. This cell was manufactured by the National Battery Company of Buffalo during the late 1890s.[62] Another such battery was the Osburn battery, which was used in some Chicago-made electric cars around

---

[60] Wade, *Secondary Batteries*, p. 441.

[61] Ibid., p. 51.

[62] Ibid., p. 448.

1900.[63] Several European inventors also tried variants of perforated sheet pasted cells.

ESB developed the first Exide battery by combining ideas developed by a number of different companies during the 1890s for the design of lightweight batteries. Among these ideas were the use of wood separators, hard rubber "protected electrodes," delicate double-grids, and punched lead sheet electrodes. This was in addition to ESB's concurrent adoption of hard rubber outer casings to replace the old wood and lead sheet boxes which had been used on the early EVC cabs.

As we saw in our earlier discussion of stationary batteries, this pattern of rapid technology transfer between the world's leading battery firms emerged during the early 1890s. By the late 1890s there still remained a considerable difference between the concepts of cell design used by various companies, but some sort of consensus had been established on certain elements of design, for example, the mix of Planté positives and pasted negatives on stationary batteries was becoming general practice towards the end of the decade. For the newer technology of transport batteries, however, considerable variation still existed near the turn of the century, with some manufacturers using all tubular plates, others all Planté, others all pasted grids, and so forth.

In the years ahead, ESB's continuing development of the Exide design had a major influence on the evolution of a worldwide consensus of transport battery design. Initially, however, the introduction of the first Exide had a fairly minimal influence. It was, after all, only a stopgap design, developed quickly to do something to replace the ESB's original electric cab cells which were driving the EVC to financial ruin. The first Exide was much lighter than the chloride-Manchester design, and it may have been more durable, but it was not very original and it does not appear to have been noticeably more durable than the cells tested at the 1900 Paris Automobile Club trials. For example, as late as 1906 one authority estimated the renewal expenses on an electric delivery truck at one set of negatives and two sets of positives per annum.[64]

The Paris tests, however, along with a number of other such trials held at the same period, gave battery designers a number of signals about the design of effective automobile batteries, and ESB was quick to pick up these signals. The chloride cells had all done poorly in the trials, but three plate designs had all

[63] Ibid., pp 449-450.

[64] Electrical Journal 3 (1906): 286.

done fairly well—Planté plates, pasted double-grids and tubular plates. All three chloride companies—ESB, TEM and Chloride—gave up the chloride process at this time. Planté plates for automobiles were a good stopgap solution for the durability problem, but it was clear that they could not be a long-range solution because of their heavy weight.

ESB, therefore, chose to concentrate on the other two successful designs—pasted and tubular plates. The first Exides appeared relatively quickly; this was probably because ESB was simply copying the ready-made technology of the thin, pasted double-grid for this cell. At the same time, however, the company also began a longer-range program to develop a really durable positive plate using the concept of the tubular electrode. The initial stimulus for this development appears to have come from the good reports appearing in France about the Phenix battery—ESB purchased the American rights to this cell shortly after the report appeared.[65]

Despite the favorable qualities of the tubular electrode, however, the Phenix design had two major drawbacks which prevented ESB from commercializing a tubular battery. First, the Phenix was expensive to make with its stacks of millimeter-thick hard rubber rings threaded on lead rods. Second—and more importantly—the hard rubber tended to crack under the expansion of the active materials. This problem was faced by all early designs of tubular electrodes, whether the hard rubber was formed into a single, integral tube or made up of many separate rings.

In our own day the selection of a strong, elastic, and inert material for the retaining tubes is much easier because of the availability of a large number of synthetic plastics. Indeed, several of these are used in modern tubular plate storage batteries. During the first decade of this century, however, these materials did not exist, and the manufacture of an effective tubular plate battery, although excellent in concept, was extremely difficult in practice. All of the metals were ruled out as retaining materials, since they were too heavy, expensive, or reactive. Ceramics and glass are too brittle and non-elastic. Celluloid is strong, but easily oxidized by PbO_2. Soft rubber also suffers from this problem, while hard rubber (ebonite) was found to be too hard and brittle. Chiefly because of this

---

[65] Private communication from the late Dr. E. Lighton.

lack of a suitable material, the ESB engineers required about ten years to develop a satisfactory tubular plate battery.

The technique for making practical tubular plates was worked out by ESB engineer Edward W. Smith in collaboration with one of the company's components suppliers, the American Hard Rubber Company, which provided ESB with its supply of ebonite battery boxes. But this company also possessed skill in manufacturing many other types of ebonite articles. One of its specialties was a process for extruding hard-rubber tubing for the casings of fountain pens. This was just the technology ESB needed for the economical, mass-production of tubular plate elements. The type of hard rubber used for the pens was still too hard and brittle to be used for this purpose, but Smith, working in collaboration with American Hard Rubber, was eventually able to produce a material with more elasticity and less brittleness than the standard fountain pen ebonite.[66]

The tubes were given the required porosity by sawing a series of grooves of fine thickness across them. These were then grouped together to form a plate in the configuration which Stanley Currie had patented in 1890-1891.[67] ESB called this the Ironclad plate and used it to replace the flat, pasted double-grid of the original Exide. These new tubular positives were advertised as having 2½ to 3 times the life of the old Exide positives.[68] For the negatives, the punched-sheet design was abandoned, probably because it was subject to rapid disintegration, and the pasted double-grid design used for it. This battery, called the Ironclad Exide, was introduced in 1911.

This design soon became a favorite throughout the world. Chloride soon adopted it, using the name Exide. AFA produced its own identical version, and adopted the trade name Panzerbatterie. Other companies responded similarly. Until the early 1950s this design remained basically unchanged for storage batteries used in electric vehicles, and the design was a world standard for this type of service. The Ironclad design, more durable than ordinary flat, pasted double-grids, is also more expensive, and most manufacturers, including ESB, re-

---

[66] Ibid.

[67] Wade, *Secondary Batteries*, p. 61.

[68] *Electrical Review* (New York), 7 January 1911: 51.

tained the flat grid for the more inherently durable negative.
The slotted tubes of the positive grid look thus: [69]

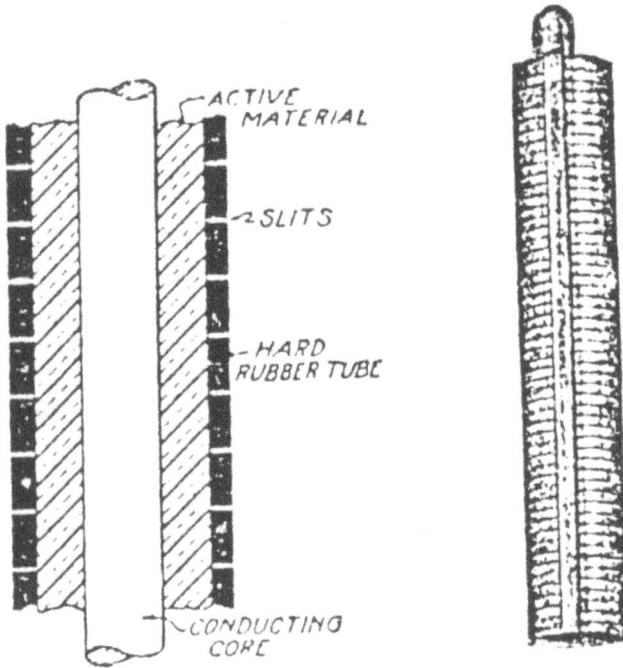

This is the pattern still in use today. Moreover, most modern
starting batteries, as contrasted with electric vehicle cells, are
not made with tubular grids, but use the flat grids for both
electrodes. This is because a starting battery is not usually
called upon for "deep cycling" use. For electric vehicles, however, the ability of the tubular positive to resist deterioration
over many cycles of charge and discharge was an advantage
which fully justified the extra cost and which assured ESB of
dominance over the American electric vehicle market.

2. The Electric Car: The Economics

William C. Whitney would not have been surprised at ESB's
economic and technological dominance over the American automotive battery industry. That was part of his great plan; ESB
was to supply the batteries for the EVC, which in turn was to
be the General Motors of the electric car industry. In fact, since
EVC controlled the basic Selden patent also, Whitney's idea
had been to make the EVC the control center for the entire

---

[69] Ibid.

automotive industry. But none of this was to be. EVC filed bankruptcy in 1907 after a number of years of decline.[1]

Nonetheless, the collapse of EVC was not quite so rapid as the editor of *Horseless Age* had predicted it would be in early 1900. During the early years of the century the company was still attempting to expand its operations and to make a success of the electric car. At the start of 1901, for example, the Boston franchise of the company was operating a line of ten battery buses in the Back Bay area. Significantly, they replaced a still existing horsecar line, and were therefore being used in the same way that new crosstown battery streetcars were in Manhattan at the same period.[2] Simultaneously, the EVC was still trying to create Whitney's dream of a worldwide electric vehicle empire; the frames and electrical equipment of Columbia vehicles were being shipped to London and Paris, where locally-produced bodies were fitted to them.[3] The EVC was setting up its garages in these cities also—a sixty-car garage on the Avenue Montaigne and another planned near the Arc de Triomphe. The fate of these European operations is unknown.

In addition, from the illustrated articles appearing in the technical press during the early years of the century it is clear that the New York cab operation of the EVC was a very serious and impressive undertaking, and not simply the stock-jobbing scheme denounced by *Horseless Age*. The battery-charging room, for example, was 300 by 40 feet and had cranes and hydraulic apparatus for the total mechanical handling of the cells. The cabs moved around the garage on overhead trolley conductors when the batteries were removed.[4]

EVC also entered the business of selling vehicles directly to consumers, as most of the electric car manufacturers were doing. Despite all these efforts, however, the EVC continued its slow slide towards bankruptcy. One might have expected that the ESB would have lost interest in automobile battery design as the EVC declined, but this did not happen. As the EVC went downhill, the ESB increased its commitment to automobile applications. In addition to its long-term project of developing the Exide tubular plate cell, the ESB also continued its policy of consolidation of the American battery industry. In 1902 it bought the Willard company.[5]

[1] For details of the Selden patent struggle and EVC's last years, see: Flink, *America Adopts*; and, Rae, *American Automobile*.

[2] *Electrical Review* (New York), 26 January 1901.

[3] Ibid., 19 January 1901: 110.

[4] Ibid., 4 May 1901: 546-547.

[5] Barak, "History," p. R6.

Willard was a Cleveland firm begun in the late 1880s to meet the demand for electric lighting system batteries. During the 1880s, Theodore Willard oriented the company towards the development of transport batteries, concentrating on lightly-built Planté-type cells. During the early years of the century, Willard, like ESB, was involved in research to develop effective hard rubber envelopes for its cells.[6] Following its takeover by ESB, the design of Willard batteries converged toward that of ESB.

The reason for ESB's increased commitment to automobile batteries despite the decline of EVC was that many more manufacturers of electric vehicles still continued to exist, and, as time went on, most of these companies adopted the Exide batteries as standard equipment. The 1900 prediction of *Horseless Age* did not turn out be be true; vehicles propelled by lead-acid storage batteries could be made a practical method of transportation. Nonetheless, by the end of the first decade of the twentieth century, the outlook of the electric vehicle industry had become quite different from what it had been a decade earlier when Morris, Salom, Rice, Whitney, Edison, and many others had dreamed of the triumph of the privately-owned electric pleasure vehicle. In order to properly understand the ESB's continued commitment to transport battery design, it is necessary to examine the economic situation which the company's management faced during this period.

In the United States, electric vehicles remained a common sight until the Great Depression, but this was not true of the electric pleasure automobile. From the late 1890s on, the production of American electric cars as a percentage of all car production fell continually, reaching only 4.4 percent in 1910.[7] The invention of the electric self-starter for gasoline cars in 1912, and the massive commercial application of this technology beginning in 1914, drove the last nail into the private electric car's coffin. By the start of the 1920s the private electric car had become a rarity.

On the other hand, the electric commercial vehicle continued to be popular. Indeed, in Europe it has never lost its popularity for commercial uses, even though private electric cars disappeared there about as rapidly as in the United States. Because of a series of totally unanticipated coincidences, the electric passenger vehicle and the electric commercial vehicle

---

[6] U.S. patents: #761,345—31 May 1904—filed 11 September 1903; #765,060—12 July 1904—filed 13 August 1903; See also: #617,003—3 January 1899; #617,004—3 January 1899.

[7] *Horseless Age*, April, 1914: 518.

experienced opposite patterns of public acceptance. At first, in the late 1890s and first years of the new century, enthusiasm for the electric private car was high, while little thought was given to the commercial version. As time went on, however, the electric car and taxi faded, while electric vans and trucks gained popularity. The reasons for this are not entirely clear, since the subject has not been subjected to any detailed analysis by historians of the automobile, but a number of tentative explanations can be proposed.

In its early days, the gasoline car did not have much of an actual advantage over the electric, since the speed of both cars was limited and the gasoline car's tendency to break down gave it an effective range not very much greater than the electric's. Even if we grant a slight speed and range advantage to the early gasoline car, its noise, vibration, smell, and other mechanical problems made it less desirable to many people. Moreover, before 1912-1914, the absence of an electric starter gave the electric an assured market among women, the old, and those who simply did not want the nuisance and danger of the hand crank. The crucial thing about these early advantages of the electric, however, is that they all declined with the advance of technology.

One early advantage which the electric vehicle did *not* have was that of cost. Electrics were expensive on both a first-cost and operating basis. From the 1880s until the early years of this century, the cost of electricity was quite high—a fact which reflects the relative immaturity of the technology and the point made earlier in this study that the electric light was originally designed for the well-to-do consumer. Gasoline, on the other hand, was initially cheap.

After the turn of the century, however, this relationship started to reverse itself and continued to do so for the next two decades. C.R. Erikson, for example, gives the following figures for the average nationwide cost of electricity and gasoline in 1903 and 1913: [8]

	Gasoline (¢/gallon)	Electricity (¢/kilowatt)
1903	10	23
1913	24	7

The reasons for the price fluctuations in gasoline are beyond the scope of this study. Suffice it to say that the period from 1900 to the early 1920s was one of increasing gasoline prices, particularly after 1910, whereas electric rates, because of the maturing of the technology and increasing economies of scale,

---

[8] *Electric Vehicle News*, 8/1972: 28.

generally fell. The graph below shows the wholesale price of gasoline at export from the United States and contemporary electric rates:[9]

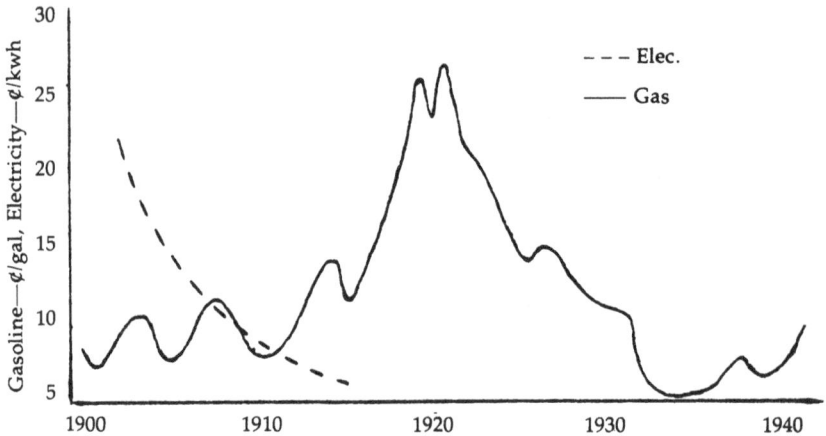

The 1913 gasoline prices quoted from Erikson are probably a bit high. Nevertheless, American gasoline prices after 1910 did enter a period of considerable inflation. The newspaper and journal articles of those days contain continual predictions of rising gasoline prices. Among other things, experts were predicting that gasoline would soon be costing many times the current prices as a result of the increasing demand and inherent shortage of crude petroleum.

Although gasoline prices did not rise very fast before 1910, electricity rates fell most rapidly during this period. Consequently, it is evident that during at least the first twenty years of this century the operating cost differential between gasoline and electric vehicles constantly and rapidly changed in a direction favorable to the latter. At precisely what time the power costs per ton mile for electric fuel fell below those for gasoline is difficult to say. But it is clear that such a point had been reached by the time of the First World War.

These operating cost differentials do not seem either to have hurt the electric car during the '90s, or to have retarded its demise later on. We can hardly have expected this—people who bought electric cars were usually well-to-do, and these

---

[9] On 15 April 1914, gasoline was reported falling from 17¢ to 14¢ per gallon retail. During 1913, however, gasoline prices had bounced between 23¢ and 14¢ per gallon. *Horseless Age*, 15 April 1914: 580.

fuel cost differentials would not have meant much. Particularly after the invention of the electric starter, they would hardly have been willing to put up with the limitations of the battery car in order to save a few dollars on fuel costs.

The electric commercial vehicle, on the other hand, confronted very different conditions. First of all, the maximum range of 35-40 miles per day was not a serious limitation for such vehicles, particularly in the less spread-out cities of those days—delivery vehicles were not needed to cover more ground than this. Low speed was also not as critical; the traffic mostly stop-and-go, and the low hourly wages of those days made rapid deliveries not as critical for holding down costs as in our day. Moreover, the electric vehicle used power only when moving, whereas the gasoline vehicle often had to be kept running during stops. This was most crucial for the delivery vehicles such as bakery trucks. Electrics were always great favorites with the bakeries, both because of the stop-and-go nature of their operation from house to house, and the light, non-battery straining nature of their merchandise.

When businessmen were shown operating statistics based on rising gasoline and falling electric prices, they increasingly tended to favor the electric truck. Of course, electric trucks were more expensive than gasoline on a first-cost basis. Moreover, the company installing a fleet of electrics had to invest in charging apparatus. Against these costs, however, were placed the much longer life of electric vehicles, their lower (non-battery) maintenance costs, and the lower fuel cost.

Commercial electric trucks appeared on the market at the same time as the first private electric cars. Statistical studies of the economics of their operation, however, did not appear in the technical press until the end of the first decade of the new century. Consequently, it is difficult to say anything definite about the early economic relationship of horse, gasoline, and electric delivery vehicles. By 1910, however, a businessman who contemplated buying delivery vehicles had at his disposal a fair amount of reliable data on which to base a decision. This development was not unlike the appearance of detailed economic data on power station stationary batteries about ten years after their initial introduction.

Most of the electric commercial vehicle studies compared the economics of the electric only with horse-drawn vehicles. This is understandable, since such comparisons were easier to make than those between electrics and gasoline trucks—in those days, when transport was still mostly by horses, the replacement of horses by either electric or gasoline vehicles was common, but the replacement of either of the two modern systems by the other was uncommon. This situation was much like that of a generation earlier, when the replacement of horsecars by either battery cars or trolleys made economic analysis of the changeover costs and receipts easy. All of the electric vehicle studies gave the electric truck a considerable cost advantage over the horse, usually between 20-30 percent based on operating expenses.[10]

Of greater interest for our purposes, however, was an extremely detailed report prepared in 1913 by a pair of MIT professors, which compared a series of equivalent horse, gasoline, and electric commercial delivery vehicles. The following tables are a condensation of the most important data from this study from the perspective of our study:[11]

I—TOTAL COSTS—*Urban Delivery Van for Parcels: 1/2 ton capacity*

	Electric (3 ¢/kwh)	Gasoline (16¢/gal)	Horse
Cost per day	$8.30	$10.10	$6.40
Cost per mile	.25	.30	.27
Cost per delivery	.042	.05	.045

II—TOTAL COSTS—*Suburban Delivery Van for Parcels: 1/2 ton capacity*

Cost per day	$8.60	$10.40	$6.50
Cost per mile	.23	.27	.23
Cost per delivery	.067	.075	.085

III—TOTAL COSTS—*Furniture Delivery Truck: 2 ton capacity*

Cost per day	$9.75	$12.50	$8.60
Cost per mile	.31	.38	.35
Cost per delivery	.31	.38	.35

IV—TOTAL COSTS—*Saloon Delivery Truck for Beer: 3 1/2 ton capacity*

Cost per day	$11.40	$14.25	$10.00
Cost per mile	.37	.43	.46
Cost per delivery	.52	.62	.66

V—TOTAL COSTS—*Coal Delivery Truck: 5 ton capacity*

Cost per day	$12.30	$16.20	$11.00
Cost per mile	.45	.51	.59
Cost per delivery	.54	.61	.71

[10] For example, *Electrical Review* (New York), 4 January 1913: 23-25; 1 June 1912: 1043-1044; 6/1914: 1245-6; 12 April 1913: 774-775.

[11] *Electrical Review* (New York), 12 April 1913: 774-775.

Note that the relative cost differentials for the electric and gasoline vehicles on a day, mile, and delivery basis always favor the electric and to about the same degree in all cases, that is, between roughly 75-85 percent cheaper for the electric. This shows that the speed of the two types of vehicle *for this particular service* was roughly equal. The cheaper cost of horse service on a *day* basis, but more expensive cost on a *delivery* basis, reveals the slower speed of the animal and the clear economic advantage of changing from the horse to either gasoline or electric power.

A breakdown of the expenses of the electric and gasoline vehicles for one of the cases listed above shows that the economic relationships of these two technologies were complicated. Consequently, a businessman trying to decide whether to adopt electric or gasoline trucks with such data was in a similar situation to the electrical engineer of the pre-1892 era trying to decide upon the relative economics of a battery versus non-battery D.C. central power station. It is not to be wondered, then, that the electric truck became more popular after such data appeared. The table below shows the breakdown for the ten-ton coal truck:

Expense per Year:	*Ten-Ton Electric*	*Ten-Ton Gasoline*	
Tires	$ 400	$ 520	(+120)
Repairs	300	700	(+400)
Battery	440		(−440)
Lubricants	15	70	(+55)
Electricity (3¢/kwh)	290		(−290)
Gasoline (16¢/gal)		430	(+430)
Garage	270	270	(0)
Driver & Helper	1210	1280	(+70)
Depreciation	400	760	(+360)
Interest	135	150	(+15)
Insurance	150	200	(+50)
	$3,610	$4,380	(+770)

The differential in favor of the electric truck was $770, yet the fuel differential was only $140. This ratio was roughly the same for all the cases analyzed by the MIT group. Most of the cost differential came from the more complicated technology of the gasoline vehicle. Nevertheless, gasoline prices were a significant factor in the increasing acceptance of the electric truck; the statistics above were taken in 1912, at the beginning of a decade of soaring gasoline costs. Moreover, it should be noted that a factor which is almost as important to businessmen as cost itself is the *predictability* of cost. In the period under discussion, electric costs were falling, while gasoline prices were rising and doing so at an erratic rate.

In July 1912 *Horseless Age* editorialized upon this theme: [12]

> The recent heavy increase in the price of gasoline has evidently been a godsend to the electric truck industry. In spite of the fact that gasoline trucks are more suited for long distance hauls and electric trucks for urban work, so that each class has a field of its own, there is a considerable rivalry between the two. There are many cases in which a gasoline or an electric truck will satisfactorily do the work, and the choice between them is largely a matter of costs—first cost and operating cost. . . . the 50% rise in the price of gasoline therefore places the gasoline truck at a great disadvantage. . . . Electric station (i.e. power companies) are very much alive to the importance to their business of continued growth of the electric truck industry and they give it every encouragement in the way of low charging rates and accommodating service.

This last sentence indicates another important factor in stimulating electric vehicle acceptance. Beginning about 1907-1908, central station managers began to recognize the potential of the electric vehicle as a new kind of load-leveling device. Storage batteries, of course, had been used for many years as load-levelers to improve the load-factor, but these cells had always been used *inside* the station or sub-station; and the power company itself had always borne the cost of buying and maintaining these cells. Now, however, electric companies saw the opportunity of using *other people's* batteries to store current during non-peak hours. Not only that, but the current used to charge the batteries would be sold *at that point* to the consumer, rather than being kept within the powerhouse. Accordingly, the 20-30 percent loss of energy which occurs during the charge-discharge cycle would now be a loss to the consumer, rather than the power station. Moreover, the attendance and maintenance cost inside the powerhouse would also be saved.

The surprising thing is that power companies took so long to recognize this analogy between the in-house load-leveling battery and large numbers of electric vehicle batteries, most of them charging during the off-peak night time hours. In 1906, for example, one author vigorously pushed the idea of a close

---

[12] *Horseless Age*, 17 July 1912: 105.

relationship between central stations and electric vehicles, but indicated that few companies had so far taken up the idea:[13]

> If the electric light and power companies at present in existence do not take advantage of this active opportunity, it will certainly be done by independent capital, to their great disadvantage and they will find themselves entirely deprived of this new and important source of revenue. . . .

The reason for this reluctance may have been the gradual decline in interest in electric passenger vehicles during the early years of the century, and the as yet undeveloped state of demand for electric commercial vehicles. Be that as it may, by the end of the decade, American central stations were beginning to actively enter the field of the electric vehicle, as these comments from a January 1913 article indicate:[14]

> The real value to the central station of the electric-vehicle load (i.e. current demand) has only recently received the attention which it rightly deserves. At the beginning of the present year (1912), or but a short while before, a great many of the larger central stations in different sections of the country began to take an active and serious interest in the development of the electric vehicle load.

One symptom of this activity was the formation of the Electric Vehicle Association in 1910. This was a trade association whose roots lay in the informal meetings between battery manufacturers, electric vehicle makers, and central station managers resulted in agreements to advertise the electric vehicle in a joint basis and to coordinate operations for sales and service. The ESB played a central role in getting these meetings going.[15]

The Electric Vehicle Association began in March 1908 as a local Boston-based group for New England.[16] In 1910 the first step towards the creation of a national EVA was taken when the various commercial interests involved established a central fund for carrying out their objectives. In September 1910 a national EVA was created with 29 charter members; membership had climbed to 335 only two years later.[17]

The cooperation between the various parties in the EVA took several forms. The New York Edison Company, which was one of the earliest power companies to push electrics,

---

[13] *The Electrical Journal* 3 (1906): 280-289.

[14] *Electrical Review* (New York), 4 January 1913: 23-25.

[15] Barak, "History," p. E3.

[16] *Electrical Review* (New York), 18 May 1912: 955-957.

[17] Ibid.

began an advertising campaign for the electric vehicle in late 1908. It also introduced its own fleet of 70 electric service trucks and published the operational records of these vehicles.[18] New York Edison also established charging stations around the city, and maps of these were mailed to consumers.[19] Big city systems, such as New York Edison, were ideally suited to the electric vehicle, since the short-range electric was inherently a "city car," and, as we have discussed, such central city areas were supplied with D.C. during the early decades of this century; thus no rectification apparatus was needed at charging stations.

The New York Edison system granted special rate contracts for vehicle charging, ranging from 3-6¢ per kilowatt hour, depending upon the rate of consumption.[20] These rates were granted on condition that the charging be done during trough-load periods. This low-rate structure for vehicle charging was perhaps the most important factor of all in the upsurge of commercial electric truck use. Central stations varied, of course, in the reduced rates they allowed, but something between 3¢ and 6¢ per kilowatt hours was the national average, with the rate going down as consumption went up. There were even a few instances of power companies supplying current free, but this was no doubt for advertising.[21]

The low rates given for high consumption was the reason why companies operating big fleets of trucks often favored electrics. One New York list from 1912 included such large businesses as: Adams Express (250 trucks); Ward Bread (230); Gimbel Brothers (76); American Express (86); and Jacob Ruppert Brewery (72).[22]

Central station managements showed sophistication in the promotion of electric vehicle sales. By 1912 the New York Edison system operated an Automobile Bureau, which maintained a card file on every electric owner in the city. The card listed the owner's occupation, residence, garage, make of car, and so forth. The information was culled from the State License Bureau. Another file was kept on businesses which still used horses; salesmen were directed to these prospects.[23]

---

[18] Ibid., 2 April 1910: 690.

[19] Ibid.

[20] Ibid.

[21] ESB Bulletin #109 4, 1908.

[22] Electrical Review (New York), 18 May 1912: 955-957.

[23] Ibid., 10 February 1912: 288-289.

In 1913, the New York Edison Company, at the urging of the Anderson Electric Car Company, appropriated $30,000 to create an electric car garage.[24] The power company undertook to charge and maintain the vehicles at a fixed monthly rate. By 1913 many electric power companies across the United States were operating such garages. One of the earliest was set up in 1907 in St. Louis by the Union Electric Light & Power Company. It charged $35 a month for all services, even including one delivery per day of the vehicle to the consumer's doorstep.[25] With the exception of this last service, these rates and services were fairly typical for most garages established at the time.

Some electric companies even became agents for the electric car manufacturers, establishing showrooms and offering garage contracts along with the sale.[26] Others hired storage battery engineers and gave them the full-time job of advising consumers of proper battery care and feeding while promoting the electric car at the same time.[27]

The Electric Vehicle Association itself provided a central headquarters for gathering and disseminating information about the technology of electric vehicles. Monthly meetings were held in New York at which technical papers were read. Most of the earlier-mentioned economic analyses of electric vehicle performance were prepared as papers to be read at meetings of the EVA. The statistics thus gathered were used to sell vehicles to businessmen, or for such specialty purposes as getting insurance rates reduced. For example, in 1912 the insurance committee of the EVA was trying to convince the insurance industry to give electrics a 25 percent differential over gasoline vehicles.[28]

In 1912 the president of the EVA, W.H. Blood, accurately summed up the short history of the electric vehicle:[29]

In the early days the electric vehicle was clearly at a disadvantage; nevertheless, it worked. . . . The early electric car was clumsy and hardly more than an adaptation of the horse-drawn vehicle; the batteries were heavy and good for comparatively short distances. The attitude of the central station toward the installation of electric vehicles was not conducive to their gen-

[24] Ibid., 12 April 1913: 754.

[25] *Electrical Storage Battery Company Bulletin #109* (Phila. ESB, 4/1908).

[26] *Electrical Review* (New York) 12 February 1910: 332-333.

[27] *Electrical World,* 25 August 1910: 437.

[28] *Electrical Review* (New York), 18 May 1912: 955-957.

[29] Ibid.

eral introduction, for I think that most managers looked upon the charging of storage batteries as a nuisance and made rates for current in accordance. Today, however, all of these conditions are changed. The modern truck has many points of superiority over the older types. The battery is lighter and more efficient and is capable of giving a mileage two or even three times that attained by earlier batteries. The central station people have awakened and they now realize that the charging of electric vehicle storage batteries is a profitable off-peak business. . . .

It may well be imagined that the ESB was confidently looking forward to the future in light of these favorable developments. There existed, however, a major cloud on the ESB lead-acid horizon during these years before and after 1910, one that more than anything else motivated the ESB to develop the Ironclad Exide. That cloud was Thomas Edison and his development of the alkaline nickel-iron storage battery.

Thomas Edison became an enthusiast for the electric car at about the same time as William C. Whitney. He, no less than Whitney, decided to try to create a monopoly over the field, but his method of creating this monopoly was technological, rather than financial. Edison had denounced the lead-acid battery during the 1880s and had left others to develop the technology. By 1900 it was much too late to enter the lead-acid field in hopes of dominating it. Domination of the industry could only have been achieved at that point by buying out the ESB, and such financial operations were of no interest to him.

Therefore, Edison decided upon a technological end-run; he developed a new kind of storage battery using neither lead nor acid. He designed it to be lighter and more durable than the lead-acid, and intended it to eliminate every existing lead battery from the market. Edison revealed these plans to the press at the same time that ESB was introducing its first Exides. The development of a successful non-lead battery would take much longer than Edison had thought, but during the next ten years he regularly announced to the press the imminent commercialization of the battery. Edison continued to denounce the lead-acid battery as a worthless piece of junk, but he was well aware of the improvements being made. He knew, moreover, that most of these improvements were coming from ESB; Edison had industrial spies at ESB and they kept him informed of details.

ESB, of course, also kept a weather-eye on Edison. It is unknown if they had spies also, but they nonetheless kept close tabs on the "Wizard's" progress. ESB publicly denounced the Edison battery, but they were equally aware of the progress being made by their famous competitor, and of the danger which lay therein. The first decade of the twentieth century, therefore, saw a technological race between the two interests. The word *race* is accurate, since the time element here was crucial; whoever was first to produce a really durable electric vehicle battery could establish a secure position in the business. This results from the different charging characteristics of the lead-acid and nickel-iron systems; once a consumer is convinced of the good properties of one type of battery and buys charging equipment for it, he cannot switch to the other system without considerable repurchase of charging apparatus.

The financial stakes involved in this competition were considerable. In 1914, *Horseless Age* published a list of electric pleasure and commercial vehicles then being sold in the United States. Although the list is not exhaustive, it includes all the major manufacturers. The following list shows the brands of batteries which these manufacturers used as standard equipment on their vehicles. By the time this list was published, both the Ironclad Exide and the Edison Nickel-Iron had been on the market 3 to 4 years, and it is clear that they were the two chief competitors:

Pleasure Vehicles	*Commercial Vehicles*
31 Total Makes	11 Total Makes
12 Specified Exides standard	6 Specified either Edisons or Lead-Acids, with no brand specified (however, Exide probable)
8 Specified Edisons, but of these, 6 also used unspecified lead-acids as well	2 Specified Exides
11 Specified lead-acids, but did not indicate brand; it is probable that at least some of these did actually specify Exides	2 Specified Philadelphia lead-acid (later, the Philco Co.)
	1 Specified W.T.X. lead-acid

If we make the assumption that half of the companies which did not specify which kind of lead-acids they used actually used Exides, the figures give an indication of ESB's domination of the lead-acid market. Unfortunately, precise statistics on the numbers of batteries actually sold for these purposes are unavailable. Nevertheless, a comparison between the private and commercial columns is revealing. Only one quarter of the pleasure vehicle models in 1914 were normally equipped with the Edisons, and even for these, most of the manufacturers also offered their customers a choice of the cheaper lead-acids. For the commercial vehicles, however, over half the models were equipped with Edisons, although all manufacturers also offered lead-acids here.

The Edison battery had few real advantages over the Exide. In size and weight it was roughly equal to the lead-acid. It cost more and operated less efficiently in cold weather. But it did have one advantage over lead-acids which appealed particularly to the commercial user—the Edison had a much longer life, and the businessman who invested in electric trucks was interested in long-term savings from reduced operating costs. The Edison battery had this appeal. This is the reason for the large percentage of truck makers specifying the nickel-iron cell. For the pleasure car, on the other hand, durability has never been an overriding selling point, whether for electrics or for gasoline cars.

The rising importance of the commercial electric vehicle against the private vehicle, therefore, put a premium on the durability of storage batteries, and this gave impetus to the race between the ESB and Edison designers. The race, however, turned out to be in vain; the technology which both these batteries sought to dominate was doomed by SLI (Starting-Lighting-Ignition).

3. Starting-Lighting-Ignition

The doom of the electric car was sealed on 17 February 1911. On that day Charles Franklin Kettering delivered his electric car starter to the Cadillac Motor Company in Detroit.[1] It worked faultlessly. From its introduction in the 1912 Cadillac, the electric starter was rapidly adopted throughout the automobile industry. By 1913 there were 44 makers of SLI equipment on the market.[2] SLI technology eliminated most of the remaining props from the private electric vehicle market—women, older people, and those who either feared the dangerous "kick" of the starter handle, or who simply disliked the inconvenience.

---

[1] T.A. Boyd, *Professional Amateur—The Biography of Charles Franklin Kettering* (New York, E.P. Dutton, 1957) p. 70.
[2] Ibid., p. 82.

The introduction of SLI technology was not the first time that electric batteries had been used on internal combustion vehicles. The ignition systems of many early cars were powered by dry cells because magneto ignition was often unreliable.[3] Moreover, the lighting of some of the earlier, more expensive gasoline cars was electric—the power being supplied from storage batteries. Indeed, in view of these early applications of electricity in gasoline cars and the great amount of inventive talent applied to the design of such cars since 1895, it is surprising that the invention of electric starting should have been so long delayed. One reason for this was the belief that a sufficiently powerful starting motor would be too large and bulky; Kettering's unique insight was that the motor could be made small, since its required torque, although great, was only needed for a brief, non-straining spurt.[4]

Another reason for the delayed appearance of SLI technology may have been that the attention of electrically-minded inventors was concentrated on the electric vehicle. It is worth noting, for example, that no storage battery company ever worked on the idea until Kettering developed it; Kettering had no prior connection with the battery or electric car business.

Nevertheless, battery makers were quick to grasp the importance of Kettering's invention. ESB was the first to become involved; in 1910 Kettering asked ESB to develop a special battery for SLI purposes and they sent an engineer to work with him. The battery which they developed reflects both the dearth of transportation battery technology at this time, as well as the undeveloped state of the SLI electrical system. It weighed 67 pounds![5]

Within twenty years of this invention, the American storage battery industry was turning out ten million SLI batteries per year. Yet when Charles Kettering's initial letter arrived at ESB, offering to purchase only ten thousand batteries, the reaction of the company was one of stunned disbelief; no one had ever wanted anywhere near so many batteries before. As O. Lee

---

[3] Harold Cross, *Automobile Batteries* (London, 1919) p. 3; ESB, Lead, its use in Storage Batteries, p. 3.

[4] Boyd, *Professional Amateur.*, pp. 68-69.

[5] Part of the problem was the need for a large number of individual cells. The cranking motor required 24 volts, but the lighting and ignition functions only needed 6 volts. Accordingly, the battery was designed with 12 cells, each holding 3 pairs of plates—72 individual plates. To perform the series-parallel shifts required, Kettering borrowed the speed controller from trolley technology. Boyd, *Professional Amateur,* p. 69.

Harrison, the ESB representative sent to see Kettering, said: "I don't want to sell you any batteries. I just want to look at a fellow who thinks he wants to buy that many batteries."[6] Another ESB engineer was more straightforward about Kettering, saying simply: "He's nuts."[7]

The surprised attitude of these battery people is understandable considering the scale of the market for electric car batteries at the time. A year before Kettering made his offer to ESB, some 3,639 electric cars were made in the United States. In that year, 1909, the total value of all American storage battery manufacture was only about $4,250,000.[8] It is probable that most of this was for the large and expensive stationary batteries. Using 1912 prices, a standard, 68 pound Exide car battery cost about $40 complete. This price was for the largest electric car battery made by ESB at the time, which contained 21 plates.[9] The first SLI Exides probably cost more, given their more complex internal structure.

If we (broadly) assume that ESB controlled 90 percent of the United States storage battery market in 1909, and that about 1/3 of this was for vehicle batteries, and if we make the further assumption that one of Kettering's first SLI Exides cost about $50, then the Kettering request for 10,000 such batteries would have represented nearly half of ESB's yearly output of car batteries. Moreover, the battery industry was quick to see the enormous potential of the new market. In the years around 1910, when the production of electric private cars remained almost constant at about 4,000 per annum, the manufacture of gasoline cars increased as follows:[10]

YEAR	NUMBER OF GASOLINE CARS PRODUCED
1908	55,400
1909	82,000
1910	180,000
1911	215,000
1912	300,000
1913	400,000

Even though SLI equipment added to the cost of gasoline cars, it was adopted very rapidly throughout the industry. The

---

[6] Ibid., p. 72

[7] Typescript of ESB history, ESB company files, p. 15.

[8] U.S. Census.

[9] ESB company price list, 1912. Type HV Ironclad Exide.

[10] *Horseless Age,* 1 April 1914: 518.

journal *Horseless Age,* referring to the year 1913, remarked,[11] "Last year. . . . the application of electric starting and lighting systems to all classes of pleasure vehicles, with the exception of the very lowest priced ones, became almost universal. . . ."

Most battery firms, therefore, soon turned their attention to the design of SLI batteries. A number of new firms entered the battery field as a result of this new and tempting market. For example, the Prest-O-Lite Company, which had been a producer of calcium carbide vehicle lamps, saw that much of its market might soon be lost, and turned a problem into a virtue by diversifying into storage battery manufacture.[12]

The lion's share of the market, however, was taken by the large, pre-existing firms—ESB, Willard, Chloride, AFA. Fortunately, little was required in the way of new technology and what was needed had already been partly developed for the electric vehicle market. No major change in battery design resulted from the introduction of SLI's, although the scale of their manufacture was to have an influence on manufacturing methods after the First World War.

The first Kettering SLI system required a more complicated battery than that used on electric vehicles. This situation was soon simplified, however, by the invention of electrical systems which used the same battery voltages for all four automobile functions—cranking, ignition, lighting, and charging. Consequently, it was possible to reduce the number of plates in SLI cells, as well as to simplify the wiring and switching apparatus.

For the past 60 years, SLI manufacture has dominated the storage battery industry. In 1909, as Charles Kettering was preparing his invention, the total value of American storage batteries produced was a little more than $4 million. By 1914 this had already risen to nearly $11 million, and by 1919 to $57 million.[13] Statistics on the percentage of storage battery manufacture devoted to various purposes are not available before the mid 1920s, but from then until the mid 1950s SLI batteries made up about 80 percent of the dollar output of the American industry. Since then, the percentage has fallen to the 73-75 percent range, because of the greater life of SLI's and the increased use of electric vehicles. Nevertheless, the SLI remains today the dominant factor in the industry.

---

[11] 7 January 1914: 50.

[12] Boyd, *Professional Amateur*: 81.

[13] U.S. Census of Manufactures.

As a result of this dominance, the needs of the SLI battery have had the strongest influence over storage battery design during this century. One affect which the SLI had was to direct the attention of battery engineers towards the design of separators which are inert, electrically-resistive objects placed between the plates to prevent them from touching and shortcircuiting. All storage batteries need some kind of separation device, but in stationary batteries the technical problems involved were small. The space between the plates could be kept fairly large and the most common method of maintaining this space was to hang vertical glass rods between the plates.[14]

With the advent of the traction battery, however, it was necessary to save space and weight. Plate distances were reduced and the glass rods were no longer satisfactory. The rods were too brittle, of course, but more importantly the separators now had to cover the entire interplate area, since pellets of active material falling from the grids could no longer easily descend to the sludge space below the plates, but might get stuck between the plates and short-circuit them. The most usual solution to this problem was to use thin veneers of wood, the thinness and porosity of which would permit sufficient electrolyte transfer through them. The veneers were usually ribbed to give them strength:[15]

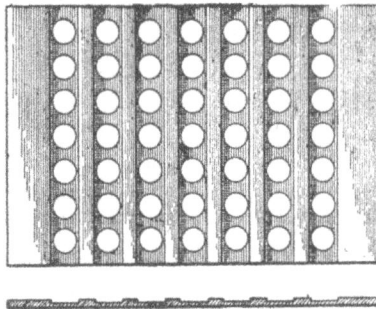

Storage battery literature reflects this changing interest in separators. The British battery expert E.J. Wade wrote a voluminous treatise on storage battery design in 1901, just at the beginning of the electric vehicle era.[16] Wade treats separators

[14] Lyndon, *Battery Engineering*, p. 191.
[15] Wade, *Secondary Batteries*, p. 49.
[16] Ibid.

not so much as insulation between the plates, than as a means of holding the active material in the grids. This, of course, was the older tradition of battery design.

Edmund Hoppe, who wrote *Die Akkumulatoren für Elektricität* three years after Wade's book, has almost no reference to separators.[17] Both Wade and Hoppe concentrate most of their attention on the chief concerns of the designers of the 1880s and '90s—grid geometry.

A significant change occurred in the 1911 edition of Lamar Lyndon's *Storage Battery Engineering*.[18] Here grid geometry, which had occupied most of Wade and Hoppe, was reduced in importance, while separators are treated in the new fashion— that is, as insulators between the plates and not as active material retainers. Lyndon indicates that ribbed wood separators were the most popular kind. This was only because of their low cost, however; he indicates that their life was short.

The short life of these wood separators is a reflection of the undeveloped state of transport battery technology in 1911, on the eve of the SLI. For example, it is obvious from Lyndon's book that not much was yet known about the preparation of wood for separator use. Lyndon makes no distinction between the various types of woods which may be used for separators, whereas George Vinal, writing the first edition of his classic storage battery textbook in 1924, is careful to alert his readers to the importance of selecting only those woods which have been found suitable for battery use.[19] Lyndon was aware that wood contains various impurities which are leached out by the electrolyte and cause trouble; he knew, for example, that the lignin is soluble in sulphuric acid and can be oxidized by it to form acetic acid, which is harmful to the plates. Lyndon also indicates that the wood must be treated to remove these impurities. But he goes on to say that the wood of separators thus treated "becomes friable and easily broken down in shreds," which indicates that the methods used around 1911 for purifying wood for separators were crude.

Vinal (1924) indicates that the treatment of wood veneers with heat, steam, caustics, and other chemicals is a tricky business and must be done with skill and experience if durable separators are to be produced. Only at the time that the SLI

---

[17] Berlin, Springer, 1898.
[18] Lyndon, *Battery Engineering*.
[19] George W. Vinal, *Storage Batteries*.

battery made its appearance was the battery industry beginning to develop that skill and experience. The industry took its
initial knowledge of wood treatment techniques from the only
industry then practicing such technology—paper making. In
paper making, the process of "digestion" uses heat and caustics to separate the lignin from the cellulose, allowing the cellulose to disintegrate into individual fibers. It is hardly surprising, therefore, that the early, treated wood separators
easily broke into shreds.

The economic importance of the SLI market stimulated the
development of more sophisticated wood treatment
techniques. The paper-making digestion process had to be
moderated so that the lignin, which is the "glue" of the wood's
structure, would only be partially removed—enough to give
the wood a microporous structure and prevent excessive acetic
acid formation, but not enough to produce an incoherent
mass. Moreover, special chemical treatments had to be invented to remove trace chemicals which, although harmless in
paper making, could cause local action in storage batteries.
George Vinal implies that much of this work had been done by
the mid 1920s, and he was able to identify a number of the
most favorable woods for separator use; by about 1920 most
American and European manufacturers had selected an Ore-
gon wood called Port Orford cedar as the most desirable
separator wood.

Although development work on wood separators took place
before the invention of the SLI, not much progress had been
made before the time of the First World War. Thereafter, progress was more rapid. There are a number of factors why this
was so. First, and most obviously, was the skyrocketing
economic importance of the SLI battery. The electric vehicle's
incentive for innovation had been much less than this. Second,
however, are a number of factors which made separators more
important for SLI batteries than they had been for electric vehicles. An SLI cell, in contrast to most of the batteries described up to this time, usually needs little of its overall storage
capacity. It gives a quick burst of power to the starting motor
and is then recharged during the run of the engine. This is
quite different from the operation of an electric car battery, or a
power house "load-leveler," which are designed for "deep
cycling" on a regular basis.

Consequently, it is both desirable and permissible to reduce
the quantity of electrolyte in an SLI cell well below that in an

electric car battery—desirable, because it saves space and weight; and permissible, since only small amounts of sulphuric acid are needed for the short discharges. On the other hand, in order to provide the high amperage bursts of power for engine cranking, the sulphuric concentration must be greater than in the electric car cell. Moreover, since the SLI battery must occasionally deliver a large portion of its total charge, the increased concentration can make up for the reduced volume. As a result of these technical changes, the practice in the design of SLI batteries became to bring the plates closer together, thus squeezing the separators and leaving only enough electrolyte to saturate the separators and fill up the small space between the ribs. The separators in the older forms of batteries had more room and were surrounded by less corrosive acid.

The result of these changes was to place considerably more strain, both chemical and mechanical, on the separators. The expansion and contraction of the plates with charge and discharge squeezed the separators at the same time that the electrolyte corroded them faster. Moreover, the more intimate contact between the wood and the lead peroxide of the positive plate created a tendency for the separator surface to oxidize.

While the separators were subjected to these more severe conditions, the evolution of cell containers was making it increasingly more difficult to replace worn separators. Stationary storage batteries had been open-topped to facilitate maintenance, and the electric vehicle batteries, although provided with covers to prevent acid spillage, could still be easily opened for servicing. From the inception of the SLI battery, however, automobile designers appear to have concluded that drivers would pay no attention to battery maintenance, since the usual practice was to hide the battery in an inaccessible place in the car, and to almost completely seal its top with molten hard rubber.

One remaining factor was the difference between the plates of electric vehicle and SLI batteries. Since the 1890s the tendency in electric car battery design was to use plates surrounded with some sort of inert sheath, usually of hard rubber. This led to the evolution of the Ironclad Exide and its various imitations. Such expensive design was permissible with electric cars, where the buyer expected to pay more both for his car and a substantial percent of the car's total cost for the battery. Since neither of these conditions applied for most gasoline cars, however, and the cost of SLI's needed to be kept low, it has been fairly universal throughout the battery industry for the past sixty years to use the flat, pasted, double-grid for both electrodes of the SLI cell. This configuration, of course, also increases the need for good separators.

As a result of these conditions, the designers of SLI batteries needed strong, durable, porous, chemically and electrically cheap, inert separators. Fortunately, they found that a number of woods, if carefully prepared, would meet these criteria and would last as long in regular service as the plates themselves. Nevertheless, from the first decade of this century numerous attempts have been made to develop substitutes for the wood separators. As we saw in chapter III, the use of perforated hard rubber electrode envelopes originated in the 1880s for the positive electrode and was the starting point for the evolution of the tubular positive. Various designs of perforated hard rubber separators also evolved. Such separators continued to be used extensively until the 1960s, when they were replaced by plastics.

Rubber separators were more expensive than wood, and, not being microporous, needed a large number of holes for circulation of electrolyte. These made them structurally weak if not done carefully, thus, hard rubber separators were used only in the more expensive SLI's.

Much work was done to give the rubber separators a kind of artificial microporosity to do away with the need for the perforations. One of the most popular of these porous rubber separators was the "threaded rubber" separator patented by Theodore Willard in 1921.[20] In this design, hundreds of thousands of fine cotton filaments were kneaded into the rubber, which was then sliced to provide separators with tiny, microscopic "wicks" for the transport of electrolyte.[21] Willard filed for this patent in February 1915; just when the battery industry started to develop a sophisticated response to the need for good separators.

Interest in fiber glass separators began around 1925. Here, however, a materials technology problem prevented fiber glass from being used until the 1950s and '60s; only when fiber glass could be impregnated with synthetic resins could it be made

---

[20] U.S. patent #1,375,763—26 April 1921—filed 26 February 1915.

[21] Arendt, *Storage Batteries*, p. 130.

strong enough to resist deformation.[22] In fact, most of the non-wood separators used before the Second World War had few advantages over wood: they were not significantly more durable than good wood separators and were more expensive. The artificial separator, however, did have one significant advantage over wood which justified the considerable amount of research investment devoted to them by battery manufacturers. This related to the economics of the distribution and marketing of storage batteries.

Since wood separators had most of their lignin "glue" leached out, they were more porous than untreated wood. Like wood, they shrink, warp, and crack if allowed to dry; only more so because of the greater porosity, hence, batteries made with wood separators must be immediately filled with electrolyte and must never be allowed to dry out. Such cells must also be kept in a full state of electrical charge until the time of sale. In the early days of the battery industry, when batteries were large, the number of individual sales small, and the manufacturer's men themselves came to install a battery at a powerhouse, or when electric vehicle owners had a new set of cells installed by their full-service garage, problems in the marketing of batteries were not large.

As the batteries became smaller and the number of sales grew much larger so did the problems. SLI batteries were one tenth the size of electric car batteries and insignificant compared with the size of a powerhouse load-leveler installation, yet the overall output of these small cells soon dwarfed that of their larger-sized cousins. They were sold, as today, in small lots to such places as service stations and department stores, where the time spent sitting idle and neglected on shelves or in crates was serious in view of the technical problems mentioned above. This was an important problem for the SLI market, and it was compounded during the 1920s by the rapid growth of commercial radio. Early radio tubes could not be operated from house current and so batteries had to be used, with storage batteries being the favorite source for the filament heating current.

To solve the problem of charge-retention in these new marketing and distribution environments, battery engineers recognized early the value of so-called "dry-charged" batteries— cells which could be shipped to the distributor sealed and bone-dry inside, and which would be in a full state of charge

---

[22] Vinal, *Storage Batteries* (1955 edition) p. 49.

as soon as electrolyte was poured in. The major problem in developing such a battery lay with the wooden separators, and this spurred such inventions as the threaded rubber separators of Willard. The first commercial dry-charged cells made their appearance in 1922, although they never became fully practical until the introduction of improved synthetic resin separators after World War II.[23]

The separator, pre-SLI component in storage battery technology, became an important part of the technology with the appearance of the SLI. The same was true of another battery component—the "expander." During the 1890s, battery engineers began to work on the problem of the tendency of the spongy lead active material to "densify" during cycling. The evolution of the box negative was a mechanical solution to this problem. Various chemical solutions were also proposed in the '90s; among the most popular chemical additives to the anode mix were lamp black and barium sulphate.

Still, before the introduction of the SLI battery, little thought was given to these substances or to their mechanism of action—they were usually conceived simply as inert bulking agents in the lead sponge, thus, the materials chosen for expanders were often chemically inert—lampblack, pumice stone. SLI technology, however, strengthened the role of expanders. First, the chemical solution to densification problems became more important with SLI's, since the use of mechanical devices such as the large and expensive box grid was impractical for SLI's. Second, the reduction in active material porosity is particularly serious in SLI's, since they are often required to provide high current outputs at low temperatures, when the electrolyte is sluggish.[24]

The third and most interesting influence upon expander innovation was the feedback effect between advance in expander design and advance in separators. During the decade immediately following Kettering's invention, some battery manufacturers who had switched from wood to hard-rubber separators detected a tendency of their negative plates to lose capacity at a faster rate than had hitherto been the case with wood separators. Some thought that there was perhaps something in the lignin, which, even if leached out only in very small quan-

---

[23] *Radio News*, 11/1922: 893. The first year in the United States that sales of dry charged storage batteries exceeded those of the wet charged type was 1961. *The Storage Battery Manufacturing Industry*—1962 Yearbook.

[24] On this point, see, for example, *Trans. Am. Electrochem. Soc.* 92 (1947): 230.

tities, might make its way to the sponge lead and prevent densification. Therefore, shortly before 1920, work began on the use of organic expanders, and, specifically, on the addition of wood powders or extracts to the negative paste.[25]

One of the first to try such techniques was the man responsible for the most popular form of early non-wood separator—Theodore Willard. The Willard branch of ESB made a big selling point of its "threaded rubber" separators during the 1920s, but the use of these separators caused a capacity problem with the negatives.[26] In 1922 Willard patented a process in which a wood powder, preferably oak, was mixed with the negative paste. This admixture, to quote from the Willard patent:[27]

> . . . . increases very materially the capacity of the negative plates. . . . It increases the porosity of the plate, and in that respect has a beneficial action, and as a result of careful experiments I believe that it liberates an energizing substance which increases or perhaps more accurately keeps up the capacity of the negative plates.

Willard's statement implies that he was not thinking of his wood expander as a bulking agent. Nonetheless, his use of untreated fine sawdust in the negative paste indicates the crude level of this technology just before the start of the 1920s. During the succeeding decades, the research laboratories of battery firms sought to isolate and extract Willard's "energizing substance," so that it could be added to the plate without any unnecessary and perhaps deleterious wood chemicals. For its source of raw materials for these extracts the battery industry turned again to the paper industry, where the waste digester fluids contain large quantities of ligno-sulphonates. Thus, the waste products of one industry became valuable additives in another industry.

Dr. M. Barak has suggested another inter-industry link in the evolution of expander technology. He suggests that the spectacular growth of the detergent industry during the first half of this century, which focused the attention of industrial chemists on the surface-active powers of complex organosulphonates in small quantities, stimulated the battery engineers to conceive of the action of the ligno-sulphonates on the fine metallic lead crystals of the negative plate in the same

---

[25] Vinal, *Storage Batteries*, (1955 edition) p. 26.

[26] Arendt, *Storage Batteries*, pp. 129-130.

[27] U.S. patent #1,432,508—17 October 1922—filed 2 June 1917.

way. This led to research into surface-active expanders, and away from the older approach of simply looking for inert bulk-ing agents.[28]

This theory seems correct since most of the pre-SLI period expander patents relate either to the use of bulking agents, for example, cotton, wool, asbestos, pumice, lampblack, or to a few inorganic agents, such as barium sulphate.[29] Patents on the use of lignin-based derivatives of the paper-making pro-cess began to appear only in the 1930s, with interest increasing sharply in the '40s. Indeed, as one writer observed as late as 1947, no technical literature on the subject of expanders then existed, except for the patent literature.[30]

The following is a list of the United States patents on lignin-based expanders, as taken from the above-mentioned 1947 article. Those marked with an asterisk apply more strictly to the simple use of wood powder, rather than an extract of wood:

Year	Number of Patent	Year	Number of Patent
1924	1,505,990; 1,508,427*	1939	2,178,680
1931	1,817,846	1940	2,217,787; 2,217,786; 2,217,814
1933	1,940,714		
1935	2,022,482; 2,030,717	1941	2,233,281; 2,251,399; 2,253,247
1936	2,079,207; 2,079,208	1942	2,290,496*
1938	2,130,103; 2,130,104*; 2,130,105*	1943	2,325,542; 2,326,689; 2,326,689; 2,365,600; 2,365,604; 2,367,453

The absence of a really concerted effort to exploit the pos-sibilities of complex wood chemicals until the late 1930s and early '40s may seem strange considering the growing impor-tance of expanders since about 1912 and the recognition of the value of wood as a source of an expander substance since the end of the First World War. The answer to this paradox lies in the growth of the science-base to storage battery technology.

---

[28] M. Barak, "The Lead-Acid Storage Battery" (Paper delivered at B.C.I. Convention, San Francisco, 5/1973) p. 4.

[29] See: *J. Am. Electrochem. Soc.* 92 (1947): 228-257.

[30] Everett J. Ritchie, "Addition Agents for Negative Plates of Lead-Acid Storage Batteries," *Trans. Am. Electrochem. Soc.* 92 (1947): 228-257. See also: Ibid., pp. 281-303.

Until the 1930s, theoretical science and its methods played a minimal role in the activities of battery manufacturers. It is true that ever since the 1880s a number of chemistry professors had researched and debated the precise nature of the lead-acid system's reaction mechanisms. However, these studies had no influence on battery design. It is also true that all of the major battery manufacturers had laboratories connected with their works, but these early labs were for quality-control purposes, such as the routine analysis of incoming lead ingots and sulphuric acid, or for purely empirical studies, such as those of Bernard Drake. Early battery company chemists had a difficult enough time simply insuring that the lead ingots did not contain massive amounts of copper, zinc, and other impurities; they had neither the time nor the training to carry out theoretical studies of the physical chemistry of battery systems.

The first significant steps towards intergrating theoretical research and commercial storage battery technology took place during the early 1930s. For example, in the early '30s the first microscopic and X-ray diffraction studies of the mechanisms of crystal growth in the active material masses took place. These studies laid the foundations for the creation of a scientific model of expander action.[31]

The studies of crystal growth in the negative plate showed that the "densification" of the lead sponge is due not to the actual contraction of the entire spongy network, but rather to the preferential growth of large lead crystals at the expense of smaller ones; a phenomenon known generally to chemists as Ostwald ripening. Lead sponge is made up of a porous mass of different-sized crystals. In the course of time, the larger crystals grow by slowly absorbing material from the smaller ones. During the 1940s, researchers theorized that this phenomenon was due to the very slight solubility of lead sulphate in electrolyte.[32]

These researchers further theorized that the wood chemical agents act by being attracted to the surface of the crystals, thus stabilizing them and preventing further growth. They also found that earlier forms of expanders, such as barium sulphate, appeared to act by a different mechanism. The researchers based this model on the similar crystal structure of barium and lead sulphate; the barium is not a surface-active

[31] Personal communication from Dr. M. Barak.

[32] For a technical discussion see: S. Uno Falk & Alvin J. Salkind, *Alkaline Storage Batteries*, p. 44.

agent, but rather acts internally, like a seed crystal, and serves to act as a center for the formation of lead sulphate crystals. Since the barium sulphate is diffused throughout the mass of lead sponge, it counteracts the formation of preferential growth centers in existing large lead crystals.

Moreover, researchers in the 1940s were able to link the microscopic and X-ray diffraction studies of negative battery plates with a much older, empirically-derived tradition from the electroplating industry. In electroplating it had long been known that the addition of certain organic agents to the plating baths helps to reduce grain size and aids in the adhesion of the plate to the base metal. Since the theoretical studies of lead-acid battery operation had shown that most densification occurs during charging, Ritchie, in 1947, drew the analogy between the reduction of grain size at the electro-plating anode caused by organic agents, and the preservation of small grain size in storage battery anodes by the addition of wood-derived organics. Thus he was able to suggest a most likely path for future applied research into new expander materials.[33]

By the late '40s and early '50s, most lead-acid battery manufacturers were using a mix of expander materials in their negative pastes. This mix, which comprised about 1 percent of the total active material, consisted of barium sulphate, lampblack, and a lignin-derived organic materials.[34] Both barium sulphate and lampblack were already being used before the First World War, and the use of wood-based organics got its start accidentally and purely empirically around 1920. Nevertheless, as the list of patents on page 298 shows, the sophisticated elaboration of this technology only began with the influence of theoretical research. It was only with the beginning of this period that the idea of using sawdust as an expander was replaced by the concept of isolating the pertinent surface-active organic from the waste liquors of paper mills. During the 1940s and '50s chemists expanded upon this basic idea and identified whole classes of organic expanders.[35]

A further connection between the evolution of theoretical research by battery firms and the growing importance of the SLI battery is shown in the growing emphasis on organic expanders in contrast to the older, inorganic types. Empirical

---

[33] Ritchie, "Addition Agents," p. 237.
[34] Ibid., p. 250.
[35] Barak, "Lead-Acid," p. 4.

studies showed that expanders such as barium sulphate and lampblack had little influence on the capacity of batteries at low temperatures, whereas lignin-based expanders did have such an influence. Also, it was found that such influence varied with the kind of lignin used.[36] Low temperature capacity had not been of great importance in the days when most storage batteries were of the stationary, indoor variety, but with the introduction of SLI technology, cold capacity increased in significance.

Another result of the growing importance of SLI technology over the past sixty years has been its influence upon grid design. Here, however, there was a major difference between the changes in grid designs we have studied so far for the period 1880-1911. In that period, changes in the geometry of the grids were often dramatic. After 1911, however, no really major changes in SLI grid geometry took place; the geometrical changes which did take place were of a more subtle and evolutionary nature. The major changes in the past sixty years have occurred in the materials of the grids, rather than in their geometry.

Because of weight and size limitations, SLI batteries are made by the pasted, rather than the Planté method. As we have seen, ever since the early 1880s, it was standard practice among manufacturers to alloy the lead for the grids with antimony. The percentage of antimony used in most early grids was high by modern standards—often up to 12 or 13 percent. The reason for this is that lead and antimony form a eutectic at a 13 percent alloy, at which point the melting point of the metal is depressed over 150°C from the pure lead.[37] The value of this eutectic mixture was of increased importance after the mid 1880s, as grid designs became more delicate and complicated—the high antimony alloy has a lower viscosity than pure lead and flows readily into thin cross-section molds and produces sharp castings. Some manufacturers even added tin to the Pb-Sb alloy to aid fluidity.[38] Pure lead not only requires considerably higher temperatures for proper casting, but the coefficient of expansion is greater for pure lead than for the alloy—a factor which creates greater casting problems, especially with the delicate, thin-walled grid molds. Since the storage battery industry before the days of the SLI had usually

---

[36] *Trans. Am. Electrochem. Soc.* 98 (1951): 325-333.
[37] Vinal, *Storage Batteries* (1955 edition) p. 18.
[38] Ibid., p. 15.

operated at a low level of capitalization, with relatively simple machinery, the use of the eutectic mixture was a considerable convenience. Moreover, the antimony both stiffens the lead of the grids, and, in the positive electrode, helps the lead resist "formation."

Antimony, however, is by no means an unmixed blessing for storage batteries, for as it helps the positive plate, it hurts the negative. During normal charge-discharge cycling, antimony is slowly dissolved out of the surface of the positive grids and migrates to the negative, where it is electrodeposited in the lead sponge. This will set up local action in the negative, causing the battery to self-discharge. The older the battery becomes, the more antimony will be deposited in the negative, and the faster will be the rate of self-discharge. The process clearly feeds upon itself, since the need to keep the battery up to full charge speeds up the process of antimony deposition. This is one of the chief reasons why older batteries tend to go "dead" by themselves faster than new ones.

A second and more serious problem is the indirect influence this process has on the positive plate. The antimony contamination of the negative lowers the charging potential of that electrode. Thus, since the potentials of the two electrodes are additive, the charging potential of the entire battery falls also. But, because of the need to maintain simplicity in automobile equipment, the charging system of an automobile's SLI equipment operates on a *constant-voltage* principle, with the voltage being set just slightly above that of a fully-charged, *new* battery. Therefore, as the (new) battery reaches a state of full charge, its own EMF rises to meet that of the generator voltage. It will consequently resist further current from the generator and will not be overcharged.

But if antimony poisoning is slowly lowering the cell's overall potential, the cell will never fully resist the charging current, and although it may be fully charged, current will continue to flow through it from the generator. This will have the effect of slowly "forming" the positive grids and destroying them, which is one of the very things antimony addition is supposed to prevent.

Other problems caused by this phenomenon are the drying out of the cells by the continual decomposition of electrolyte water, and, of course, the nuisance of a "dead" battery.

Battery manufacturers began to recognize the seriousness of this problem for car batteries in the 1920s. A series of statistical studies of the causes of SLI battery failures, begun in the 1940s, revealed that positive grid "formation" caused by overcharging was by far the major source of failure.[39] In addition to these empirical studies, theoretical research into the electrochemical mechanism of antimony poisoning was begun in the early 1930s and intensively pursued throughout the decade. By the end of the decade, battery engineers had a good understanding of the basic physical chemistry of the problem.[40]

Here we see one of the ways in which SLI technology essentially reversed the degree of seriousness of certain earlier storage battery maintenance problems and thus helped to redirect the path of technological innovation. During the 1880s, Bernard Drake identified undercharging as the chief villain of battery durability, and he considered overcharging to be of minimal harm. By the mid twentieth century, however, overcharging had been recognized as a much more significant source of battery failure than undercharging.[41]

During the late nineteenth century, storage batteries were industrial tools designed to save fuel costs, but were labor intensive in their operation. With the introduction of SLI batteries as the major economic support of the industry, however, this earlier situation was reversed. Storage batteries became consumer goods, designed not to reduce any costs, but to be a convenience—that is, to *reduce* labor. The great care which had been lavished on the maintenance of the load-leveling cells was replaced by the habit of early car manufacturers of hiding the SLI battery under the floor boards or seat, where maintenance was almost impossible. Of course, the decline in strict battery maintenance had already begun with the electric car, but, because of the much greater cost of an electric vehicle battery as compared with an SLI, and since owners of such cars often had them stored and serviced at special garages, electric car cells were never as badly ignored as SLI's.

With SLI technology, the entire *raison d'être* of the system was the saving of labor; therefore, neglect of maintenance and the proper care of the batteries became almost a de facto part of the technology. Only during the 1930s, over twenty years after the invention of electric starting, did automobile manufacturers stop installing the batteries in inaccessible places and made

---

[39] Ibid., p. 303; Barak, "Lead-Acid," p. 2.

[40] See particularly the interesting discussion following the paper by Byfield: *Trans. Am. Electrochem. Soc.* 79 (1941): 259-268. The earliest study of the deposition of Sb in the lead sponge was by Crennell and Milligan, Ibid. 27 (1931): 103-113.

[41] For example, Vinal, *Storage Batteries* (1955 edition) p. 303.

it standard practice to install them under the hood, where they could be properly looked after.[42] The problem of excessive, chronic overcharging, which was easily prevented in powerhouse battery operation, became therefore the overriding problem with SLI's.

Overcharging in a load-leveling system wastes energy, and thus money, and will therefore be carefully controlled in a powerhouse. Moreover, its presence will be detected immediately, since it is accompanied by the bubbling and hissing of the gassing cells. In an SLI battery, on the other hand, the operator's counterincentive to proper maintenance insures that steps to prevent overcharging will not be taken.

To compound this problem, the self-discharge caused by antimony poisoning is a more serious problem in SLI than in load-leveling or electric car use. This is because SLI cells normally use little of their storage capacity, whereas load-levelers are continually cycled. Consequently, self-discharge represented a small and generally unnoticed part of the normal, daily charge-discharge cycle of such cells. SLI batteries, on the other hand, if allowed to stand for even a short period of time unused, could go "dead," or become at least seriously undercharged. The danger of long-term undercharging is over-sulphation and the risk of warping and buckling the plates.

Although the evolution of SLI technology helped to focus attention on the problems of self-discharge, overcharging, and antimony-poisoning, a concurrent change in another branch of electrical engineering practice had an equivalent influence on focusing attention on this problem. This change was the growth in the use of stationary batteries for emergency standby service and the decline in "cycling" stationary batteries, such as load-levelers. A standby battery is, like an SLI cell, one which normally does not "cycle"; consequently, the effect of this change in stationary battery usage was analogous to that of the SLI.

In the early days of electric power service, some batteries were used for emergency standby service in case of generator breakdown. The main function, however, was load-leveling. The importance of maintaining absolutely uninterrupted service was not so critical in those days; many of the earliest systems were designed to operate only during certain times of the

[42] Nels E. Hehner, *Storage Battery Manufacturing Manual II* (Largo Fla, IBMA, 1976) pp. 69-70.

day. By the turn of the century this situation was changing rapidly. Electrical engineers were developing many methods of improving load factors which did not rely on batteries.[43] The use of load-leveling batteries, with their elaborate and expensive end-cell switches, boosters, and so forth, had always been an expensive proposition; with the improvement of non-battery solutions to the load-factor problem, they became an unnecessary nuisance. As one storage battery authority observed in the mid 1920s:[44] "The necessity for the storage battery in central stations, as it existed some years ago, has practically vanished. . . ." And, of course, the almost universal use of A.C. by the mid 1920s, both for transmission and consumption, made the use of battery D.C. all the more difficult.

Yet just as the function of load-levelers diminished, the role of standby cells increased. Among the most important factors in easing the load-factor problem was the increased use of electricity for more functions—streetcars, appliances, stationary motors, daytime lighting. But this also meant that society was becoming increasingly dependent upon uninterrupted electric service; the necessity for maintaining absolutely constant power for subways or elevators can hardly be exaggerated.

During the first few decades of the twentieth century, therefore, storage battery installations were designed to carry the entire load of a distribution system in case of breakdown. Some big-city systems tried to maintain such standby installations as late as the 1930s, but, of course, the ever increasing size of the load made the scale of such batteries impractically large.[45] Also, the increasing reliability of generating equipment was matched by its increasing complexity, and, more to the point, the greatly increased complexity of the control equipment. In the case of emergencies, such complex control equipment had to act automatically. But if the emergency entails a loss of power, there would be no current to operate the automatic controls.

As a consequence, the trend in power stations during the course of the present century has been to replace standby

---

[43] These methods applied both to evening the external load over the 24 hours and to making the station's equipment better able to adapt itself to varying external loads, for example, the increased ability to operate boilers under forced draft, the use of turbines instead of steam engines and the ability of generators to tolerate greater overloads for longer periods. Vinal, *Storage Batteries* (1924 edition), p. 390.

[44] Ibid.

[45] Ibid.

load-carrying batteries with batteries which will supply power only to the control devices should generator power fail. In normal operation, these cells "float"[46] across the terminals of the control devices. In case of power failure, they provide power for the control devices without even an instantaneous break in the control's current. These functions are particularly crucial in nuclear power stations, since the station's own power is continually required to prevent the reactor from destroying itself. Nuclear station storage battery setups are for this reason much more elaborate than those in conventional powerhouses, and this trend is likely to continue.[47]

The most important market for modern stationary batteries, however, is not power stations, but telephone switching stations.[48] Storage batteries began to find their way into communications use in the late 1880s, when some of the larger telegraphic stations abandoned their cumbersome and expensive regiments of primary cells in favor of dynamos. These were cheaper, but they could not be used to send signals directly on the line because of the large amount of noise in their output. The current from the dynamo was therefore filtered through lead-acid cells in a manner similar to the way in which the filaments of early incandescent bulbs were protected from generator fluctuations.

By the later 1890s, the replacement of primary batteries by storage batteries was becoming general throughout the telegraphic industry. Since the power needs of telegraphy and telephony are similar, the technology of using dynamos and batteries was transferred to the latter industry also. In the early telephone systems, however, storage batteries were not used, since the power came from small dry cells or wet Leclanché batteries contained in each subscriber's telephone box. This system was inefficient, however, and during the 1890s it was replaced by the Common Battery System, in which each phone was powered in common from a large battery in the central switchboard. At first, these too were equipped with dry cells or Leclanché's.

The large numbers of dry cells consumed in common battery operation was an expensive proposition for the phone companies, particularly considering that turn of the century dry

[46] "Float" service is that in which the battery neither charges nor discharges, but which is maintained at a constant rate of charge by a continuously applied external voltage.

[47] G. Smith, *Storage Batteries* (London, Pitman, 1964) pp 139-140.

[48] Ibid., p. 143.

cells had much lower capacity and durability than modern dry cells. Moreover, due to technical problems connected with the circuitry of switchboards, any power source which has significant internal electrical resistance will induce cross-talk between the various circuits connected to the same source. Dry cells, particularly if they polarize rapidly, have such significant resistance.

As a result of these problems, generator and storage battery current began to replace dry cells in telephone offices around the turn of the century. And despite the increasing reliability of generator operation since then, telephone offices have become ever more dependent upon storage batteries during the past 70 years. This is because the increasing complexity and automation of telephone switching systems has resulted in an increasing sensitivity to current fluctuation, so that even a momentary interruption today would blow fuses throughout the system and snarl the dial impulse circuits.[49] Standby storage cells on "float" service guard against this.

Not surprisingly, it was a telephone company which first suggested eliminating some of the storage battery's problems by doing away with antimony. In 1935 two researchers at Bell Laboratories recommended the replacement of antimony by calcium in battery grids.[50] They found that calcium, even if added in percentages of only .1 percent reduced the rate of self-discharge by a factor of ten over that of standard 9 percent antimony grid batteries. Two other Bell Labs researchers, working at the same time, investigated the difficult problem of casting the lead-calcium grids.[51]

Beginning in the late 1930s, a number of battery manufacturers began producing lead-calcium grid cells for the Bell System, and Bell standardized its own operations on this type of battery. It is interesting to note, however, that the development work in the lead-calcium batteries was undertaken outside of the battery industry. Moreover, it is clear that the battery manufacturers undertook the lead-calcium manufacture reluctantly and only because a major customer had demanded it. There were a number of reasons for this.

From the consumer's standpoint, lead-calcium grid batteries were desirable, but from the manufacturer's perspective they presented numerous headaches. As indicated earlier, lead-acid

---

[49] Ibid.

[50] *Trans. Am. Electrochem. Soc.* 68 (1935): 293-307.

[51] Ibid., pp. 309-319.

battery manufacture was a non-capital intensive business. The grid-casting, for example, was performed in simple, gravity-feed molds, with the molten metal fed from an open pot near the side of the casting machine. The antimony-lead alloy was ideal for such operation, partly for reasons already discussed, but also because the percentage of antimony in the metal did not need to be controlled between narrow limits. In a 5 percent Sb alloy, for instance, the percentage of Sb may vary up to 10 percent.[52]

The control of calcium alloying, on the other hand, is a much more complicated business; the percentage of calcium which is used in the alloy is small, and must be controlled within narrow limits—neither below .07 percent nor above .1 percent.[53] Variation outside these limits results in low battery life. To compound this problem, the maintenance of this percentage of calcium in the melt is made difficult by the readiness with which calcium oxidizes in air. Consequently, the usual open-topped melt pots, in which a layer of dross is allowed to form on top of the molten metal, present problems. The melting point of the lead is not reduced nearly as much as with the large percent antimony alloy, and the setting range of the calcium alloy is much sharper.[54]

Even though the Bell system required that all their batteries be made with calcium alloy grids, the battery industry made no attempt to use calcium in the grids of its SLI batteries. It has only been in recent years that the industry has invested in the research and development necessary to produce a calcium grid SLI battery—the so-called "lifetime," "maintenance-free," or "totally-sealed" batteries. The reluctance of manufacturers to abandon the traditional high antimony grids was due in part to the factors mentioned above, but was also due to another design objective which conflicted with that of lower antimony alloys—thinner plates.

The development of thinner electrodes had been an objective of battery designers ever since the appearance of the electric car. Such plates represented a savings in size and weight, or, for the same overall size battery, an increased amperage. At the time of the First World War, SLI grids were in the average range of 0.188 to 0.225 inches thickness. By the early 1920s, it was commercially possible to cast grids down to a thickness of

---

[52] Between 4.75 and 5.25%. Hehner, *Manufacturing Manual*, p. 4.

[53] *Trans. Am. Electrochem. Soc.* 92 (1947): 313.

[54] Hehner, *Manufacturing Manual*, pp. 72-75.

0.125 inch.[55] By the Second World War, SLI cells were being made with grids of 0.090 inches.[56] Today, some plates are cast down to 0.040 inches, although 0.070 to 0.080 inches is average.[57]

The movement towards thinner plates has received a number of different stimuli during the past seventy years. The two world wars were important, particularly the second. During World War I a number of SLI batteries were developed for aircraft which had plates in the 0.03 to 0.05 inch range. For wartime service, where cost was no issue, these were successful, but they proved to be commercial failures. Renewed attempts were made to revive these batteries for commercial production during the 1920s, but again they failed.[58] It was during the Second World War that success was achieved with very thin plate cells. For example, the Germans put a priority on the development of electric-drive torpedoes; the standard, compressed air-drive torpedo left a very noticeable wake. The German success in producing large numbers of electric torpedoes during the war was largely due to AFA's development of very thin plate cells, since these could provide the high amperage in a small space.[59]

The short life of many of the early thin plate batteries led to the common belief in the 1920s that plate thinness necessarily means low durability.[60] This attitude changed during the '30s and '40s as the first thin but durable plates were developed. The industrial intelligence reports on German industry which followed the war, describing what had been done in the way of successful thin grid development, helped stimulate post-war research in this area in other countries.[61] Today, in contrast to the attitude of many battery designers of the 1920s, the use of thinner plates is credited with *increasing* the life of batteries, since the additional surface area of the greater number of plates creates less strain per unit area during charging.[62]

The evolution of thin grid technology was primarily dependent upon the development of suitable casting machinery. For example, one of the factors which allows the mass casting of

---

[55] *Trans. Am. Electrochem. Soc.* 39 (1922): 497-505.

[56] Private communication from Dr. M. Barak.

[57] Barak, "Lead-Acid"; Hehner, *Manufacturing Manual*, p. 7.

[58] *Trans. Am. Electrochem. Soc.* 39 (1922): 497-505.

[59] Private communication from Dr. M. Barak; BIOS Report #1129 (1947).

[60] *Trans. Am. Electrochem. Soc.* 39 (1922): 498.

[61] Private communication from Dr. M. Barak.

[62] Hehner, *Manufacturing Manual*, p. 70.

thin SLI grids today is the construction of grid molds with electric heating coils built in. These coils keep the metal at proper fluidity as it fills the delicate mold channels, where a cool mold would very rapidly conduct heat away from the molten metal. The coils are located at the opposite end of the mold from the sprue; their purpose is to insure that the metal which flows to the farthest part of the mold cools and solidifies at the same rate as that which fills the portion right by the sprue.[63] Grid casting machines built a half century ago did not have such features. Consequently, when an attempt was made in the 1920s to make grids of less than 0.125 inches in thickness, they had to be formed by punching rather than casting techniques.

Today, most SLI grids are produced by highly automated, capital intensive casting machinery. The old, labor-intensive, capital non-intensive manufacturing methods have been gradually replaced over the course of the past sixty years in the SLI industry by processes in which the casting, pasting, curing, drying, of the plates is performed and controlled by automation, with little human labor apart from supervision. The manufacture of stationary batteries, on the other hand, or those for submarines, or for other relatively small-scale markets, is still carried on with a considerable amount of hand labor and control, much as it was done a half century ago. The automation of SLI manufacturing was made possible by the enormous scale of this sector of the battery market; in general, the automation of SLI production has reflected the automation of the manufacture of most automobile parts.

The scale of SLI manufacture has made it feasible for battery producers to invest in very expensive automated machinery. This, in turn, has provided the economic incentive for a number of process equipment manufacturing companies to specialize exclusively in the design of storage battery manufacturing equipment. Until recently, these companies were all American; the lack of mass production of automobiles in Europe prevented any machinery firms there from specializing in battery-making machines. American battery-making machines were always better than the European and were, until recently, largely made only in this country.[64]

The equipment designers were therefore able to specialize in the design of the complex and expensive features necessary to cast very thin grids quickly and with precision. Thus, a feed-

---

[63] Ibid., pp. 8-9.
[64] Private communication from Mr. Nels Hehner and Mr. Edgar Oldham.

back relationship existed between the evolution of thin plate SLI batteries and the automation of that technology. Increasing demand for SLI batteries has encouraged manufacturers to invest in increasingly more expensive and elaborate production machinery, which has permitted the manufacture of thinner grids on an inexpensive per unit basis. The plates of stationary batteries remain today relatively thick; they neither need to be thin, nor could the economics of their market permit them to be made that way. For SLI's the situation is, fortunately, the exact reverse.

Beneficial though the reduction in grid thickness was, however, it discouraged attempts to reduce the antimony content in the grid alloy. Although antimony poisoning was known as a serious limiting factor in battery durability since the late '20s, the excellent casting properties of antimony alloy kept its content in storage batteries high until the 1940s. At that point, however, the needs of World War II had the first serious effect in reducing antimony content.

Until World War II, the percentage of antimony in SLI plates varied between 5 and 12 percent, with the majority of manufacturers probably using about 9 percent.[65] In all the countries involved in the war, however, planning committees feared shortages of antimony. Although no actual shortage ever materialized in the United States, the War Production Board limited American battery producers to a 7 percent antimony alloy for the grids. It is an illustration of the poorly developed state of grid casting technology at that time that some American manufacturers found they could not cast proper grids with this alloy and obtained special government permission to use more antimony.[66] Today, 7 percent antimony would be considered a high concentration by all manufacturers.

In addition to antimony supply problems, a number of the traditional technical problems of antimony use were aggravated by various kinds of war service. For example, one of the serious battery problems on submarines is the evolution of hydrogen gas during charging. Antimony poisoning of the negatives, by lowering the hydrogen overvoltage of the plate, increases hydrogen evolution and thereby greatly increases the danger of explosion. Wartime submarine battery development thus put a premium on eliminating antimony, so that by the end of the war some submarine cells were made with as low as 2 percent antimony.[67]

---

[65] Vinal, *Storage Batteries* (1940 edition) p. 14.

[66] Hehner, *Manufacturing Manual*, p. 3.

[67] Private communication from Dr. M. Barak.

By the end of the war, SLI grids with 6-7 percent antimony were fairly standard. Stimulated by the wartime progress, battery manufacturers during the post-war period continued the development of lower antimony alloys. More fundamentally, this continued effort was a reflection of the great post-war investment in research and development. In contrast to the pre-war failure of manufacturers to develop alternatives to high antimony alloys, a number of alternatives appeared after the war.

ESB invented a grid alloy called Silvium, which was a silver-arsenic-lead mixture. It had a brief vogue in the 1950s, but was dropped because of cost.[68] Other manufacturers tried reducing the antimony content to 4 percent, or even 2 percent, while making up for the antimony with arsenic, tellurium, or cadmium.[69] The so-called "lifetime" batteries which have become popular during the past decade have been made possible by the use of a new ternary alloy of lead-calcium-tin. The tin makes the grid stronger and aids in molding. The fluidity of the Pb-Ca-Sn alloy is crucial, since it aids in speeding up the rate of casting by modern machinery—without the complexity of such machines it would be very difficult to cast the new alloy, yet the complexity is expensive, and only the high volume of castings produced makes their expense tolerable.

A second major change in battery materials stimulated by World War II was the movement toward synthetic materials for separators, outer cases, and the grids themselves. Before the war, separators were made of wood or natural rubber. Rubber, of course, was very important during the war, and was tightly rationed. Since the most popular separator woods came into short supply, particularly in Europe, a number of attempts were made to develop microporous separators with plastics.[70]

Despite this work, no practical synthetic separators were produced commercially until late 1950s. This was chiefly because of the initial high cost of acid-resistant plastics such as polyvinylchloride (PVC).[71] But when synthetic resins were reduced in price, they freed the storage battery engineers from the technical restraints of half a century. The wood separator, with all the complex and tricky technology that it required, was retired in the early 1960s. It was replaced with cheap separators made from PVC and blotting paper. These plastic-

[68] Ibid.

[69] Vinal, *Storage Batteries,* (1955 edition) p. 20.

[70] Ibid., p. 59.

[71] Private communication from Dr. M. Barak.

paper separators essentially reversed the technology of the wood separator—in the wood, one began with a complex, non-porous object and treated it to make it simpler and more porous. With paper, the object was to convert a radically-altered form of wood back into a more complex, more coherent form. A PVC-solvent mixture is used to saturate the blotting paper. This is then stamped to form the ribs and the solvent evaporated out to form a micro-porous structure.

In other techniques, PVC powder is sintered to produce the microporous structure, or it may be cast with a finely-divided filler, such as starch, which may then be dissolved out to create microporosity.[72] More recent developments of the 1970s have stressed the use of non-chloride containing plastics, for example, polyolefins instead of the vinyl chlorides.[73] The advantage here is economic and is related to the efficient operation of high automation industry; much of the raw material for SLI batteries today comes from recycled cells and it is costly if each old cell has to be disassembled before being melted down. Yet the chlorine in PVC is an undesirable contaminant in lead and the separators must therefore be separated from the plates before melting.

Plastic separators have had a number of fundamental influences on the technology of the storage battery. The rapid growth in production of dry-charged SLI's has been one result; plastic separators have cheapened the dry-charge technique. Another influence has been the reduction in separator thickness. The old wood separators were no thinner than 1 mm, whereas the paper and plastic separators can be made down to 0.5 mm, and the more recent plastic film separators are made down to 0.1 mm thickness.[74] This has permitted SLI manufacturers to increase the number of plates in the same sized battery box, and thus to increase battery life.

The post-war development of plastics has had one final influence on storage battery design; this concerns the plates themselves. The strength and oxidation-resistance of many plastics have been used to augment the durability of batteries in ways that were impossible before the war. For example, in some modern paste mixes, fine fibers of Dynel are added to either the positive or negative paste to increase the cohesion of the mass. In effect, this is an attempt to do on a macro scale what is going on at the same time in the paste on a microscopic

---

[72] Barak, "Lead-Acid," p. 4.

[73] Y. Miyake & A. Kozawa (eds.) *Rechargeable Batteries in Japan* (J.E.C. Press, 1977).

[74] Ibid., p. 50.

scale; as the PbO of the paste is converted to $PbSO_4$ during "setting," the lead sulphate crystals assume a filament-like structure themselves. As we have seen earlier, the idea of adding threads of fibers to the paste is as old as the pasted plate itself, but the designers had never before had materials possessing the necessary properties to provide good cohesion over long periods.

Plastics have also allowed manufacturers to make the first major changes in the design of tubular electrode plates since the invention of the Ironclad Exide in 1911. Although the Exide slotted hard rubber tube design was copied by several European, Japanese, and other storage battery companies during the second and third decades of the century, the technology remained essentially unaltered until the mid 1950s. Then, in 1953, ESB introduced the first microporous synthetic tubular positive—fiber glass threads were braided to form square cross-section tubes. These tubes were then given rigidity and stability by soaking them in polyethylene solution and evaporating out the solvent. The figure below shows the present day ESB design of this plate: [75]

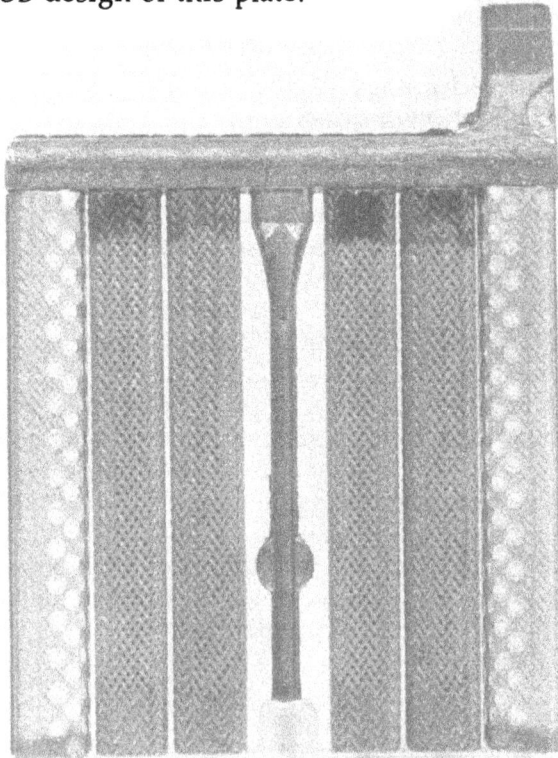

---

[75] Photo by author; sample grid courtesy of ESB.

The idea of braided fiber tubes was as old as the concept of the tubular positive itself; ESB pioneered the concept at the beginning of the 1890s. Without plastics, however, the idea could not be reduced to practice. The modern concept of the braided tubular electrode appears to have come from the Italian Tudor Company; one of the many spinoff firms of AFA. In the early 1950s, an engineer at the Italian firm patented the idea of woven tubes made of plastic fibers; his inspiration came from the braided weave pattern of military webbing belts.[76] This patent was sold to Chloride, which, of course, entitled ESB to use it.

Meanwhile, Chloride went ahead with its own development program. Instead of ESB's fiber glass and plastic mix, Chloride developed a simpler technique: two sheets of woven or felted Dacron are pressed against a set of parallel, cylindrical mandrels, and a multiple-needle sewing machine sews the two sheets together between the mandrels. The mandrels are removed, lead alloy support bars are inserted to take their place, and the annular space is filled with active material paste:[77]

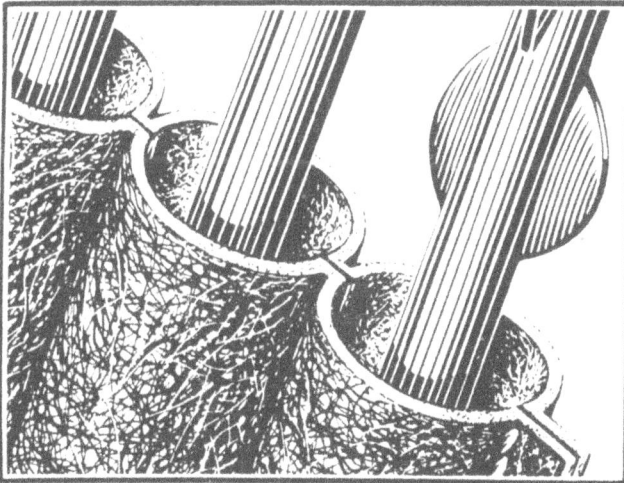

This technique was developed during the early '60s. Both the ESB and Chloride processes continued to be used, as well as a number of variant patterns developed by other companies. Much development work is still being carried on with this

---

[76] Private communication from Dr. Eric Sundberg.

[77] Barak, "Lead-Acid," fig. 11.

system as new plastics and weaving-braiding techniques are invented. The replacement of the old slotted rubber tubes by the microporous plastic ones has produced batteries with greatly increased electrolyte permeability and an energy density about 40 percent greater than the older cells.[78] The life of such batteries is between 3 to 5 times greater than contemporary SLI batteries, in which flat, pasted double-grids are used for both electrodes.[79] Because of their expense, however, they are limited to electric vehicle use, but they will probably make up an increasingly larger share of total battery production in the years ahead as gasoline prices continue to rise.

One final aspect of the materials revolution in SLI technology remains to be discussed. This has to do with the raw materials used in preparing the active pastes. We have indicated that during the 1880s battery manufacturers discovered that two lead oxides were satisfactory for the production of coherent pasted plates—litharge (PbO) and red lead (Pb_3O_4). When mixed with sulphuric acid, these oxides form the sulphate and thus become hard and coherent masses, not unlike plaster of Paris. Since the red leads takes a less rigid "set" than litharge, but it is easier to "form" than the other oxide, a mix of the two materials was often used, with different mixes used for different purposes.

The source of these materials was the lead and lead pigments industry. Even though the battery manufacturers bought metallic lead for their grids and could therefore have manufactured the oxides themselves, they chose to purchase them ready made; there is no record of a pre-SLI period manufacturer making his own oxides. The reason for this is connected with the relatively small scale and capital non-intensive nature of the pre-SLI battery industry; manufacture of the oxides required large reverberatory furnaces and skilled workmen. It would not have paid the early battery makers to have made their own oxides.

The growth of the SLI industry, however, made it feasible for battery manufacturers to begin thinking in terms of cost savings through vertical integration. The initial stimulus for this came from the Japanese storage battery industry with the invention of the Shimazu ball-mill process for oxide manufacture.

---

[78] Ibid., p. 7.
[79] Miyake & Kozawa, *Rechargeable*, p. 62.

Genzo Shimazu was the creator of the Japanese storage battery industry. He was an industrialist involved in a number of early aspects of electric power technology. Shimazu had been interested in the use of batteries for electric lighting use since the 1890s, and in 1910 he established the first commercial manufacture of lead-acid batteries in Japan. This manufacture was at first carried out in a section of one of his existing factories, in much the same way that ESB and Chloride began.[80] In 1917 this operation split off as an independent company; the Japan Storage Battery Company.

Initially, Shimazu produced a Planté-type battery based on AFA technology; this was logical given the company's original orientation towards stationary batteries. During World War I, however, Shimazu began developing a pasted grid cell. He investigated existing French technology, but decided against it on economic grounds. Nevertheless, the First World War provided the incentive for Japanese innovation in storage battery technology, since shipments from European producers were cut off at the time when Japanese demand increased rapidly. Two more Japanese battery companies were formed during the war—Yuasa and Furukawa Battery Companies.[81]

Shimazu developed a technique whereby litharge was manufactured less expensively than by the traditional methods. The savings were achieved by cutting down greatly on the fuel and skilled labor required. The process was essentially a small modification of the well-known ball milling technique, in which steel balls are rolled inside a rotation drum along with the substance to be ground. In the Shimazu process, the balls were made of lead, and it was they themselves which were ground down by mutual impact. A current of heated air was passed over the balls as they rolled; when minute lead particles were ground from them, they were immediately oxidized and carried from the drum by the hot air stream. Outside the drum, the lead dust was collected by standard cyclone-type dust collectors.

The beauty of this process is that it consumes little energy compared with the older reverberatory furnaces, and the process equipment used—air blowers, rotating ball mills, cyclonic dust collectors—was all standard industrial process equip-

---

[80] Ibid., p. 26.
[81] Ibid.

ment. Nothing new had to be built; the engineer simply assembled standard, commercially-available components.
Shimazu applied for a Japanese patent in 1920 and shortly
thereafter in the United States and Europe.[82] By the mid 1920s
battery companies in these countries were beginning to take
notice of the technique. In 1924, for example, the famous German chemist Fritz Haber visited Shimazu and helped to get
out the German patent soon after. It is impossible to say precisely when the various European and American firms began
shifting to commercial manufacture of the new oxides, but the
years 1930-35 were the major transitional period.[83]

It is interesting to note that this major change in technology
occurred at a time when one would not ordinarily expect much
innovation activity and investment—at the depth of the Great
Depression. The reason for this was economic—ball-milled
oxides were cheaper than the older litharge and red lead. There
were a number of reasons for this; it was a less expensive
process to carry out, and its use involved a less expensive
materials shipping system. For the old oxides, the lead pigments makers shipped the materials to the battery firms in
expensive oak barrels, the lead ingots, however, could be
shipped on cheaper pallets.[84]

The battery manufacturers themselves were not the only
ones interested in the new oxidation technology. The lead and
pigments producers also studied the economics of the system
and were consequently led to adopt the method. In several
instances, a battery manufacturer's adoption of the new oxides
was a two-stage process: he was first offered the cheaper material from his regular supplier of lead and lead oxides, but then
later decided to buy a ball-mill plant himself. Indeed, the lead
producers often sold the manufacturing equipment for oxides
to the battery companies.[85] Some battery firms, on the other
hand, continued to buy the new oxides from the lead companies and did not set up their own plants.[86] By the 1970s
about 75 percent of all United States-made battery oxide was
made by the battery firms themselves.[87] Normally, the size of

---

[82] U.S. patents: #1,584,150—11 May 1926; #1,985,465—25 December 1934.

[83] Nels E. Hehner, *Lead Oxides* (Largo Fla, IBMA, 1974) p. 5.

[84] Private communication from Mr. Nels Hehner.

[85] Hehner, *Oxides*, p. 6.

[86] Private communication from Mr. Edgar Oldham.

[87] Hehner, *Oxides*, p. 6.

the battery manufacturing operation will determine if the manufacturer of batteries makes or buys his oxide—small size favors the latter option.

The success of the Shimazu process encouraged the development of similar techniques. The most popular of the alternatives to the ball mill method was the Barton process. Beginning in 1899, George Barton in England suggested making lead oxide by stirring molten lead in a pot and blowing air through it.[88] The process was not originally designed for the battery industry, but it was nonetheless picked up in the 1930s and soon made up about an equal share of oxide manufacture with the ball mill technique. In the modern Barton process, molten lead is squirted from nozzles as a fine spray into a chamber containing rapidly whirling blades. The blades further disperse the lead droplets, which are quickly oxidized and solidify as a fine dust.

The price of the new oxides was low, but technologically, they presented a serious problem for the industry—they were not what the battery makers were used to. The older oxides—PbO or Pb_3O_4—were 100 percent oxide; there was no admixture of metallic lead. Indeed, most early manufacturers specified that this be so, since they did not believe that proper plates could be made with high metallic content litharge or red lead.[89] Because of the low temperature at which the new methods were carried out, however, some of the lead always remains unconverted. Therefore, modern oxides are between 15-35 percent metallic lead depending on the process used. In battery industry jargon, these are called "leady" or "gray" oxides; the oxide is always a form of litharge (PbO), since these processes cannot produce red lead.

The problem with the leady oxides was that active material pastes made from them would not "set" the same way that pure oxide pastes did. With the older oxides it had been possible to paste the grids and then lay them aside to dry under ambient conditions; the setting would be completed by the time the plate was dry. This was not the case with the leady oxides; if pasted grids were simply allowed to dry, the material would shrink, crack, and not adhere to the grid bars.[90] Thus, a

---

[88] U.S. patents: #633,533—19 September 1899; #988,963—11 April 1911; #1,060,153—29 April 1913.

[89] Hehner, *Oxides*, p. 5.

[90] Private communication from Mr. Edgar Oldham.

way had to be found to control the rate of drying of the paste. This led to extensive work during the 1930s on the development of factory-floor techniques for the proper setting of leady oxides.

This work was not connected with any theoretical or experimental research such as the battery companies were beginning to conduct at this time; the production engineers knew very well what the problem was and were confronted with a question of production economics. The amount of metallic lead in the paste had to be reduced to no more than 5 percent by the time the plate dries out, or the high percentage of metallic lead would interfere with the formation of the coherent, fibrous structure of the paste as PbO is converted to $PbSO_4$ by the sulphuric acid. Since the oxidation of the lead to PbO is relatively slow, the drying of the paste must be slowed. The question for the process engineer was not whether this could be done, but how to do it without adding greater costs than the technique was worth.

To compound this problem, the leady oxides were introduced to the battery industry at the same time that it was beginning its conversion to automated manufacturing methods. For example, the American SLI industry began introducing automatic grid casting and pasting machines in the mid 1920s, with the Europeans adopting the American machines and methods from about 1930 onward.[91] Therefore, at just the time that the industry was coming to rely more on the speed of machines and was adapting its operations to the replacement of human control over its products, it was introducing a new raw material which slowed down part of the process and added quality control difficulties as well. The case history of one company's adaptation to the leady oxides will demonstrate these problems.

Oldham & Son is a British manufacturer of lead-acid batteries.[92] Like the Shimazu company in Japan, Oldham's entered the battery business during the First World War, having diversified into this new field from its prior involvement in the manufacture of battery-powered lamps. Oldham's entered the SLI and home radio battery market in the early 1920s, and began introducing mass production equipment in the early

---

[91] Hehner, *Oxides*, pp. 5-6.

[92] *Vintage—Centenary Issue of the Grid* (House Organ, Oldham & Son) (Manchester, Oldham & Son, 1965).

'30s; in 1932 they purchased their first grid pasting machine and in 1935 the first grid caster.[93] At just this time (1934), they began the changeover to the leady oxides.

At first, the Oldham engineers tried to speed up the setting time of the new oxides by some rather heroic methods. The freshly pasted plates were placed in giant pressure cylinders and subjected to super-heated steam at 90 pounds pressure. This technique, however, proved to be unreliable, somewhat dangerous (from the steam), and counter-productive from the point of view of introducing automation into battery manufacture, since it was labor intensive.

Next, the company adopted a more moderate approach, which involved placing the plates in stoves at 100 percent humidity and 140°F. When this didn't work because it was too difficult to maintain proper humidity at this temperature, the company tried an even more moderate approach: cartloads of plates were simply wheeled into rooms whose walls and floors were kept wet. This technique worked and it is basically the same method used today throughout the industry—the plates are kept at 70-90 percent humidity and 80-90°F for about 48 hours and are then allowed to dry at ambient conditions.

Oldham's use of large, wet rooms reduced costs, but the method was still labor-intensive, since the fresh plates had to be individually stacked on the curing carts (this involved six men) but the casting and pasting machines reduced labor to a minimum. The solution to this problem was "flash drying"— as the freshly pasted grids leave the pasting machine they are given a quick, very hot drying, which dries a superficial layer at the surface of the plate. Flash-dried plates will not stick to each other and consequently they can be simply piled one atop the other. Oldham's found that they could reduce both labor time and stacking space in the curing rooms by this technique.

The critical factor in this technique is proper adjustment of the flash drying; too much and the plates crack, too little and they stick together. The design of flash driers has therefore been of considerable importance in the maintenance of efficient mass production. Moreover, as thinner grids were developed in the 1940s and '50s, the problems of proper flash drying were increased.

The combination of flash drying and delayed setting of leady oxide plates is known as the "hydroset" process. The details of

---

[93] Private communication from Mr. Edgar Oldham.

this process were evolved during the 1930s by the various American and European firms which adopted the new oxides; the American firms appear to have been in the lead here, transferring their ideas to the Europeans. The hydroset techniques have been almost universal throughout the industry since the '40s. Once standard techniques were developed, it was found that the new oxides actually yielded more durable plates than those made with 100 percent litharge. As the wet plates cure, oxidation of remaining metallic lead to PbO raises the temperature of the mass, since the reaction and drying of the plates occurred at the same time. As the lead is oxidized, it expands and fills some of the remaining void spaces, making the mass more coherent. This does not happen with pure litharge.

V. The Alkaline Storage Battery

1. Origins

The lead-acid storage battery dwarfs all competitive energy storage systems economically, and there are a number of basic reasons for this. The lead-acid's materials of construction are relatively cheap and the methods of manufacture are uncomplicated and easily automated. Moreover, the physical nature of the lead-acid system lends itself to a spectrum of manufacturing methods. Durable and efficient lead-acids can be made and the setting and curing properties of the lead pastes allow SLI batteries to be made in simple and inexpensive configurations.

No other electrochemical system is so accommodating to the battery engineer. Every alternative to lead-acid has some property which either raises raw material costs or which requires that the design and manufacturing techniques be complicated and expensive. Gaston Planté was the first to observe this and no one to this day has proven him wrong. Still, the lead-acid system is far from ideal. It's most obvious drawback is weight. Less obvious, although more significant for stationary uses, are the system's durability limitations and its inability to hold a charge as well as some alternative systems. Ever since the 1870s, inventors have sought to solve these problems by using some alternative to the $Pb-H_2SO_4-PbO_2$ system. Most of these efforts have failed, and the inventors have either given up or turned back to lead-acids as the only practical method.

The reasons for these failures are multiple. Some substances, like silver, have excellent electrochemical properties, but are too expensive for most battery applications. Other metals, like zinc, possess excellent energy-to-weight ratios, but they dissolve into the electrolyte on discharge, and no one has yet found an acceptable way to redeposit the zinc on charge. Ever since the beginning of battery history, zinc has been the favorite metal for *primary* battery anodes; many storage battery engineers over the years have experimented with a $Zn-H_2SO_4-PbO_2$ system. This system makes an excellent primary battery, but it will not recharge well. Still other materials, like manganese and nickel, have oxides with very high electrical resistance, and therefore batteries made from these materials need a complicated and expensive structure to provide internal conductivity.

Only one approach to the problem of designing a non-lead-acid system has so far succeeded. This is to use electrodes which are insoluble in the electrolyte and to use an electrolyte which does not vary in concentration during cycling.

The alkaline storage battery had its origins in the design of alkaline primary batteries in the 1870s. The 1870s marked a crucial watershed in the evolution of primary battery technology. Before this time, primary electric batteries were essentially a *capital good*—that is, they were used by industry as tools in the production of consumer goods or services. The industries in which the cells were used were electroplating and telegraphy. In addition, the batteries were used as tools in scientific research. In both of these applications, the batteries were handled by professionals who looked upon these devices as tools of their trade and who therefore used them properly and with skill.

In the 1870s, however, the primary battery evolved into a *consumer good*. At first only for doorbell and annunciator operation, but then for such things as telephones, burglar alarms, fans, and sewing machines, the primary battery found an increasing market as a household device. As dynamos made their way into electroplating factories and telegraphic offices, the primary battery ceased to be a tool for professionals, and evolved into a labor-saving convenience for home or office.

This evolution caused a major change in primary battery design and manufacture. The older forms of primary cell were not well suited for a consumer market. One problem was in the simplicity of the cells—to reduce costs, some of the early primary batteries were reduced to the absolute minimum of physical complexity. Such was the extremely popular gravity Daniell cell, which, except for its glass jar, contained nothing but two simple copper and zinc electrodes and the copper and zinc sulphate solutions. This simplicity not only reduced initial cost, but also simplified maintenance steps and permitted easy replacement of components. In contrast with primary batteries of the present day, in which the entire assembly must be thrown away if one or more components are exhausted, gravity cells were refreshed by simply adding more copper sulphate crystals and a new zinc anode rod. The anode was hung on the glass jar with wire hooks. The simplicity also reduced manufacturing costs—a perfectly respectable gravity cell could be made in a kitchen.

Simple battery designs, however, usually lack durability. For example, the chemicals in gravity cells self-discharge within days if the cells are left unattended. This was no problem for the professional telegrapher, since he used the cells continuously and maintained them carefully. For the consumer battery, however, this was a contradiction, since it was purchased as a labor-saving device and was used only occasionally. The new consumer battery manufacturers had to build complexity into their cells. As a result, they also had to add additional costs to the batteries.

One of the ways in which both complexity and cost rose was in the choice of the depolarizing system. Most of the early primary cells employed a liquid depolarizer: nitric acid in the Grove cell; aqueous copper sulphate in the Daniell; aqueous chromates in the Fuller cells. Liquid depolarizers lend themselves to simplicity of construction if they are used in constantly-maintained batteries. Since they are highly dissociated in solution, these depolarizers provide excellent electrical conductivity within the cells, eliminating the need to add special means of improving electrode conductivity.

In a consumer battery, however, liquid depolarizers are unsuitable. On open-circuit the ions diffuse to the anode and cause self-discharge. This can be retarded by using a diaphragm between anode and cathode. The problem with this solution, however, is that a diaphragm resistant enough to ion flow to give a battery long shelf-life would also have impractically high electrical resistance. No modern consumer batteries use liquid depolarizers.

Liquid depolarization can be replaced with mechanical depolarization. For example, the cathode can be constantly vibrated, thus shaking loose the hydrogen, or the surface area of the cathode can be expanded by perforation, granulation, shape-change, and so forth. These methods are all of minimal use, however, and gaseous depolarization is impractically expensive, except in the so-called air cell, in which a large, sponge-like cathode built of inert material brings atmospheric oxygen into contact with the nascent hydrogen. Because of their size, however, air cells are both expensive and clumsy.

Consequently, the only method of depolarization for consumer needs is solid depolarization. The lead acid storage battery, for example, uses a solid depolarizer (PbO_2), as does the modern dry cell (MnO_2), and the increasingly popular mercury batteries for electronics (HgO). Beginning in the 1860s, and increasing through the next decades, battery designers invented large numbers of solid-depolarized primary battery designs. At first, such cells were intended for use on low-traffic telegraphic circuits, where the volume of the traffic could not justify proper maintenance of the more standard telegraphic cells. Increasingly, however, such solid-depolarized batteries were intended as fool-proof, maintenance-free cells for consumer use.

Manganese dioxide in ammonium chloride solution was a popular depolarizing system, since the raw materials were cheap. The high electrical resistance of the MnO_2, however, limited its use. Lead peroxide was popular as a primary battery material as well as for storage batteries. It was cheap, had good electrical conductivity, and its high electropositivity made for a powerful battery when a PbO_2 cathode was opposed to a zinc anode in sulphuric solution. A number of PbO_2-H_2SO_4-Zn primary batteries were marketed extensively during the late 1870s, '80s and early '90s. One such was the Roberts cell:[1]

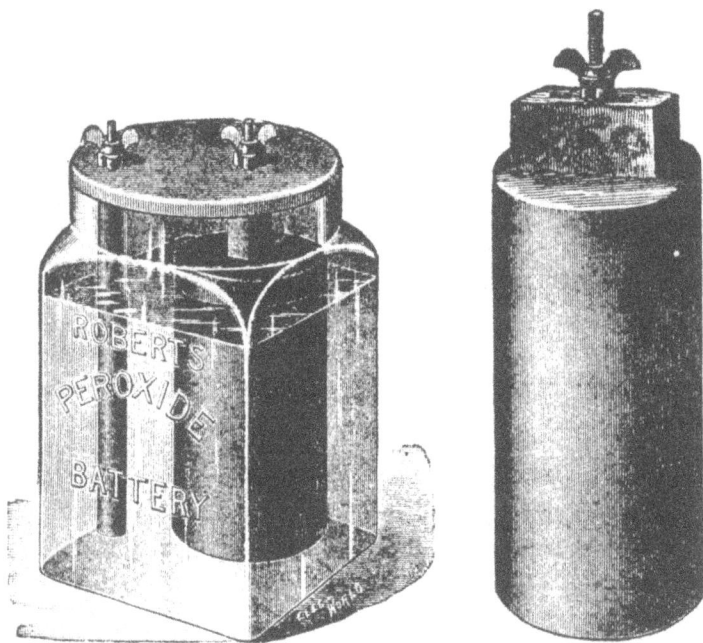

The Roberts cell was designed to supply a strong current

---

[1] Photograph from: Donato Tommasi, *Traite des Piles Electriques* (Paris, George Carre, 1889) p. 264; see also, Henry S. Carhart, *Primary Batteries* (Boston, Allyn & Bacon, 1891) p. 74; Alfred Niaudet, *Elementary Treatise on Electric Batteries* (New York, Wiley, 1890) p. 179.

with maximum durability and maintenance-free operation. These are difficult characteristics to combine in a single battery. A strong current requires powerful chemicals and a minimum of internal resistance. Solid depolarizers must be some form of oxide and oxides are usually bad conductors, thus leading to high internal resistance. Yet solid depolarizers are essential for high primary battery durability. Strong chemicals, such as sulphuric and nitric acids, increase the strength of the current, but they also dissolve most metallic oxides, thus preventing durability.

Fortunately for the early consumer battery makers, strength of current was not of great importance for most household applications. Consequently, the battery designers could sacrifice current strength and constancy to obtain durability. For this reason the most popular primary battery design for the consumer market was the Leclanché. This system consists of an electrolyte of ammonium chloride, and an MnO_2 depolarizer. Zinc is used for the anode. In nineteenth-century Leclanchés, the MnO_2 was frequently omitted, since, for such purposes as ringing doorbells, chemical depolarization could be dispensed with. Batteries such as these did not have much power, but they could be connected to a circuit and then left for months or even years on end with no attention.

Consumer batteries such as the Roberts, however, were designed for more energy-intensive uses: driving fans, operating sewing machines, or even supplying the current for incandescent lamps. The development of high intensity primary batteries for incandescent lights attracted a good deal of attention during the 1880s and early '90s, since many inventors believed that homes outside of central city areas could never be economically supplied from central stations. Consequently, many designs for gargantuan home lighting primary batteries appeared. In such batteries, it was popular to use caustic alkalies for the electrolytes instead of the hitherto more popular strong mineral acids. Alkali solutions have an advantage over the acids; they do not dissolve as do many potentially useful depolarizing agents. Use of alkaline electrolytes broadened the battery designer's option and became increasingly more popular in consumer battery designs.

The use of alkaline electrolytes was not a completely new idea; they had been suggested in the early nineteenth century. They were not used, however, for telegraphic or electroplating use, since liquid depolarizers were satisfactory for this service and allowed cheaper construction and usually provided stronger current. NaOH and KOH electrolytes, therefore, offered no advantages over acids for commercial uses, and they were more expensive and harder to come by than the strong mineral acids in the nineteenth century.

The first oxide to be used as the depolarizer in a commercial alkaline battery was copper oxide: CuO. This substance was both inexpensive and a good conductor. Moreover, since the metallic ion was the same as that used in the depolarizer of the Daniell cell, that is, copper, the same weight differentials existed between the cell's two cations (Cu^{++} and Zn^{++}) as existed in the Daniell. This fact permitted use of the density separation principle of the gravity Daniell, and was the primary reason for the choice of this oxide.

Copper oxide was first proposed as a depolarizer in 1870. The French inventor Denys suggested that CuO be substituted for $CuSO_4$ in the standard acid gravity Daniell cell.[2] Copper oxide is less soluble in an aqueous acid solution than the sulphate and it was hoped that this would produce a battery less subject to self-discharge than the standard gravity cell. The Denys cell never caught on, however, because the Leclanché cell solved the same problem more satisfactorily.

The first inventors to use CuO in an alkaline primary battery were Felix de Lalande and Georges Chaperon in 1881. These two, aided by D'Arsonval, were trying to "constituer un élément voltaique de longue durée, a montage permanent, et cependant susceptible d'un debit important et soutenu."[3] The early patents reveal that Lalande and Chaperon had given little attention to a special physical design for their batteries; the drawings show standard porous pot and gravity Daniell telegraphic cells of the time. The figure on p. 329 shows the gravity version:[4]

Here c is the CuO powder; d a KOH solution; b the zinc anode; a' a metallic copper current collector; and e a glass jar. Except for the CuO, all these components were standard items of sale in contemporary telegraphic supply stores. Nevertheless, this cell stood little chance of competing in the telegraphic market; its EMF was less than a volt and the copper oxide bed gave the cell a higher internal resistance than most telegraphic batteries.

During the 1880s, Lalande and Chaperon turned towards the

[2] W.R. Cooper, *Primary Batteries* (London, Penn Bros., 1920) p. 172; Arthur V. Abbott, *Telephony* (1904) p. 321.

[3] *La Lumiere Electrique* 12 (1884): 260-264.

[4] U.S. patent #274,110—20 March 1883.

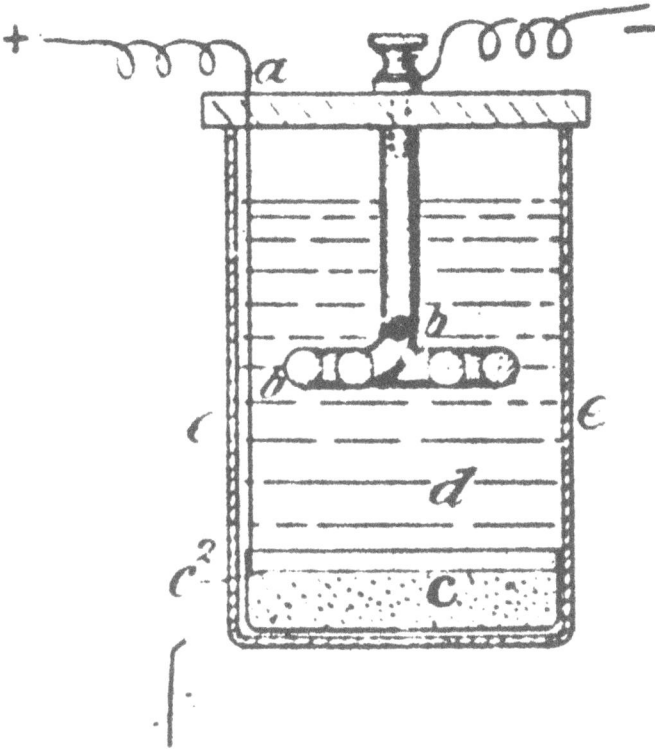

consumer market and developed larger, sealed versions of
their cell for home use. Throughout the '80s various forms of
the Lalande-Chaperon, as well as other alkaline batteries, enjoyed a brief vogue for home incandescent lighting. For example, in 1884 a British chemical firm introduced the first of a
series of Lalande-Chaperon designs intended not only to provide current for home lighting, but also to make a profit for the
consumer who used it. The consumer was to install a series of
cast iron tanks in his cellar, each of which held 15 gallons of
potassium hydroxide electrolyte apiece.

Each tank had six electrodes—two, twenty-pound zinc
anode plates, and four, ten-pound copper oxide cathode plates.
The cathodes were made coherent by squeezing a heated mixture of CuO powder, metallic copper chips, and magnesium
chloride into a folded iron mesh current collector. During discharge, the cathodes were reduced to metallic copper, while
the zinc anodes oxidized and dissolved into the alkaline solution. Regeneration of the spent cells was described by the inventors as being both easy and profitable. The cathodes were
air dried, and then placed in a furnace, which reoxidized

them. The electrolyte was then drained into a separate tank, where carbon dioxide was bubbled through it. This precipitated zinc white, ZnO. Since metallic zinc could be brought at £15 a ton, but zinc white pigment could be sold at £20 a ton, these batteries not only provided light, but also made a profit for their owners.[5]

Although this idea of running a small chemical plant in the consumer's basement may seem absurdly impractical to us, the method actually derived from the standard operating procedure at the larger telegraphic offices, where spent zinc and copper sulphate from the Daniell cells was recycled by the office's battery man. This practice had been common for many years by the 1880s.[6] The large number of variants on this scheme which appeared in the 1880s and early '90s is another reflection of the tendency of early lighting engineers to carry over into lighting the experiences of their early training in telegraphy.

The copper oxide system was not the only alkaline battery system to be proposed for consumer use in the 1880s. In 1882, for example, Alfred R. Bennett created the Electromotive Force Company in Glasgow to manufacture his patent Fe-KOH-Zn battery. The metallic iron cathode consisted of a sponge of iron turnings from machine shop scrap; depolarization came from atmospheric air rather than by the chemicals of the battery. Bennett proposed that the zinc oxide produced be sold to pigment manufacturers.[7] In 1886 the German Aron proposed substituting HgO for the CuO of the Lalande-Chaperon. The system which he proposed: HgO-KOH-HgO, became of great commercial value after the Second World War with the evolution of the transistor, but was too expensive for the nineteenth century.[8]

Despite the existence of these variants, however, the Lalande-Chaperon CuO system was the only alkaline primary battery system to achieve any popularity in the late nineteenth-early twentieth century period. The most success-

---

[5] *Electrical Engineer* (London), 11/1887: 417-419; *The Lalande-Spence Primary Battery* (advertising pamphlet) (London, 1885).

[6] Another example of this system was proposed by Thomas Slater in 1880. Slater suggested a battery working on the system: Ni-NiSO$_4$-(diaphragm)-H$_2$SO$_4$-C. The nickel salts thus generated would be sold to nickel platers for more than the cost of the original metal. *The Electrician* (London), 29 May 1880: 20.

[7] *Bennett's Patent Iron Battery* (Glasgow, the Electromotive Force Co., 1882).

[8] German patent #38,220—1886.

ful attempt to commercialize this sytem began in 1889 when Lalande licensed the American rights to the battery to Thomas Edison. On September 30, 1889, Edison created the Primary Battery Division of the Edison Manufacturing Company to produce these batteries.[9] This cell has been continuously manufactured ever since.

Edison's interest in the Lalande-Chaperon derived from his development of the phonograph. From the spring of 1887 to mid-1888 he was preparing the phonograph for commercial manufacture. He saw that the original clockwork drive was a nuisance and he therefore devised an electric motor drive, but then, the availability of the current became a problem. Edison originally conceived his phonograph as a business machine, with the entertainment function evolving later. Therefore, the machine was designed to be light, portable, and convenient, without the need for cords. Edison reasoned that if the machine were to have maximum appeal to businessmen, it would need a small, durable, foolproof battery. Today we use dry cells for this purpose, but in 1889 the recently-invented dry cell was much too weak and inconstant a power source for this purpose. Consequently, Lalande redesigned his battery to fit the dictating machine market. The loose CuO powder of the original designs was replaced by porous copper oxide wafers. These were made by mixing CuO powder with tar, forming the mixture into sheets, and burning out the tar at 500-600°C, leaving a coherent but porous structure.[10]

These wafers were then inserted into a hinged copper electrode holder and placed in the glass jar along with zinc anode plates. Replacement of wafers and zincs was a simple operation (see figure, p. 332).[11]

Edison hoped to sell these batteries in large volume; unfortunately the consumers liked his phonographs but not his batteries. By 1894 he was complaining to Lalande that people preferred to use lead-acid storage batteries to run his phonographs.[12] The Lalande-Edison batteries continued to be made, however, and found a market in fans, sewing machines, and other household appliances.

---

[9] *Seventy Five Years of Packaged Power* (Bloomfield, N.J., McGraw-Edison, 1964).

[10] U.S. Patent #479,887—2 August 1892—filed 6/2/1891, patented France #205,592—28 May 1890.

[11] W.R. Cooper, *Primary Batteries, Their Theory and Use*, p. 166.

[12] Draft of letter, Edison to Felix de Lalande, 1 March 1894, *ENHSA*.

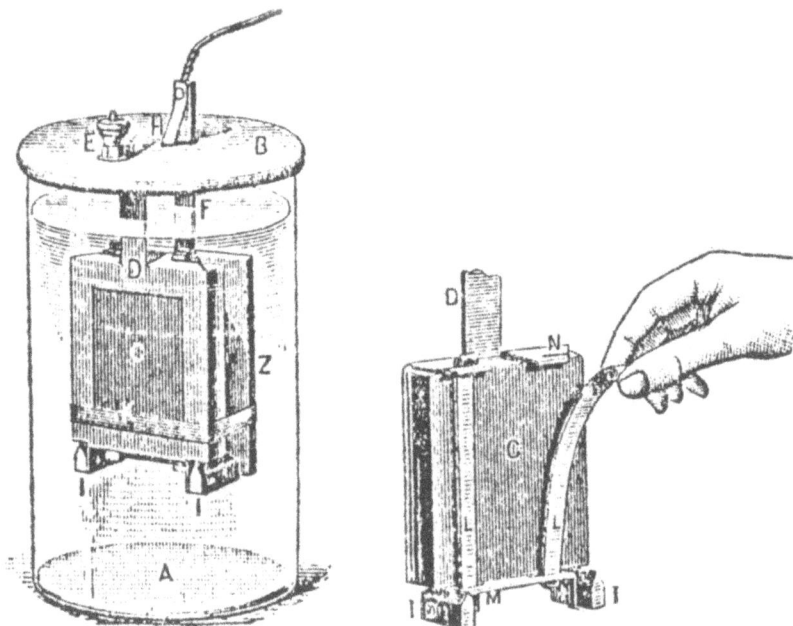

The earliest patent of Lalande and Chaperon appeared in 1881, the same year as the first Faure patent. Storage battery excitement was intense at the time and the two inventors did not fail to point out the storage battery potential of their cell when they applied for an American patent in 1882:[13] ". . . the piles are reversible, that is to say, capable of being recharged by an electric current of sufficient electromotive force."

This idea was similar to the 1880 patent of Thomson and Houston for a rechargeable Daniell cell, mentioned in Chapter I. It is doubtful if Lalande and Chaperon got any further than Thomson and Houston in commercializing this idea. Lalande did take out two French patents in 1882-1883 specifically limited to secondary forms of his battery, but these patents cover no details of construction.[14] The reason for Lalande's failure to follow up on this work was that he was not involved with the one technology for which the alkaline storage battery seemed to have a clear advantage over the lead-acid; namely, electric transportation.

---

[13] U.S. patent #274,110—20 March 1883—filed 11 July 1882.

[14] French patent #150,454—3 August 1882; #158,945—3 December 1883; Niaudet, *Elementary Treatise*, p. 121.

The alkaline electrolyte enabled the battery designer to dispense with the heavy lead for both electrodes; the use of a zinc anode in the regular lead-acid storage battery reduced the weight of only one electrode. If the designer chose to gain the double advantage of weight reduction on both electrodes, he was then faced with two very difficult technical problems—the difficulty of getting the dissolved zinc to replate on the anode during charging, and the problem of making a coherent cathode from active materials which did not possess the "setting" properties of the lead oxides. Nevertheless, a number of battery transportation enthusiasts took up the technical challenge of designing one or another of these alternatives to the lead-acid cell.

Among those enthusiasts of battery transportation who took up the design of the PbO_2-H_2SO_4-Zn battery was Jules Julien, who worked on battery transportation in the late 1890s after his father's death. Only one patent appeared on this type of cell, however, and most of Julien's work seems to have been concentrated in the more traditional all-lead design.[15] This was a typical pattern for the half dozen or so European and American inventors who worked on the zinc-lead system in the 1880s and '90s; most of them had only one or two patents, indicating that little was done on this design beyond the experimental stage.[16] There are no records of streetcars actually being propelled by zinc-lead storage batteries.

As we saw in the discussion of the battery streetcar, interest in motive-power lead-acid batteries picked up in the later 1880s and reached its peak by the mid '90s. Interest in a motive power version of the Lalande-Chaperon cell followed roughly the same chronological sequence. The first inventors to work seriously on a rechargeable CuO-alkaline-Zn battery were three Frenchmen: Alexandre de Virloy, Edmond Commelin and Gabriel Bailhache. In an 1884 French patent they described a set of flat cathodes.

Each of these electrode assemblies, which the patentees described as having the form of "a very shallow box," were in-

---

[15] U.S. patent #600,693—15 March 1898—filed 26 September 1896—patent Belgium #120,281—10 March 1896.

[16] For example: U.S. patents: #227,445—15 May 1883—filed 1 May 1882 (Emil Boettcher); #543,680—30 July 1895—filed 5 January 1895 (Ludwig Epstein); #615,246—6 December 1898—filed 15 March 1896 (Owen Bugg); #615,476—12 June 1900—filed 11 July 1899 (Owen Bugg); #695,707—18 March 1902—filed 19 July 1900 (Owen Bugg); #736,420—18 August 1903—filed 22 November 1899 (Joseph Middleby); #1,321,947—18 November 1919—filed 24 September 1915 (Elmer Sperry).

serted in a long wooden box filled with an alkaline electrolyte saturated with zinc salt. The battery was charged by external current, which caused the zinc from the solution to plate out on the fine wires of the copper gauze:[17]

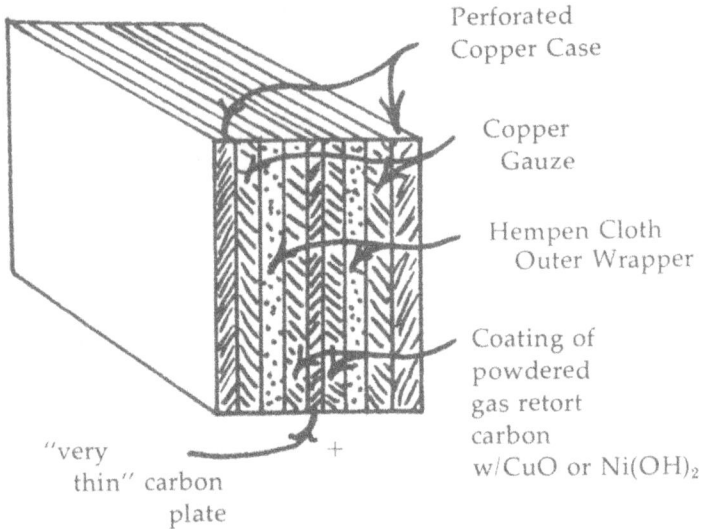

Perforated
Copper Case

Copper
Gauze

Hempen Cloth
Outer Wrapper

Coating of
powdered
gas retort
carbon
w/CuO or Ni(OH)$_2$

"very
thin" carbon
plate

The use of the gas retort carbon current collector for the cathode was typical of many early attempts at storage battery design, both for the standard lead-acid cells and alkaline designs. This, again, was another example of the transfer of standard telegraphic battery manufacturing technique into a new field of electrical engineering. Since it proved to be an unsuitable technological transfer, however, the carbon set up local action with the depolarizing material and produced self-discharge. In 1886 the three inventors substituted a screen of pure metallic copper for the central carbon plate.[18]

Another problem remained, however—the lack of coherence of the copper oxide mass. Copper oxide does not "set" like litharge or red lead, and it tended to fall away from the cathode. The inventors, therefore, developed a series of techniques for producing coherence. Instead of beginning with a copper salt, a copper-mercury amalgam was cast on the copper screen, and the mercury was evaporated out to leave a porous structure. Alternatively, copper was electrolytically deposited in the screen and then squeezed tightly into place

[17] U.S. patent #345,124—6 July 1886—filed 27 March 1885; French patent #164,681—9 October 1884; see also: Wade, *Secondary Batteries*, p. 129.

[18] Ibid., p. 129.

under pressure. Neither of these techniques seems to have been satisfactory.

In 1886, Camille Desmazures took over this storage battery project from de Virlog, Commelin, and Bailhache. He patented a final form of this particular variant on the Lalande-Chaperon system. Desmazures prepared two, millimeter thick sheets of pure porous copper by subjecting chemically-precipitated metallic powder to a pressure of 4-8 tons per square inch. These sheets were then wrapped in parchment envelopes and opposed to tinned iron negative current collectors which received the zinc plating during the charging operation.[19]

The Desmazures-de Virloy-Commelin-Bailhache battery had one brief moment of fame when it was used to power the French submarine *Le Gymnote* in 1889. *Le Gymnote,* as its name implies, was a pioneering attempt to propel a submarine by electricity. The ministry of marine commissioned the naval architect Gustave Zédé to build a boat of positive bouyancy; previous subs had submerged by becoming denser than water through the filling of ballast tanks, whereas the *Gymnote* submerged by the thrust of her electric motors against downward-sloping vanes.[20]

Gymnote was launched in 1889. It was a large ship for its time—59 feet long by 6 foot beam, with 30 ton displacement:[21]

---

[19] U.S. patent #402,006—23 April 1889—filed 31 March 1887; British patent #7966—15 June 1886.

[20] Maurice Gaget, *La Navigation,* pp. 315-323.

[21] Martin & Sachs, *Electrical Boats,* p. 67.

A number of different batteries were tried on the *Gymnote*. The first was a chromic acid primary battery designed by the famous balloonist, Renard. The chromic acid battery had appeared in the 1840s, but was largely ignored until the 1870s because of the expense of its chemicals. The chromic came into vogue in the '70s because its high current output made it useful for the more energy-intensive pieces of electrical apparatus then coming into use. For such a thing as a submarine, however, even the chromic acid primary battery cells were too weak.[22] They were therefore removed and the Desmazures cells substituted.

The Desmazures installation was large—564 individual cells weighing a total of nearly ten tons. Emile Reynier, the French lead-acid expert, said in 1889 that the Desmazures battery possessed no advantages over its lead-acid rivals as a transportation power source; he claimed that it was as heavy as a comparable lead-acid, was bulkier, and cost more.[23] Other accounts, however, gave the Desmazures a clear weight advantage—an energy output of 11.8 watt hours per pound.[24] By contrast, the early Exide car battery of 1905 gave only about 8.2 watt hours per pound.[25] Nevertheless, the Desmazures batteries were a failure. They were soon removed from the *Gymnote* and replaced with TEM lead-acids. French submarines immediately after *Gymnote* were also TEM equipped.[26]

The disappearance of the Desmazures batteries after the *Gymnote* trials was not so much the result of their poor performance, as of the concurrent death of their promoter in 1889. This is not to say that the batteries did not have serious durability problems, but they also possessed considerable advantages over lead-acids, and the *Gymnote* trials, although a failure in themselves, stimulated others to take up where Desmazures left off. The advantages of the copper-zinc storage battery stemmed mainly from the higher energy densities of both these materials over lead. Thus, even though the system has an EMF of .85 volt, as compared with 2 volts for lead-acids, it still possessed an overall potential weight advantage.

The chief problems with copper zinc come from their solubility. Copper oxide is not really soluble in alkaline solution,

---

[22] Gaget, *La Navigation*, p. 316.

[23] *The Electrician*, 18 January 1889: 302; Martin and Sachs, *Electrical Boats*, p. 66.

[24] *Electrical World*, 10 January 1891: 19.

[25] *Cyclopedia of Applied Electricity* (Part II) p. 11.

[26] Gaget, *La Navigation*, p. 316; for example: the *Gustave Zede* and the *Morse*, both 1893. The TEM cells were also reported to work badly.

but it does slowly form a colloidal suspension which diffuses to the anode, plates out there, and ruins the cell. Zinc would be a superb anode material for storage batteries if it could only be made to replate out of solution properly. Unfortunately, the metal does not plate out evenly over the surface of the anode, but tends to favor certain spots. Over the course of several cycles, such sites become increasingly favored, until "trees" are formed—branched growths leading from one electrode to the other, short-circuiting the cell internally.

Desmazures was the last French inventor to work on the Lalande-Chaperon storage battery. From 1890 on a series of German inventors took up the idea. None of these German efforts appear to have gotten beyond the experimental stage, however, and the few German patents went little beyond the level already reached by Desmazures: Boettcher (1890); Schoop (1893); Oppermann (1894); Schmidt (1895).[27] Some minor American work on copper alkaline electrodes also went on at this time, for example, the 1891 patent of Marmaduke Slattery.[28] This work also was little more than a copy of Desmazures.

These efforts tapered off after the mid '90s, although as late as 1899 Walther Nernst had a graduate student trying again to solve the copper-zinc reversibility problem.[29] One German inventor in 1894 noted that the upsurge in European interest in battery streetcars had produced new concern for the alkaline cell: "In neuerer Zeit werden die zuerst von Lalande angegebenen Elemente wieder mehr beachtet, nachdem dieselben lange Zeit ganz in Vergessenheit geraten waren."[30]

Dr. Paul Schoop was a noted German storage battery expert with connections to AFA. His 1894 article entitled: "Hat der alkalische Zinc-Kupfer Akkumulator Aussicht auf baldige praktische Verwendung im Trambetrieb?" came up with a mildly pessimistic answer to this question, but indicated that recent developments were encouraging.[31] Schoop began by indicating that battery streetcars had proved technically successful but economically unworkable. This failure was caused by the high weight of their lead-acid cells and their excessive

---

[27] German patent #57,118; British patent #7711; *Elektrochemische Zeit.* 10 (1894): p. 191; 13 (1895): 224.

[28] U.S. patent #466,104—10 February 1891—filed 6 April 1889.

[29] Woolsey Mc A. Johnson, *Trans. Am. Electrochem. Soc.* 1 (1902): 187.

[30] *Elektrochemische Zeit.* 10 (1894): 191.

[31] *Zeit. f. Elektrotechnik u. Elektrochemie* (1894) p. 131.

rate of depreciation—50 to 100 percent per annum. Schoop then revealed the temptation of the zinc-copper cell—it can deliver between 20 and 27 horsepower hours per ton (7.46 to 10.07 watt hours per pound), depending upon the discharge rate, whereas the best a lead battery could do is 5 to 7½ horsepower hours per ton. The data for the lead battery seem unrealistically low, unless Schoop was assuming that only Planté-type batteries had sufficient durability for streetcar drive. Even granting this, however, the difference clearly shows the advantage of the alkaline cell.

Schoop then turned his attention to the question of the copper-zinc battery's reliability and durability. "Und hier beginnin die Zweifel," he says with considerable understatement. In fact, at the time Schoop was writing these words, the only company ever to undertake the commercial development of the alkaline copper-zinc storage battery was being driven into bankruptcy. This was the Waddell-Entz Electric Company of Bridgeport, Connecticut.

Montgomery Waddell and Justus B. Entz got their start in the electrical field as employees of the Edison Machine Works in Schenectady. Entz was the technical brains of the team.[32] He entered Edison's employ in 1887 at age 20 and within two years was head electrician of the Schenectady plant. The Waddell-Entz Electrical Company was created about 1889. It was typical of many small and ephemeral electric machinery companies which sprang up in the late 1880s to make dynamos, motors, and similar equipment. For a few years the company specialized in installing electric lighting plants in the New York Area.[33]

Like so many other companies of the period, Waddell-Entz was interested in capturing part of the electric streetcar market. Entz designed a special motor for streetcar drive, and in 1889 he and William A. Phillips, an English engineer still in the employ of the Schenectady works, began work on a new form of the Desmazures cell for use in driving the planned streetcars. The inventors alleviated the manufacturing problems of the flat-plate Desmazures cells by wrapping copper wires with copper mesh, impregnating this with copper oxide and wrapping this with a cotton cover. The copper oxide was made

---

[32] Application by Justus B. Entz for membership in Edison Pioneers, 1936, *ENHSA*; Falk & Salkind, *Alkaline Batteries*, p. 6; Thomas C. Martin & Stephen L. Coles, *The Story of Electricity* (New York, Story of Electricity Co., 1919) p. 219.

[33] *Electrical Engineer* (London) 5 June 1891: 556-558.

coherent by mixing it with molten sulphur, applying the paste to the mesh and then heating to drive off the sulphur. The patentees stated that these electrode tubes could be mass produced and bent to make any desired shape or size electrode.[34]

It may be purely coincidental that the same factory where Entz and Phillips worked in Schenectady was where the phonograph development work went on and where the Edison-Lalande work was also carried out. It is known, however, that Edison himself carried out some unsucceful attempts to see if his battery was reversible. It is probable that both the Desmazures battery and the Edison-Lalande work had a motivating influence on the creation of the Waddell-Entz battery.[35]

In practice, the cathode tubes were gathered into a flat-plate configuration and a thin, perforated sheet of tin-plated iron was bent around this to serve as the anode support for the plated zinc.:[36]

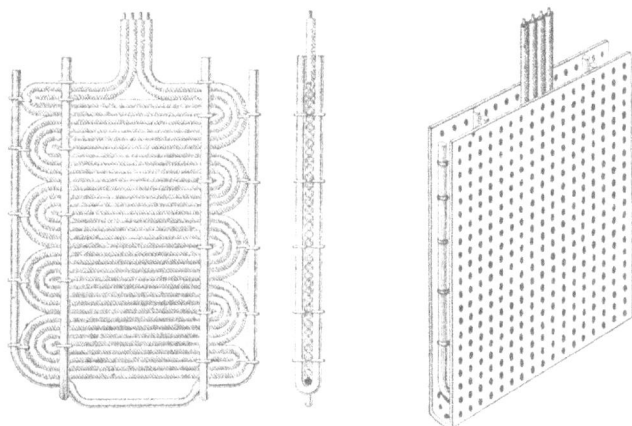

Waddell and Entz set up a factory in Bridgeport to manufacture the component parts for their cars. In 1891 a few cars were run experimentally in Coney Island and Philadelphia.[37] During

---

[34] U.S. patents: #421,916—25 February 1890—filed 9 October 1889; #440,023—4 November 1890; #440,024—4 November 1890.

[35] The name Waddell-Entz battery was always given to this cell, although from the patents it is clear that the name Entz-Phillips would have been more appropriate; Waddell had little to do with the actual design.

[36] From: Wade, *Secondary Batteries*, p. 131; U.S. patents: #461,823—27 October 1891—filed 31 October 1890; #461,858—27 October 1891—filed 31 October 1890; #518,996—1 May 1894—filed 20 June 1893.

[37] *Electrical World,* 16 May 1891: 354-355; *New York Times,* 16 October 1892: 14.

the years 1894-1895 a number of cars ran on commercial service in Germany and Vienna under an agreement between Waddell-Entz and AFA.[38] The big test of the system, however, was in Manhattan. This was the conclusive trial for the alkaline zinc-copper storage battery, since it was a full-scale commercial operation with a large number of cars and a specially outfitted and designed powerhouse.

The Whitney syndicate gave Waddell-Entz a contract in mid 1892 to operate cars on 2nd Avenue; this was three years before the first Manhattan experiments with ESB batteries on the Madison Avenue line.[39] The line was finished by June 1893; it ran from 96th to 127th Streets with a charging station on 127th. Initially, only three cars were run; this was increased to six by November, ten by early 1894, and eighteen by December 1894, by which time the battery cars had replaced all horsecars on the line.[40] For a while, therefore, the Waddell-Entz experiment looked like a success—but then so did almost every other battery streetcar line for a year or so. Costs per car mile were reported at less than 10¢.[41]

Problems soon appeared, the Waddell-Entz batteries proved to have a durability no greater than lead-acids. Moreover, they often proved incapable of being recharged if discharged at too high a rate, and they had a tendency to self-discharge on open-circuit. The zinc replating problem had not been solved, since the formation of zinc "trees" steadily ruined the cells. Finally, the cells proved to be only slightly lighter than contemporary lead-acids designed for the same service; a full complement for a car weighed over two tons.[42] The inventors attempted to solve some of these problems by means of special charging apparatus.

These techniques could not save the venture from failure. The 2nd Avenue Line was discontinued in 1895 and the plant and stock of the Bridgeport factory sold at auction in May. The company's patents and the equipment of the Manhattan line

---

[38] *Electrical Engineer* (London), 30 November 1894; 622; 4 January 1895 2; *The Electrician* (London), 1 February 1895: 387.

[39] *New York Times*, 16 October 1892, p. 14.

[40] *Electrical World* 21 (1893): 327; *The Electrician*, 17 November 1893: 72-73; 8 December 1893: 135.

[41] *The Electrician* (London), 17 November 1893: 72-73. The actual figure was 9.32¢.

[42] Oscar T. Crosby & Louis Bell, *The Electric Railway in Theory & Practice* (New York, 1896) p. 244.

were also sold off.[43] Justus Entz, having failed to lick the lead-acid battery, decided to join it—he entered the employ of ESB after his own firm went bankrupt. He eventually rose to the position of chief engineer in ESB and was instrumental in the transfer of European storage battery technology to ESB, and thus to the American industry in general. For example, he spent some time at the AFA plant in the late 1890s learning their techniques.[44] Entz was also responsible for elaborating the design of the differential booster. As we saw in the last chapter this was one of the innovations which allowed such companies as ESB to create a substantial market in the trolley industry.[45]

Entz was a highly respected member of the first generation of electrical engineers. The later reputation which he established in the lead-acid industry shows him to have been an excellent designer of electrical systems. And yet, his pioneering attempt to commercialize an alkaline storage battery collapsed very rapidly. We emphasized in an earlier chapter that storage battery companies tended to succeed in direct proportion to the amount of financial support they were given by larger corporations; regardless of how good the initial design of the batteries was or how competent the engineering skill, each new company needed enough capital to carry it over the half dozen or so years during which it developed a product which was really worth buying. Waddell-Entz was simply too small a company to afford the vast number of false starts which are inevitable in the evolution of a new storage battery design.

---

[43] *New York Times*, 23 May 1895: 3.
[44] *Electrical World*, 11 January 1896: 54-55.
[45] Martin & Coles, *Story*, p. 219.

2.   Early Commercialization: Jungner and Edison

The Waddell-Entz battery was the first and last commercial form of the alkaline zinc-copper battery. Its failure did not stop others from experimenting with the system, but the bankruptcy of the Waddell-Entz company warned others not to stake their fortunes on this particular electro-chemical system. Consequently, when several commercially successful alkaline storage batteries were developed during the first decade of the twentieth century, they were designed to avoid the chief source of trouble in the Waddell-Entz cell; that is, they were designed with electrode materials which were perfectly insoluble in the electrolyte.

For the cathode, a number of insoluble oxides had been proposed since the 1880s, but nickel oxides had been the most popular. In 1887 Dun and Hasslacher took out German and British patents which specified the use of nickel, cobalt, bismuth or silver oxide cathodes in combination with zinc anodes in alkaline electrolyte.[1] Desmazures had suggested nickel oxide as an alternative to copper oxide in 1886, although he does not seem ever to have employed it in actual practice.[2] In 1896 Louis Krieger, the director of the French Thomson-Houston company and a promoter of both battery streetcars and battery cabs, invented a storage battery which was perhaps stimulated by the Desmazures patent—it worked on the system: Ni_2O_3-KOH-Zn.[3] The most persistent of this group of nickel cathode people were a pair of Austrian engineers—Stanislaw Laszczynski and Titus von Michalowski—who worked on the problem of using nickel cathodes in connection with zinc anodes from 1889 until the early years of the new century.[4] Moreover, whereas all of the earlier patentees who specified nickel also mentioned the use of alternative cathode depolarizing agents, Laszczynski and Michalowski were the first to concentrate solely on nickel. As they pointed out in a 1902 American patent: "Of all the metallic oxids (sic) hitherto employed as depolarizers for the zinc-alkali storage batteries, nickel oxid (Ni_2O_3) is, as is well known, the most suitable."[5]

The reasons for nickel's popularity as a cathode material are clear. Nickel, like copper, was a material which practical electrochemists of the late nineteenth century knew well; just as copper was the most popular depolarizing substance in various commercial primary cells, so nickel was the most commercially important of the electroplater's metals. Engineers tend to build new systems from preexisting components; indeed, this is the most rational way to proceed. It is not surprising, therefore, that the earliest designers of alkaline storage batteries

---

[1] German patent #38,383; British patent #1862.

[2] U.S. patent #345,124—6 July 1886—filed 27 March 1885.

[3] Rudolphe Herold, *Etude theorique et pratique de l'Accumulateur Fer-Nickel* (Belgium, 1924) p. 4; H.B. Coho, "An Historical Review of the Storage Battery," *Trans. Am. Electrochem. Soc.* 3 (1903): 159-168.

[4] Coho, "Historical," p. 165.

[5] U.S. patent #714,201—25 November 1902—filed 23 February 1901; British patents: #15,370—1899; #17,569—1900; German patent #112,351—1899.

should have chosen the chemical components they were most familiar with from earlier forms of electrical technology.

Moreover, the use of large quantities of nickel by electroplaters ever since the 1860s had made this metal a common commodity by the late nineteenth century. The table shows the increase in world production of nickel:[6]

YEAR	SHORT TONS	YEAR	SHORT TONS	YEAR	SHORT TONS
1850	7-8	1880	60	1892	460
1860	15	1885	60	1893	480
1870	50	1890	450	1894	540
1875	56	1891	680	1896	460

Metals such as cadmium or cobalt were suggested as alternatives to nickel, but in the late nineteenth century, sufficiently large commercial supplies of these metals were not available. Nickel had the disadvantage over copper of much lower electrical conductivity in the oxide form, and this was probably the reason for the greater initial popularity of copper oxide alkaline experiments. Nevertheless, nickel oxide possessed the greater advantage of complete insolubility in the electrolyte.

As far as the anode was concerned, there does not appear to have been as much stress upon one particular metallic oxide, although iron was popular. The first recorded attempt to design an alkaline storage battery, both of whose electrodes were insoluble, was by the German Dr. Georg Leuchs in 1883. In an article he proposed a series of possible variations, some of which he claimed to have tried experimentally, including cadmium-manganese, cadmium-iron, and nickel-iron.[7]

The failure of the Desmazures copper-zinc cells on the *Gymnote* in 1889 stimulated the first French effort to construct an alkaline cell with insoluble electrodes. Georges Darrieus was the naval officer commanding the *Gymnote,* and its successor, the *Gustave Zédé.* As a result of his experiences in these submarines, Darrieus became convinced that neither lead-acids nor alkaline soluble electrode batteries were practical, and he set to work designing his own storage battery. In 1893 he took out a French patent in which he proposed either a bismuthcopper oxide or a cadmium-copper oxide couple.[8] The influence of Desmazures's work is obvious here in Darrieus's use of

---

[6] Robert C. Stanley, *Nickel, Past & Present* (Toronto, 1927).

[7] *Centralblatt f. Elektrotechnik* 5 (1883): 499; also, letter, Frank Dyer to Brandon Bros., Paris, 7 December 1903; and, manuscript, "A History of the Alkaline Storage Battery," ENHSA.

[8] French patent #233,083—1893.

the copper oxide cathode, which was, of course, not really insoluble. Nevertheless, this was a crucial patent, since it was the first to specify the importance of using insoluble active materials. After criticizing the lead-acid and zinc-alkaline storage batteries, Darrieus says:

> My new accumulators with an alkaline electrolyte are, on the contrary (i.e. different from other storage batteries), based on the following principle: to constitute the electrodes by means of spongy metals giving rise as well during charge as during discharge to the formation of bodies which are practically insoluble. . . . everything goes on as if there were simply decomposition of water and the simple transport of oxygen and hydrogen from one pole to the other.

During the final years of the century a few patents were granted in the United States and Europe which incorporated the basic concept which Darrieus specified. W.L. Merrin, for example, took out a United States patent for an alkaline lead-iron battery in 1896, and in 1898 Karl Pollak in Germany patented an iron-iron cell (metallic iron as anode and ferric oxide as cathode).[9] None of this work, however, reached the commercial stage. The first successful commercial alkaline storage batteries were developed during the first decade of the twentieth century by two inventors working independently: Thomas Edison in the United States and the Swedish chemist, Waldemar Jungner. The batteries which these men invented are still manufactured today. Jungner was responsible for pioneering the nickel-cadmium storage battery, and Edison, the nickel-iron battery. These men succeeded since they had the resources to push their research beyond the point at which their predecessors failed.

We have already seen how Edison's initial involvement with alkaline batteries stemmed from his interest in the phonograph; Waldemar Jungner's initial interest came from his work on fire alarms. Jungner's entrance into the field of electrical inventing was not unlike that of his fellow chemists, Camille Faure or Charles Brush. Although trained in chemistry, Jungner began inventing electrical gadgets at an early age. Jungner was a 19-year-old college student when he started work on his fire alarm system; at the same age Faure began his first work in telegraphy, which was a closely-allied technology.[10] As a result

---

[9] U.S. patent #882,573; German patent #107,727—1898.
[10] Falk & Salkind, *Alkaline Batteries*, pp. 12-13.

of this work, Jungner became dissatisfied with the current obtained from dry cells, which were very crude at the time (1888), and with the rapid self-discharge of lead-acid cells.

Jungner's basic fire alarm concept, which he patented in 1894, was to replace the standard fusible metal plug alarm with one that would be more sensitive and thus faster acting, he designed a solid-state alarm based on the thermocouple principle. A series of copper-iron couples strung end-to-end would produce a current under the influence of the fire's heat. This device would generate signals as soon as the heat reached it, in contrast to the fusible plug device, in which a plug of metal holding the contacts of a switch open had to melt before the signal was given.[11] The thermocouple current itself would not be powerful enough to ring a gong or turn on sprinklers, but it could be used to close a delicate switch which could cut in a more potent power source. It was for this more powerful current source that Jungner found the contemporary dry cell too weak and the lead-acid battery too unreliable, so he began work on a secondary form of alkaline cell in the early 1890s. Not surprisingly, his starting point was the electrochemical system which had received the most attention up to that time; the alkaline zinc-copper oxide cell. Jungner developed a form of reversible dry cell by gelatinizing the alkaline electrolyte. This modification of the wet batteries of Desmazures and Waddell-Entz made sense, since Jungner was designing a consumer battery. It is not entirely clear if Jungner really intended these cells to act as storage batteries, or merely like the earlier-mentioned Roberts cells. At any rate, the company which Jungner set up in Stockholm in 1897 to manufacture these cells—the Dry Accumulator Company—soon failed.[12] From this point Jungner gave up the idea of zinc electrodes.

After his failure with the zinc anodes, Jungner adopted the approach which Darrieus had put forward several years previously; namely, to develop an accumulator whose both electrodes were insoluble in the electrolyte. It is certain that Jungner was aiming at the design of a rechargeable cell by the late '90s, since his interest had shifted to electric automobile batteries. During the four years between 1897 and 1901 he experimented with about half a dozen different electrochemical combinations. In his early experiments, copper oxide and

---

[11] U.S. patent #521,168—12 June 1894—filed 9 March 1894.

[12] Sven A. Hansson, "Waldemar Jungner och Jungnerakkumulatorn," *Daedalus, Tekniska Museets Arsbok* (1963) pp. 78-91; British patent #15,880—23 August 1895.

silver oxide (Ag_2O) played the most prominent role as depolarizing agents. This is understandable, since both substances have high energy/weight ratios. Indeed, the energy density of silver as a depolarizer is higher than for almost any other material.

By 1899 Jungner appears to have been seriously considering manufacturing some kind of silver oxide-depolarized storage battery for electric vehicle drive. In that year he tried silver oxide in combination with a series of different anode materials—iron, copper, and cadmium.[13] In fact, in 1899 Jungner patented the idea of using either iron or cadmium as an anode material; both are good for this function since they form hydroxides on discharge which are electrically conductive, porous, and insoluble in the electrolyte.[14] Nevertheless, it is apparent that in 1899 Jungner was still far from certain about the most desirable electrochemical combination to use for his intended transportation battery. This can be seen from the wide diversity of different systems he cited in his 1899 patents: copper oxide-copper; silver oxide-copper; manganous hydroxide-ferrous hydroxide; silver oxide-cadmium.[15] The patents do not mention it, but he was doubtlessly considering silver oxide-iron and copper oxide-iron.

In January 1901, Jungner took out his basic patent for a nickel oxide-depolarized-iron storage battery and from this point on concentrated his attention either on this cell, or upon a variant using cadmium in place of iron. Jungner never abandoned his belief that the silver oxide-cadmium system would someday be developed into a practical storage battery.[16] His abandonment of it in 1900, however, was the result of two problems: cost, and the tendency of the silver to form a colloid, diffuse to the anode, and ruin the battery. This was the same problem which made copper oxide unsuitable, and Jungner abandoned both these potential cathode materials at about the same time.

Jungner's decision to concentrate on a nickel depolarizer after 1901 was also the result of a number of factors. His decision to abandon silver and copper left him with few other alternatives. Moreover, in 1897 or 98, while he was doing some preliminary experimentation with nickel, Jungner discovered

---

[13] Falk & Salkind, *Alkaline Batteries*, p. 14.

[14] German patent #114,905.

[15] German patent #110,210—1899; British patent #7892—1899.

[16] Falk & Salkind, *Alkaline Batteries*, p. 26.

a new hydroxide of the metal which, when used as a depolarizer, gave an excellent energy/weight ratio.[17] Still, Jungner's initial reluctance to concentrate on nickel and his preference for silver and copper no doubt stemmed from the high electrical resistance of the nickel chemicals, which made the physical construction of the cathode difficult.

In 1899 Jungner put forward his basic philosophy of storage battery design in a German and a British patent. The ideas presented therein almost inevitably led him to the choice of a nickel-iron or nickel-cadmium system as the basis of his commercial cells:[18]

> The faults of electrical batteries, whether primary or secondary, . . . may be said to be originally due to chemical alterations which occur in the electrolyte or during passage of the electric current. The chemical changes alter the relation of the materials of the liquid to the active mass or electrodes. The lead accumulator or Lalande's secondary element . . . form exceptions only so far as the chemical quality of the electrolyte is concerned, not as regards the quantity (i.e. the concentration). . . .

> The present invention is intended to produce an electrical element . . . in which on discharging or charging the electrolyte remains throughout the same both in quality and quantity (i.e. in which the ions remain the same and in the same concentration). The electrolyte does not here take part in the electrical action but rather fills the role of conductor of secondary order between the electrodes and its relative weight can thus be reduced to a minimum. . . .

The concepts presented here are identical to those in the Darrieus patent quoted earlier, but Jungner stresses at the end of his statement the importance of such a battery design for transportation batteries: if the electrolyte undergoes no changes, either through the dissolution of the electrodes or its own reaction with the active materials, then the weight and volume of electrolyte can be reduced to a minimum sufficient to maintain electrical conductivity. Indeed, what Jungner was suggesting here was something that contemporary designers of lead-acid cells were starting to do. As we saw in the last chapter, the generous inter-electrode spacing in stationary lead-acid batteries was continually reduced by the designers

---

[17] Ibid., p. 16.
[18] German patent: #110,210—1899; British patent #7892—1899.

of electric vehicle and SLI batteries. In lead-acid batteries, however, some compensating factor had to be added as the electrolyte volume was reduced, since the sulphuric acid concentration varies with the charge. As we saw, one such compensating factor was to increase the acid concentration as the total volume was decreased. For Jungner, however, the attraction of the alkaline battery was that no such compensation was needed.

As soon as Jungner realized this, he abandoned the concept of the zinc electrode, and advanced one step over the earlier design thinking of such men as Desmazures and Waddell-Entz. What he failed to appreciate initially, however, was the more subtle problem of the colloid formation from such materials as copper and silver oxide. The formation of colloid sludges in the batteries was a slower and thus a less easily noticed problem than the oxidation and ionic dissolution of the zinc. In a sense, the concentration of early alkaline battery designers on the problems of zinc solution and their relative disregard of the problems caused by the cathode colloid formation was analogous to the tendency of early lead-acid designers to concentrate all their attention on the durability problems of the PbO_2 cathode, and to ignore the more subtle problems of the anode. Nevertheless, the abandonment of the zinc electrode allowed the plates to be brought closer together, with the reduction in the quantity of electrolyte, and this naturally augmented the colloid problem. Since colloid formation is not ionic, there was no formation of actual "trees" between the plates, as was the situation with the zinc anodes. Nevertheless, colloid formation in batteries can lead to the formation of conducting ribbons of sludge between the plates, which can be almost as serious a source of internal short-circuits.[19]

With hindsight, we can understand how Jungner's thinking on alkaline battery design questions matured: the serious and obvious problems of using zinc electrodes led to a concentration on insoluble active materials, which in turn permitted the designer to reduce electrolyte volume to a minimum. This increased the problems caused by formation of colloid sludges from the cathode materials which had been believed to be "insoluble," which finally forced the designer to concentrate on only those active materials which were truly insoluble. Thus, Jungner was led by the early years of the new century to

---

[19] Falk & Salkind, *Alkaline Batteries*, p. 26.

select the oxides or hydroxides of three metals which were both insoluble and yet also possessed the electro-chemical properties to produce a sufficiently powerful automobile battery, namely, nickel oxide for the cathode, and iron or cadmium hydroxide for the anode.

Jungner's eventual selection of a nickel-iron or nickel-cadmium cell may sound fairly straightforward. In practice, however, the direction of the research must have seemed anything but obvious. No precise reconstruction of this research can be given, but it is worth noting that Jungner took four years of intensive work to reach his conclusions. It is not surprising that men such as Waddell and Entz, who attempted not only to select an electrochemical system, but to commercialize a physical battery based on that system in a much shorter time, should have failed.

At the same time that he was investigating possible electrochemical combinations for his automobile batteries, Waldemar Jungner was also working out their mechanical construction, and his studies of electrode support materials were to have a crucial feedback effect on his choice of an electrochemical system. Jungner began from the groundwork laid by Desmazures and Waddell-Entz; in his initial mechanical designs the active oxides were packed into wire mesh current collectors and surrounded with cloth wrappings. As might be expected, these constructions did not have good durability. Thus, during 1897 Jungner turned his attention towards electrode supports of greater rigidity. He developed a series of narrow pockets made of thin, perforated metal sheets. The active oxides were packed into these pockets under pressure.[20] This work was analogous to contemporary work going on in transportation lead-acid battery design, as discussed in the last two chapters—the design of celluloid and hard rubber envelopes and tubular shells to increase the durability of the PbO_2 cathodes. The lead-acid design work began before Jungner's work and probably inspired it; soon the lead-acid tubular electrode design was also copied into alkaline storage battery construction.

Such rigid, shell-type electrodes were more important for alkaline storage batteries than for lead-acids. This was for two reasons: first, because the active materials have no inherent "setting" characteristics; and second, because the active materials usually have high electrical resistance and therefore must

---

[20] British patent #16,361—1897.

be surrounded with conducting materials. Jungner solved this by mixing the active material with graphite flakes and packing the mixture into metallic envelopes.

Since the batteries were intended for transport, the envelopes had to be light. Yet at the same time they had to be strong and resistant against attack by powerful alkalies. In 1897-1898, Jungner tested a variety of metals to see if they would withstand anodic oxidation.[21] Platinum, silver, cadmium, iron, and bismuth all showed considerable oxidation within three months. Nickel-plated steel sheet, however, proved both strong and resistant to chemical attack. Moreover, Jungner later discovered that thin flakes of pure nickel mixed with the active material powder worked better as a conducting medium than graphite.

In his attempts to electrolytically oxidize nickel, however, Jungner did discover a way to attack it by using a dilute alkaline solution. The hydroxide of nickel thus produced was the one mentioned earlier which Jungner found to be an excellent depolarizing agent. It was this substance that he then used in the cathode of his commercial batteries. Whether or not Jungner would have eventually chosen nickel for his cathodes anyway is problematical; it is clear, however, that the earlier technology of nickel-plating influenced Jungner's decision.

In 1897 at the yearly banquet of the Association of Edison Illuminating Companies, Thomas A. Edison sat talking with the chief engineer of the Detroit Edison Company. After listening to the engineer discuss a new project, Edison enthusiastically responded: "Young man, that's the thing! You have it!—the self-contained unit carrying its own fuel with it. Keep at it."[22] The young man was Henry Ford and his project was the gasoline automobile.

This is a famous anecdote often quoted in histories of the automobile, but there took place another discussion at that same meeting which is not so famous, but which is also important for the history of automotive technology. The Association of Edison Illuminating Companies was a trade association of electric utilities using the Edison system of distribution. Like most such associations, one of its functions was to encourage

---

[21] Falk & Salkind, *Alkaline Batteries*, p. 14.

[22] Matthew Josephson, *Edison—A Biography* (New York, McGraw-Hill, 1959) p. 406.

the free flow of technical information and the discussion of mutual technical problems. The yearly meetings helped accomplish this. At this particular meeting, Thomas Edison and the managers had been discussing the possible influence of electric automobile charging on the improvement of central station load-factors.[23]

Thomas Edison was impressed by the potential of electric vehicles and it was not long after this discussion took place that he actively entered the field. Moreover, Edison's notebooks and letters reveal that he was keeping a close eye on the development of electric vehicles and their batteries during the last years of the century. He was aware of the durability and weight problems of the lead-acid cells, and he certainly knew of the difficulties faced by such ventures as the various Manhattan battery streetcar lines and the Lead Cab Trust. When Edison entered the field of automobile battery design, therefore, he was aware of both the problems as well as the potential of this technology.

As far as the potential of the electric car was concerned, Thomas Edison's interest was more complex than that of William C. Whitney and the promoters of EVC. The EVC was interested in the commercial potential of its vehicles and batteries. Of course, this was true for Edison as well, but he also had an economic as well as an emotional attachment to the system of electrical distribution which he had pioneered and which was represented by the nation-wide system of Edison companies. That system was D.C., and it was under severe challenge from A.C. at the time. By the later 1890s Edison was fighting a rear-guard action to retain D.C. in the urban areas where it had first been introduced in the 1880s. The replacement of D.C. in city centers had not yet gotten underway in the late 1890s, but the handwriting was clearly on the wall.

The development of the electric car into a major form of urban transport, and, consequently, into a major consumer of urban-generated electric current, would slow the conversion of urban electric systems to A.C., since A.C. would require the installation of expensive and energy-consuming conversion apparatus in each consumer's premises. Edison was aware of the limited range and speed of the electric car and therefore realized from the start that it was restricted to urban, or at most suburban, transport. But he was also aware by the late '90s that D.C. transmission was a short distance technology limited to urban areas.

---

[23] Ibid., pp. 405-406.

Furthermore, one of the important although lesser advantages of A.C. was its ability to improve load-factors because of the extent of its coverage; one A.C. generating station could cover a number of areas having different consumption patterns, and this would tend to smooth out the effect of sharp peak periods in each of the areas. Therefore, the load-leveling function of the electric vehicle was a not unimportant factor in these later skirmishes of the Battle of the Systems. As was noted in the last chapter, the Edison system utilities were among the earliest and strongest supporters of the Electric Vehicle Association.

Edison began designing an alkaline storage battery towards the end of 1898,[24] but it took eleven years before he succeeded in producing a satisfactory commercial battery. For two reasons we will analyze in some detail the steps which Edison took: first, the record of his researches lucidly reveals the way in which his thoughts evolved; and second, because the pattern of this research refutes the stereotyped image of Edison's methods as described in most of his biographies.

According to most of his biographers, Edison's methodology was brash, independent, individualistic, and energetic—the apotheosis of the Yankee tinkerer. In fact, it is perhaps incorrect to call Edison's supposed method a methodology at all, since the word implies a form of disciplined, step-by-step approach to problem solving. According to the biographies, when Edison began a project, he had a firm idea only of his goal, and had little or no idea of the path he would follow to that goal. Eschewing the well-worn paths of past inventors, and ignoring the writings of the theorists, Edison locked himself in his laboratory and applied himself tirelessly to trying every conceivable approach to the problem. The Edison method was anti-method; the Yankee inventor, freed from the shackles of theory and preconceived ideas, was able to try everything until something worked. This image, of course, is romantic, and at times Edison is portrayed with vaguely superhuman powers; for example, there are the famous stories of the round-the-clock inventing punctuated by five minute catnaps in the roll-top desk.

The basic Edison biography from which most of the later ones were copied is Dyer and Martin's 1929 study: *Edison—His Life and Inventions.*[25] Martin was a well-known writer on electrical topics and Dyer was Edison's patent attor-

---

[24] *Trans. Am. Electrochem. Soc.* 6 (1904): 135.

[25] Frank L. Dyer & Thomas C. Martin, *Edison—His Life & Inventions* (New York, Harper & Bros., 1929) (2 vols).

ney. The book's account of Edison's alkaline storage battery development is typical of the romantic treatment given most of his inventive effort. First comes the image of the brash pioneer:[26]

> It should be borne in mind that from the very outset Edison had disclaimed any intention of following in the *only tracks then known* (my italics) by employing lead and sulphuric acid as the components of a successful storage battery . . . he determined boldly at the start that he would devise a battery without lead.

As he began his researches, Edison knew only that he would use an alkaline electrolyte. He went into his laboratory and began trying every chemical combination he could think of. It was only after: ". . . he had examined and tested practically all the known elements in numerous chemical combinations, the electrical action he sought for had been obtained, thus affording him the first inkling of the secret he had industriously tried to wrest from Nature."

This romantic image of Edison's methods was not a creation of his biographers alone; they merely copied it from the image Edison himself created over the years. For example, the image of the Yankee tinkerer blazing new paths and contemptuous of established authority was carefully nurtured by the use of mild profanity and the judicious gob of tobacco juice on the floor while giving interviews to newspaper reporters. Thus, Edison himself created the alkaline battery pioneer image:[27] ". . . I don't think Nature would be so *unkind* as to withhold the secret of a good storage battery, if a real earnest hunt were made for it. I'm going to hunt."

The hunt began in late 1898. Edison assigned a group of assistants to review and abstract everything that had been published on the subject of non-acid batteries, both primary and secondary, and both in the scientific and engineering literatures. All applicable French, German, British, and American patents and journal articles going back to the start of the nineteenth century were abstracted.[28] Edison even sent some of his people to France because: ". . . in France there was a period for years when the government did not publish its patents; yet notwithstanding this I had the written records searched."[29]

---

[26] Ibid.

[27] Josephson, *Edison*, p. 407.

[28] Assorted notebooks, *ENHSA*.

[29] Letter, Edison to Harjes, 16 January 1905, *ENHSA*.

Nevertheless, Edison's notebooks confirm one point made by his biographers; at no time during his automobile battery work did Edison contemplate working on the traditional lead-acid battery. This decision was determined by a number of factors. First, as an experienced commercial inventor, Edison knew the danger of entering a field where somebody else already controlled many of the potential innovative paths with masses of patents. The "somebody else" here, of course, was ESB. It is clear from what will be said later that ESB would have launched a massive legal campaign against Edison if he had attempted any lead-acid option.

Second, it simply was not Edison's style to enter a field where the technology was already well developed. Edison's style was not that of W.W. Gibbs and the founders of ESB, who bought their way into an already commercialized technology and then made profits from selling the products of that technology. Edison's technique was to take pre-existing, but as yet non-commercialized technological ideas and to reduce these ideas to a set of commercially-viable patents. The Edison laboratory would then sell these patents to new companies created for the purpose of attracting capital to buy the patents. The money obtained from the patent sale would then be used to generate new patents in the laboratory, and thus keep the cycle going.

Here we see the major difference between the operation of Edison's laboratories and a modern R&D center in industry. The modern research center, obtaining its support from the normal economic operation of a large manufacturing company, can engage in a lot of long-term research and investigate new ideas which are far from commercial exploitation. The Edison labs, on the other hand, did not have this economic support; they were supported solely from the sale of patents. Research on brand-new ideas was a risky pursuit for Edison, and most of his efforts, therefore, were directed towards producing the crucial, commercializing patents for technologies which others had already pushed fairly well along before Edison began his work.

Thomas Hughes has recently pointed this out with regard to Edison's development of the incandescent light bulb: [30]

It was Edison's custom to read the available technical literature and patents in the area he intended to enter because, as an

---

[30] Thomas Park Hughes, Thomas Edison—*Professional Inventor* (London, Science Museum, 1976) pp. 21-22.

experienced inventor, he knew that major inventions had often resulted from small but critical improvements in existing devices that had not proved commercially successful or technologically feasible. History records no less than twenty inventors . . . who had experimented with incandescent lamps before Edison . . . and (this) earlier effort with incandescent lamps indicated the technological possibility of a less intense interior light.

We might well paraphrase the last sentence to read: history records no less than a dozen inventors who had experimented with alkaline storage batteries before Edison, and this earlier effort with storage batteries indicated the technological possibility of an electric vehicle battery lighter more durable than the lead-acid. Far from being the lone pioneer of his biographies, Edison followed very closely in the tracks of his predecessors.

The first laboratory work was begun in the summer of 1899, after about nine months study of the literature. This early research was devoted mainly to a study of the Desmazures-Waddell-Entz type batteries. In June, Edison ordered stocks of zinc, copper, and copper oxide for his West Orange laboratory, and for the next thirteen months the laboratory notebooks were filled with experiments in the CuO-KOH-Zn system.[31] Variants on the basic Lalande-Chaperon system were also tried, some involving different copper compounds. Various additives were mixed with the zinc and copper compounds, apparently with the hope of improving the replating or preventing colloid formation. Experiments were also carried out on electrode supports for the active materials; these were also derivative of the earlier work; for example, the materials were packed into fine nickel wire mesh, into perforated sheet metal pockets, or into porous carbon blocks.[32]

About six months after starting the copper-zinc work, Edison began some experiments on silver oxide depolarization. If we may judge from the number of entries in the lab notebooks, this work was apparently considered of less importance than that involving copper oxide. Nevertheless, experiments with Ag_2O-KOH-Zn and Ag_2O-KOH-Cu cells were carried on from December 1899 until September 1900 when both these and the CuO-KOH-Zn tests ceased.[33]

---

[31] Letters, Edison to Matthiesen & Hegler Zinc Co., 26 June 1899 and 12 September 1900, *ENHSA*.

[32] Lab notebook, entries dated: 17 August 1899, 15 December 1899, 12/1899, 21 August 1900, 22 June 1900, 15 December 1899, 13 August 1900, *ENHSA*.

[33] Lab notebook, entries dated: 15 December 1899 and 24 August 1900, *ENHSA*.

At this point it may be well to stop and ask if Thomas Edison was not reading over Waldemar Jungner's shoulder. This is possible, since Jungner began his work well before Edison. It is more likely, however, that both men, by following the "state of the art" of alkaline storage battery research, were thereby led along very similar, convergent paths. This is not to say that the direction of Edison's research effort was completely independent of Jungner, since Edison's correspondence indicates that he was keeping tabs on what his Swedish rival was doing. Nevertheless, the Edison laboratory notebooks indicate similar patterns of thought on the part of these two inventors, rather than a case of copying.

A two-page spread from one of the notebooks written in Edison's hand during mid-summer 1900 demonstrates this point. The following is a copy: [34]

Alkalies

Copper	Slightly Soluable	Nickel	*Insoluable*
Aluminum	Very Soluable	Niobium	Soluable
Bismuth	*Insoluable*	Osmium	Soluable
Boron	Soluable	Palladium	Slightly Soluable
Antimony	Soluable	Phosphorus	Soluable
Arsenic	Soluable	Platinum	Somewhat Soluable
Barium	Soluable	Rhodium	Soluable
Berryllium	Soluable	Rubidium	Soluable
Cadmium	*Insoluable*	Ruthinium	Soluable
Calcium	Soluable	Samarium	
Cerium	*Insoluable*	Scandium	
Chromium	Soluable	Selenium	Soluable
Cobalt	Slightly if at all Soluable	Silicon	Soluable
		Silver	Slightly Soluable
Didymium	*Insoluable*	Shontium	Soluable
Erbium	Probably Insoluable	Sulphur	Soluable
Gallium	Soluable	Tantalum	Soluable
Germanium	Soluable	Thallium	Soluable
Gold	Soluable	Thorium	*Insoluable*
Indium	Soluable	Thulium	
Iridium	Soluable	Tin	Soluable
Iron	Soluable	Titanium	Soluable
Lanthanum	*Insoluable*	Tungsten	Soluable
Lead	Soluable	Uranium	Soluable
Lithium	Soluable	Vanadium	Soluable
Magnesium	*Insoluable*	Ytterbium	
Manganese	Soluable	Yltria	*Insoluable*
Mercury ?	Slightly Soluable	Zinc	Soluable
Molybdenum	Soluable	Zirconium	Soluable

---

[34] Lab notebook entry dated 1 January 1900, *ENHSA*.

No comments accompany this table, but its meaning is clear nonetheless. From mid-1889 to mid-1900 Edison experimented with the soluble zinc anode, but was eventually defeated by the same problems that had beset Waddell-Entz and others. In a lab notebook entry dated 6 September 1900, Edison admitted this and announced that he was taking a new approach:[35]

Last few days have been experimenting on the negative plate— instead of plating zinc out of solution I have sought for a finely-divided (i.e. porous) plate of a metal which would oxidize (he must mean *reduce*, since the anode is reduced to the metallic state during charging) to the metal by the current. Have proposed metals reduced in hydrogen in tube by heat of (sic) Molybdenum, Tungsten, Iron, Cerium and Cadmium.

By mid-1900, therefore, Edison had come to the same conclusion that Darrieus and Jungner had already come to—the only practical alkaline battery design was that using insoluble electrodes. Since he knew of the Darrieus patent, and probably had a copy of Jungner's 1899 patent by this time, it is doubtful if this conclusion was arrived at entirely independently. The list reproduced above, which was doubtlessly prepared from a literature search, represents Edison's attempt to define all those metals for possible use in storage batteries. The underlined metals (excluding the rare ones) are: Bismuth, Cadmium, Magnesium, and Nickel. All except magnesium had played a major part in Jungner's research. This demonstrates the tendency of battery design ideas to converge, even though directed by only a single concept such as insolubility.

Edison became disenchanted with zinc from June 1900 on. Like Jungner, however, he did not so readily perceive the problem with the use of the copper oxide cathode. Thus, in June, Edison began running tests on a CuO-KOH-Cd battery.[36] Apparently the copper-cadmium work went well; by September Edison was saying:[37] "If this cadmium works O.K. it will be a good solution to the light storage battery."

By the end of October Edison filed an application for his first patent on a copper-cadmium storage battery and by the beginning of December 1900 he had his agents out searching for large-scale cadmium ore deposits to buy.[38]

---

[35] Lab notebook, entry dated 6 September 1900, *ENHSA*.

[36] Lab notebook, entry dated 22 June 1900, *ENHSA*.

[37] Lab notebook, entry dated 6 September 1900, *ENHSA*.

[38] U.S. patent #871,214—19 November 1907; Letter, Mallory to Darling, 8 December 1900, *ENHSA*.

Towards the end of December, however, hopes for the new cell started to dim. Edison realized that if his battery was going to compete with lead-acids, the raw material costs would have to be fairly low—lead cost only 4¢ a pound,[39] but a pound of cadmium cost, at least $1.20.[40] Consequently, cadmium experiments disappeared from the notebooks around the beginning of 1901.

At the same time that he was abandoning cadmium for the anode, Edison was also giving up on copper oxide for the cathode. Here, unfortunately, the notebooks give no clue to the reason, but it is probable that, like Jungner, he had discovered the copper's tendency to migrate to the anode and poison it. A final patent application was filed for a copper-cadmium battery in March, but serious efforts to commercialize the system had ended three months before.[41]

Having abandoned both cadmium and copper, Edison turned his attention in December 1900 to a nickel or cobalt oxide depolarizer in combination with an iron anode. This occurred at precisely the time that Jungner was making a similar decision to concentrate either on a nickel-iron or a nickel-cadmium cell. Again, although it is possible that Edison was following Jungner, this seems very unlikely, and it is more likely that both men came to almost identical systems because they were following the same innovative path. Edison's adoption of iron for the anode is curious, since he had earlier mistakenly identified it as being soluble in alkaline electrolyte (see table on p. 356). Edison, like Jungner, had always used nickel as a grid material, but, again like Jungner, he seems initially to have rejected it as a depolarizing material. This is not surprising, considering nickel oxide's lower electrical conductivity compared to copper oxide, as well as its much lower theoretical energy density: 40.6 versus 19.5 ampere minutes per gram for CuO and NiO_2 respectively.[42] Nonetheless, the necessary rejection of copper oxide forced both men to turn to nickel.

In early 1901 Edison committed himself to the commercialization of either a nickel-iron or a cobalt-iron battery. Cobalt did not play a very important role in Jungner's experiments, but Edison toyed with the idea of using cobalt to replace nickel from time to time as his development work went on. He never

---

[39] Stanley, *Nickel*, Chart #3.

[40] Letter, Edison to (illeg.), 18 September 1900, *ENHSA*.

[41] U.S. patent #678,722—16 July 1901—filed 1 March 1901.

[42] George W. Heise & N. Corey Cahoon, *The Primary Battery* (New York, Wiley, 1971) (Vol. I). p. 176.

took the idea to the stage of commercialization, mostly because of cobalt's high cost.[43]

At the start of March Edison filed the basic nickel-iron/cobalt-iron patents, and by mid April a chemical plant was being set up at Silver Lake, New Jersey, for intensive work on improving the electrochemical activity of the nickel and iron active materials.[44] On 25 May 1901 the Edison Storage Battery Company was formed and the patents were sold to this enterprise. By July the first attempts were underway to secure large-scale sources of nickel.

Edison spent two years selecting a suitable electrochemical system for his storage battery. Although a great number of experiments were performed, these experiments were designed to choose between no more than about half a dozen possible systems: CuO-Zn; Ag_2O-Zn; CuO-Cd; CuO-Fe; $Ni(OH)_2$-Fe; and $Co(OH)_2$-Fe (with some speculation about the use of magnesium). Edison selected these systems either because others had already worked on them, or because the general design ideas put forward by their prior inventors pointed in the direction of these systems. Each of the systems, therefore, was chosen with a good deal of forethought and reading of the literature, and not, as stated in the Edison biographies, by a wide-ranging process of elimination resulting from laboratory searches. The laboratory notebooks give no indication of such searches. Indeed, it is clear from these records that the Edison lab never studied more than two possible systems at any one time.

If Edison were alive today and attempting to invent a new battery, he might adopt the pattern of basic laboratory research ascribed to him by his early biographers. But it is anachronistic to imagine that such a process could have been tried at the turn of the century. Edison's Menlo Park and West Orange labs did not have the resources to carry out large-scale exploratory researches. They were invention mills, or, perhaps more precisely put, they were patent mills. The organization was finely tuned by Edison to mass-produce the small but significant innovations necessary to convert an "almost technology" into a commercial process.

When Edison created his first laboratory in the 1870s, he established that pattern and it never changed over the years. The first lab was designed specifically to mass-produce patents

---

[43] Lab notebook, entry dated 4 April 1901, *ENHSA*.

[44] U.S. patents: #678,722—16 July 1901—filed 1 March 1901; #701,804—3 June 1902—filed 1 March 1901; #704,304—8 July 1902—filed 1 March 1901; #700,135—13 May 1902—filed 5 March 1901; #700,136—13 May 1902—filed 5 March 1901.

for the "almost technology" of multiplex telegraphy. A good deal of work had already been done on this technology and it was just entering the commercial stage when Edison set up his Newark patent mill. Edison was able to establish dominance over emerging technologies by his careful reading of the most recent patents, his concentration upon the paths which seemed most promising, and his subdivision of the work along these paths among separate groups of workmen. This was the pattern of Edison's work on the telegraph, the electric light work, and the storage battery.

After his selection of an electrochemical system, Edison set to work designing a mechanical construction for his battery. With his usual flamboyant manner, he was soon announcing to the press the imminent introduction of the new battery on the market, and the consequent death of the lead-acid battery. Nonetheless, it took Edison three years, from 1901 to 1904, to develop his first commercial battery. This model, called the Type E, was almost identical in construction to the Jungner cell. It consisted of a set of electrodes made of small, thin, flat pockets of perforated, nickel-plated steel sheet. The pockets for the anode and cathode were identical. The active material in the anode pockets was a carefully purified, metallic iron powder. The cathode pockets held NiO_2 powder. The cell reactions are complex, but can be represented simply:

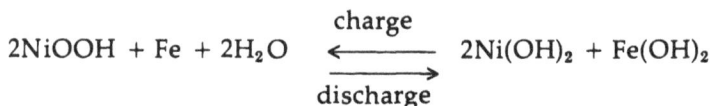

$$2NiOOH + Fe + 2H_2O \xrightleftharpoons[discharge]{charge} 2Ni(OH)_2 + Fe(OH)_2$$

The conductivity of these powders was increased by admixture of graphite flakes. The electrode pockets appear as in Figure in p. 361.[45]

The completed pocket assemblies were placed in the same configuration as that of a standard lead-acid cell, and inserted in the container, which was a sealed, welded can of nickel-plated steel.

At first the Edison batteries sold well. A number of electric vehicle manufacturers, including the EVC, designed their cars to hold the new battery.[46] Then, however, the problems started. Hairline cracks appeared in the outer steel can and the caustic electrolyte seeped out. In addition, after a number of

---

[45] Wade, *Secondary Batteries*, p. 138.

[46] Letter, Edison to (illeg.), 11 January 1904, *ENHSA*. Baker, National and Studebaker also adapted some of their cars to the Edison cell.

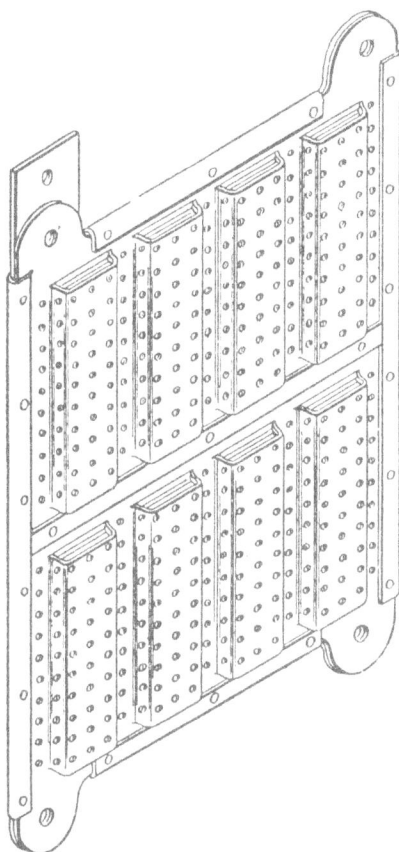

discharges the capacity of the batteries fell off. Edison realized that he had placed his battery on the market without really thorough testing; this may have been as a result of his fear that ESB or the other lead companies might have obtained control of the market before he could come on the scene. At any rate, Edison decided to shut down the factory, buy back all the bad cells, and return to the laboratory.

This time, however, Edison had no prior technological tradition in alkaline cell work upon which to base his design; he was faced with what turned out to be a hard, five-year push to develop a truly practical battery. The "d____d leaky can problem" proved to be fairly straightforward; by 29 December 1904, Edison revealed that he had the can welding technique perfected.[47] The loss of capacity, however, was a more fundamen-

---

[47] Letter, Edison to Parshall, 29 December 1904, *ENHSA*.

tal problem concerning the physical properties of the electrochemical system itself. It soon became apparent that the loss of capacity was centered on the cathode; in fact, the problem was analogous to the difficulty which the designers of lead-acid batteries for transport purposes had faced with the PbO_2 cathode during the later 1880s.

During the charge and discharge, the cathodes of the alkaline batteries underwent the same patterns of swelling and shrinking as in lead-acids. Consequently, the contact of the nickel oxide with the metallic current-conducting shells of the pockets continually changed. Given the very high electrical resistance of the active material, the influence of these contact changes increased the total resistance of the battery. Edison became so discouraged about solving the nickel cathode problem that he again contemplated giving up nickel and substituting cobalt, despite its high price and unavailability in commercial amounts. By the summer of 1905 Edison was conducting extensive experiments with a mixture of nickel and cobalt in the cathode, and this seemed to help the capacity problem somewhat.[48]

Consequently, in the summer and fall of 1905 Edison again had his agents out looking for a cobalt mine; in November he reported that he had just purchased "one of the best Canadian cobalt mines."[49] As in the case of cadmium, however, Edison found that the price of the cobalt was too high. In 1908 he stated that he had been ready to use 150 tons of cobalt a year in his batteries, but was deterred by cost. Bismuth also was found to be a useful additive for preserving the capacity of the nickel, but it again was found to cost too much.[50] Whatever happened to the Thomas A. Edison Cobalt Mine is unknown—his last comment on the subject of cobalt was: "I spent $128,000 in mines and chemical works and I am so thoroughly disgusted, that I shall never use it again."[51]

The solution of the cathode problem turned out to be mechanical rather than chemical. Just as the basic cathode problem which Edison faced was similar to the problem facing the lead-acid designers, so the solution also turned out to be similar. Edison developed tubular electrodes to replace the flat

[48] Lab notebook, entry dated 5 April 1905, ENHSA.

[49] Letter, Edison to Portland Cement Co., 17 November 1905, ENHSA; Edison to (illeg.) 18 January 1906, ENHSA.

[50] Letter, Edison to E.V. Machette, 10 January 1908, ENHSA.

[51] Letter, Edison to E.P. Earle, 17 January 1908, ENHSA.

pockets of the original cathodes. These were assembled into plates very similar to the configuration used in the Ironclad Exide six years later: [52]

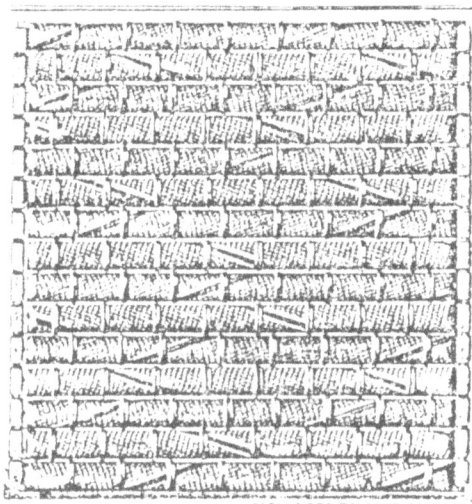

It has been suggested that the design of the Ironclad Exide was inspired by the Edison tubular positive. [53] This is certainly possible, although no direct proof exists. A more probable interpretation, however, is that both these two early twentieth century cathode designs were part of a developing tradition of cathode design which began in the late 1880s in the lead-acid industry.

The Edison tubular electrodes were patented and tested during early 1905. [54] Yet the development of the tubular electrode was only one part, and perhaps the less difficult part, of the solution. The innovation which was to make Edison's NiO_2 electrode a success was the replacement of the original graphite mixture with nickel flakes. This innovation was a characteristically Edisonian one—brilliant both in its scope and simplicity, highly ingenious, and yet rooted solidly in the technical traditions of the nineteenth century. The evolution of

---

[52] Smith, *Storage Batteries*, p. 154.

[53] Private communication from Dr. M. Barak.

[54] U.S. patents: #860,195—16 July 1907—filed 28 April 1905; #862,145—16 August 1907—filed 28 April 1905; #976,791—22 November 1910—filed 28 April 1905; #880,979—3 March 1908—filed 2 November 1905; #940,635—16 November 1909—filed 2 November 1905.

this technique shows the Edison engineering style at its best and most characteristic.

Edison felt that by switching from graphite to metallic flakes he could improve the conductivity inside the NiO_2 mass. But the problem was to make the flakes thin enough so that the grid pockets would not be filled up with so much nickel metal as to reduce the battery's capacity by the displacement of nickel hydroxide. This became more serious in the tubular grids, since they allowed for less internal volume than the older flat pockets. At first Edison tried rolling nickel wire into thin leaf, but this could not produce sufficiently thin material. Finally, during 1906 and early 1907, Edison developed a technique for producing nickel flakes so fine they would actually float on air: [55]

The photograph on page 365 shows an electroplating plant. The shiny cylinders are cathodes which receive the metallic deposits. Below the cylinders are tanks of copper and nickel sulphate solution, each tank holding alternately the copper or nickel solution. Small electric motors mounted atop each shiny cylinder kept them continually spinning. In operation, the cylinders were dipped alternately into the two plating solutions, remaining in each tank for only a few seconds before being raised, shifted horizontally to the next tank, dipped, raised, shifted back, and so forth. During each complete cycle drums acquired a microscopically-thin layer of each metal, one surrounding the other like the rings on a tree. The rotation of the cylinders insured a perfectly smooth layer. After a hundred complete cycles, the cylinders acquired a paper-thin deposit of two hundred concentric rings. This deposit was then sliced off the cylinders, unrolled to form a sheet, cut by dies into squares 1.6 mm on a side, and immersed in an acid bath. This leached out the copper, leaving the incredibly delicate nickel flakes for the electrodes. The basic patents for the process were filed January 1907.[56]

The nickel flakes were not simply mixed into the active material powder; Edison developed special machinery to tamp alternate layers of nickel flake and nickel oxide into the tubular pockets.

---

[55] Photo from glass negative, *ENHSA*.

[56] U.S. patents: #936,525—12/10/1909—filed 13 January 1907; #865,687—10/9/1907—filed 19 January 1907; #865,688—10/9/1907—filed 19 January 1907.

By 1908 Edison was confident of having solved all the problems, but it was not until July 1909 that testing was completed and full-scale commercial manufacture of the "A" cell had begun.

A number of historians have demonstrated how Thomas Edison used his studies of the economics of gas lighting to tailor the technology of electric lighting to fit these commercial realities. Although the records are not as elaborate for Edison's planning for the storage battery design, it is clear that he followed much the same procedure here, and made at least some of his decisions with an eye on his competitor's activities.

Edison perceived that he had two major competitors to worry about; Jungner and the Electric Storage Battery Company. Jungner was the more immediate threat, since his company dealt with the same technology and therefore with the same patents. ESB, on the other hand, was a more serious threat on a long-range basis, given its capital and patent base, its innovative leadership in car battery design, and its links to European firms who would immediately adopt any improvements ESB made. It is interesting to note that Edison did not see the internal combustion car as the major threat to his dream of creating a major electric vehicle battery market, nor did he anticipate the idea of the electric starter; Edison's method of inventing provided him with a very acute vision of technologies he was close to, but he often failed to see the broader technological context.

Of his two potential competitors, Edison found Jungner easier to deal with. Jungner began building a commercial organization in the spring of 1900, when he established the Jungner Accumulator Company of Sweden.[57] Shortly thereafter, Jungner came to an agreement with AFA, whereby they were to operate his patents in Germany. AFA had been interested in alkaline storage batteries for some time, as shown by their earlier licensing of the Waddell-Entz patents on the continent. Jungner was not a direct threat to Edison in this country, since he apparently contemplated no American manufacture of his cells. Nevertheless, AFA's agreement with Jungner, and the cartel between AFA and ESB, certainly indirectly threatened Edison, should the major lead-acid manufacturers have decided to enter the alkaline field in a big way.

The Jungner-Edison struggle began on 3 March 1903 when

---

[57] Falk & Salkind, *Alkaline Batteries*, p. 16.

Edison's basic nickel-iron patent was granted in Germany.[58] In October 1904 Sigmund Bergmann established the German Edison Accumulator Company in Berlin to work the Edison patents. Bergmann was an associate of Edison's from the early incandescent bulb days. During the 1880s he had manufactured fittings for the Edison system in a New York factory. Bergmann returned to his native Germany in 1890, where he founded the firm which became the giant Bergmann Electricitätswerke AG.

With the formation of the German Edison Accumulator Company, AFA[59] started a press campaign to discredit the Edison battery. For example, the previously mentioned Dr. Schoop, who was allied with AFA in 1904, wrote an article for *L'Éclairage Électrique* which attacked Edison for copying the Jungner battery. Among other things, Schoop claimed that Jungner had conceived the nickel-iron cell as early as 1893![60] In December 1904 one of Edison's Paris correspondents wrote him that:[61] "As soon as the German company was formed, the newspapers came out with articles, stating that the Edison battery infringed the Jungner patent, and practically intimated that the whole enterprise was simply an American enterprise to fleece the German public." Edison pretended not to be worried about the Jungner battery, and denounced the Swede's work as "worthless."[62] Nevertheless, he was concerned about the situation; in October 1902 a letter from a European correspondent informed Edison that General Electric had an agent in Europe who was bargaining for the Jungner patents.[63]

Although such information shows much about the heated nature of the Edison-Jungner competition, most of it appears to have been unsubstantiated rumor. For example, in late 1903 Edison heard that Bergmann had made a secret deal with AFA to manufacture the Jungner battery, to which the choleric Bergmann responded:[64]

---

[58] Brandon Bros., Paris to Edison 12/6/1904; Bergman, Berlin to Edison 8 December 1903, *ENHSA*.

[59] Actually, the licensee of the Jungner patent in Germany was Gottfried Hagen, and not specifically AFA. AFA, however, was allied on this project with Hagen.

[60] Not long thereafter Schoop was to have a falling out with AFA and joined Bergmann. Vol. 40 (17 September 1904): 441-449.

[61] Letter, Brandon Bros. to Edison, 6 December 1904, *ENHSA*.

[62] Letter, Edison to Stewart, 15 July 1901, *ENHSA*.

[63] Letter, (illeg.) to Edison, 8 October 1902, *ENHSA*.

[64] Letter, Bergmann to Edison, 8 December 1903, *ENHSA*.

That I should have gone to Hagen to acquire an interest on the
Jungner patent is simply nothing but a big lie and after you have
read the translation on the Jungner battery (i.e. the patent
#110,210). . . . you will not think that I am a big enough fool to
waste my money on such a *thing*. I think no doubt the Jungner
people could make great capital out of it, if I would take an
interest in their, I think, hopeless case. Now . . . just let us start
in and manufacture the Edison battery and we will shut them up
d—n quick.

A lengthy patent fight ensued in the German courts. Finally,
in January 1906 the court overthrew the basic Jungner patents
on grounds of non-originality, but upheld the Edison
patents.[65]

After this defeat, a number of tenders were made by the
Swedish Jungner firm to Edison to come to an agreement. For
example, in 1909 the former Edison employee John Roos, then
living in Stockholm, wrote Edison that:[66]

Being in a friendly way connected with both parties I propose to
you as I have to the leading men of the Jungner Co., that holders
of the Edison and Jungner rights should try to come to an understanding that would put an end to this fight. I have reason to
know, that the Jungner people would like to partake in such an
arrangement.

Edison replied in his typically blunt fashion: "My dear Roos;
Under no circumstances will I have anything to do with Jungner. Edison."

In 1906 Frank Dyer wrote to Dr. L. Sell at the German Edison
Accumulator Company:[67]

Now that the Jungner patent is finally cancelled, I should imagine that the Kölner Akkumulatoren Werke Gottfried Hagen and
the main company Akkumulatoren Fabrik Aktiengesellschaft
Berlin-Hagen (AFA) would be entirely willing to make a reasonable arrangement under which they would agree to withdraw
from the manufacture of alkaline batteries. . . .

That is not what happened, however; in conformity with
battery industry's tradition of buying up one's competitors,
AFA purchased DEAC (Deutsche Edison Akkumulator Com-

---

[65] Letter, Moffert and Sell to Dyer, 10 January 1906, *ENHSA*.

[66] Letter, John D. Roos to Edison, 1 February 1909, *ENHSA*, from microfilm.

[67] Letter, Dyer to Sell, 26 January 1906, *ENHSA*.

panie) in 1913. It has remained a part of VARTA ever since.[68]

Edison found the challenge of the lead-acid battery harder to overcome than that of Jungner. For one thing, there were no patents here that he could attack. The lead-acid manufacturers of the United States and Europe were making advances in the designs of their vehicular batteries, and these advances were certainly not "based on theory." The correspondence between Sigmund Bergmann and Edison reveals that Bergmann's chief fear was not the Jungner battery, but that Edison's lengthy development work on his improved battery would allow the lead-acid people to get an economic stranglehold on the market which would prove increasingly harder to break. In January 1903 Bergmann wrote:

> Please tell Edison not to wait until he gets the last improvement on the battery, but give me a battery to start manufacturing. . . .
> If we wait too long, I am afraid we are losing valuable time, as the Other Storage Battery people—I tell you—are not lying still and idle.[69]

And in November 1906, Bergmann wrote to Edison,[70] . . . the Lead Storage Battery people have made such improvements and their prices are so low, . . . that as the battery stands today, we cannot do business without a big loss.

After Edison gave the go-ahead to restart manufacture, Bergmann wrote to him in September 1909 "The lead people are doing everything in their power to prevent us from coming to the surface, but I think we are in a position to put lead cells out of business."[71]

Before 1909 Edison had a difficult time holding his German associate back. Bergmann had invested in machinery and plant and was being prodded by the shareholders to manufacture despite Edison's qualms. Edison, however, realizing the difficulty of challenging the lead-acid industry with an imperfect battery, had to restrain his impatient partner:[72] "Now Bergmann there is no use going ahead with manufacturing until I find the trouble otherwise you will get into a world of trouble. . . . you will have to meet a very serious loss and ruin of your future prospects."

---

[68] Treue & Nadolny, *Varta*, p. 58-60.

[69] 22 January 1903, *ENHSA*.

[70] 19 November 1906, *ENHSA*.

[71] 1 September 1909, *ENHSA*.

[72] Letter, Edison to Bergmann, 1/1904, *ENHSA*.

In the United States the chief potential rival of the Edison alkaline battery was the Exide of ESB, and, Edison studied the performance characteristics of the Exides while designing his own cells. The records of the West Orange laboratory contain ESB data and documents which Edison used for this purpose. These documents were obtained by Edison through an intermediary who had contacts at ESB; the correspondence between the intermediary and Edison describes one of these contacts as: "One of our friends, who is in a position to act intelligently. . . ."[73] Edison kept track of such diverse data as ESB stock prices, the number of workers at the plant and the volume of sales.

The most important information which Edison collected, however, was data on the ESB research effort. Edison studied the data from ESB and had Exide batteries tested regularly by his own staff. It is difficult to tell from the records precisely how Edison planned his own design efforts to compete with those of ESB. It does appear, however, that when he first began the storage battery project in late 1898 and early 1899, Edison based his concept of the best lead-acid vehicle battery on the ESB's chloride negative-Planté positive design, which was then being used in the Morris-Salom cabs. Indeed, that is all he could have done at the time, since ESB had not yet developed the Exide.

This explains Edison's initial underestimation of the difficulty of designing an alkaline battery which would replace the lead-acid. In 1899-1900 Edison was predicting to the press that he would soon have a battery which would far outclass the lead-acid. Given what we have said in the last chapter about the early batteries of the Lead Cab Trust, it would appear that Edison was being candid and not unreasonably optimistic in making these predictions. But what Edison apparently failed to take into account was the rapid pace of storage battery innovation during the late 1890s. Edison's contempt for the lead-acid battery concept was real, and this may have been the reason for his failure to anticipate the rapid advance of vehicular battery design.

For example, in early 1901 Edison had some tests run on the latest ESB vehicle batteries. These were the early Exides. When he received the data from his staff, he was confused, and wrote back:[74]

---

[73] Letter, Pilling & Crane to Edison, 15 February 1901, *ENHSA*.

[74] Memorandum, Edison to Barstow, 24 April 1901, *ENHSA*.

There must be something wrong in the data you gave me as to the weight per horse power hour (i.e. energy density ratio). The catalogue of the Battery Company (i.e. ESB) calculated on their largest and best cell gives 164 pounds total weight per horsepower hour (i.e. 4.54 watt hours/lb). Their lightest automobile *chloride* (my italics) plate gives 188 lbs per hp hour (i.e. 3.96 watt hours/lb.) at normal discharge. Where is the error? You state 67 lbs. (i.e. 11.1 watt hours/lb.) at normal discharge.

Edison, therefore, as late as 1901, was basing his estimate of ESB technology on the chloride process, which they had abandoned a year before. This is understandable, however, since the Exide was only introduced in 1900 and it appears to have entered full-scale manufacture rather slowly. For example, two thirds of EVC's cabs still remained equipped with the old chloride-Manchester batteries as late as 1902.[75] Nonetheless, it is confusing that Edison should not have been more up-to-date on Exide progress; perhaps he was being fed false data by his ESB spies.

At any rate, Edison was partly right; the 11.1 watt hours per pound figure quoted by his lab was too high. Contemporary Exides gave about 8.2-8.3 watt hours at normal discharge rates, but this is still about *twice* the capacity quoted by Edison.[76] Edison's figures are correct for the pre-Exide ESB cells.

Between 1901 and 1904 Edison designed three models of the nickel iron cell —the "C", "D" and "E". These models had energy/weight ratios of 11.7, 11.5 and 10.2 watt hours per pound respectively.[77] In other words, their capacities were only equal to or just slightly better than contemporary Exides. Moreover, even before the type E was commercialized in 1904, the tendency of the battery to lose capacity with use had been noted, although it was not realized how serious the problem was until the customers started to complain. In addition, the slight weight advantage of the Edison cell had to be set against its considerably smaller energy efficiency—the energy efficiencies (that is energy out/energy in) of the early Exides and Edisons at normal discharge rates were 73 percent and 50 percent respectively.[78] Finally, the cost of the complicated Edison cell was a problem; between two to three times more expensive than a comparable Exide.

---

[75] *Electrical World*, 39 (12 April 1902): 643.

[76] *ENHSA*, microfilm Lab report, unlabeled page, 1910.

[77] *Trans. Am. Electrochem. Soc.* 1904: 135-151.

[78] Letter, Atlanta office of ESB to A.R. Whitman, 3 May 1910, *ENHSA*.

It is no wonder, then, that Edison stopped the manufacture of his type E cells in 1904 and spent the next five years improving their durability and capacity. By 1904 he must surely have been aware of the rapid progress being made in the durability of lead-acid vehicular batteries; failure to improve his own cells, therefore, would have left him with a battery of no greater capacity than the lead-acid, and probably not much greater durability, for several times the price. Except for the minor and easily-solved problem of the leaky cans, the 1904 Edison cell was not a bad battery. Even with its loss of capacity it was no worse than contemporary lead-acids. But Thomas Edison was well aware of the crucial differences between a technology which works and a technology which people will buy. By 1909, Edison had raised the capacity of his cells to over 14 watt hours per pound, and could guarantee that they would hold this capacity several times longer than the best lead-acids could hold their lower capacities.

In 1898-1899, when Edison first determined to develop an alkaline storage battery, he based his decision in part upon a misconception which most of the early alkaline battery pioneers had also made. This misconception is based on the heavy weight of the lead; building an alkaline battery without lead enabled the designer to use active materials whose theoretical energy densities were far higher than lead. It *should* have been possible to make an alkaline storage battery much lighter than a comparable lead-acid. Edison's comment that he did not believe that Nature would be so unkind as to withhold the secret of a good alkaline storage battery is a sign of this preconception. Unfortunately, Nature is unkind; alkaline battery active materials are lighter than lead active materials, but the complexity of the electrode designs required for alkaline cells cuts down on the percentage of active material weight to total weight to such a degree as to just about cancel the energy density weight advantage. In other words, what you gain electrochemically, you lose mechanically. And while it is true that practical alkaline cells have often a small weight advantage over lead-acids, this is usually not worth the extra cost of these more complex cells.

3. Consumer Alkalines: The Sintered Plate

The development of the Edison alkaline battery was a process of careful, meticulous engineering. This is particularly clear in the details of the battery's construction and in the machinery used in the manufacture. The complex techniques for making the nickel flake have already been discussed. Equally complex were the tubular-electrode forming techniques. The tubes were made from a stock of thin steel ribbon. This is perforated by being run under rotating pin wheels. Next it is nickel-plated and passed through rolls which form it into a tube. This process is not unlike the technique used to make the inner cardboard tubes for holding paper towels, except that the seams here are interlocked instead of being glued.[1]

The most detailed part of the operation, however, was the filling of the tubes. Edison developed tube-filling machines which perform 960 steps to fill a single ten centimeter long tube. Seen in cross-section, the filled tubes show alternate layers of flake and nickel hydroxide—each layer is less than 1/3 of a millimeter in depth and is prepared by a three step process—the machine drops a small quantity of hydroxide in the tube, then some flake, and finally tamps it. This is repeated 320 times. Finally, eight tiny steel rings are slipped on the outside of the tube to keep it from unraveling, and a series of the tubes are mounted in a grid.[2]

The thoroughness and attention to detail which Edison lavished on this project reveal the mechanical difficulties he faced in designing a battery to compete with the lead-acids. Yet it was just this detailed craftsmanship which doomed the Edison battery to play a minor role in the twentieth century battery market. The SLI market was captured almost completely by lead batteries of simple and cheap construction. The industrial electric truck market was likewise captured chiefly by lead-acids, although here the more durable, complicated, and expensive Ironclad design was usually chosen.

Nevertheless, interest in alkaline storage battery design did not end with Edison. Edison's contributions to the technology ended essentially with World War I. Edison encountered a number of disappointments with his battery during the war, and these may have soured his previous enthusiasm for it. For example, the United States Navy rejected the nickel-iron battery for use on submarines because it generated hydrogen gas

---

[1] Falk & Salkind, *Alkaline Batteries*, p. 101.
[2] Ibid., p. 102.

at a much higher rate than lead-acids.[3] Edison believed that one of the most popular uses for his batteries would be on submarines, as he wrote Bergmann in 1910:[4] "We are assured by the government that they will not renew any more lead batteries in the submarines and that all renewals will be made with our battery."

Of course, the appearance of SLI technology and the rapid decline of electric vehicles were also disappointing. Nevertheless, the failure of the electric vehicle in the United States did not eliminate the manufacture of Edison batteries in this country. They have been made ever since Edison restarted manufacture in 1909, although since 1960 the company has operated as a branch of ESB. Nonetheless, the American market for the nickel-iron batteries has always been fairly small. This explains why innovation at the American Edison company has been minimal since Edison's day. Until ESB took over the firm in 1960, the machinery, processes, and battery designs used by the Edison Storage Battery Company remained virtually unchanged from those introduced by Edison.[5] Innovations have been introduced since 1960; a fact which reflects increased American interest in alkaline cells during the last decades.[6]

The renewal of American interest in alkaline storage batteries since the Second World War has been chiefly due to the earlier, interwar innovative activity of European firms. In Europe, interest in alkaline storage batteries continued strongly after the First World War. One reason for this strong European interest was the much less rapid decline of interest in electric vehicles in Europe, produced by a number of factors such as the greater cost of gasoline there. Another factor of major importance was the Second World War, which stimulated the commercial production of the sintered plate alkaline cell. The sintered plate was one of the two main innovations produced by the European battery industry between the wars.

The other major innovation was the substitution of cadmium for iron as the negative active material. Cadmium had several advantages over iron. Nickel-iron batteries tend to

---

[3] "A Streak of Luck," Robert Conot, p. 390. Edison Battery received a trial in USSE-2 in 1915, Explosion 15 January 1916; Edison-Matthew Josephson p. 422-423; T.A.Edison Album p. 138 *Technical World* February 1915.

[4] Letter, Edison to Bergmann, 12/24/1910, *ENHSA.*

[5] Private communication from the late Dr. E. Lighton.

[6] Ibid., Falk & Salkind, *Alkaline Batteries,* pp. 96-97.

self-discharge after a time, while nickel-cadmium cells hold their charge longer. At low temperatures, with high rates of discharge, iron can become passive. In addition, cadmium oxide has better electrical conductivity than iron oxide—always a factor of critical importance in alkaline battery design.[7] Finally, manufacture of the iron active material is difficult, since the anode material is in the reduced form. The active material must be in a finely-divided state, yet if reduced iron is prepared in this state it quickly oxidizes and is not easily reduced. Cadmium does not oxidize as rapidly.

As indicated earlier, both Edison and Jungner had experimented with cadmium; it was the Swedish Jungner company, however, which commercialized the technique. The defeat of the Jungner nickel-iron patent by DEAC in 1906 did not end the plans of the Jungner company to market a battery. As with most such patent struggles, the contest was more one of economic attrition than of the outright exclusion of one or the other of the competitors—between 1905 and 1910 Jungner's company was twice forced into liquidation and was twice reorganized.[8] Throughout all this, however, the design of alkaline batteries continued.

From 1909 onward the Jungner company concentrated on a nickel-cadmium system. Actually, the system should be called a nickel-cadmium and iron cell, since the anode mixture consists of both these materials, but with the cadmium predominating. The reason for this mix is that the Jungner company found that cadmium and iron both possess undesirable properties which can be cancelled by the other metal. Thus, for example, if finely-divided iron is co-precipitated with cadmium from solution, the resulting alloy powder will protect the iron component from oxidation. The iron is necessary since pure cadmium in a finely-divided state has a tendency to "densify," in a manner analogous to the anodes of lead-acid batteries—a large number of small crystals form into a smaller number of large ones.

In 1909 Axel Estelle, president of the reorganized Jungner Company, patented a process for the co-precipitation of cadmium and iron.[9] Initially, Estelle was chiefly interested in this

---

[7] *Trans. Am. Electrochem. Soc.* 9/1939: 435-452; C. Drucker & A. Finkelstein, *Galvanische Elemente und Akkumulatoren* (Leipzig, 1932) p. 232.

[8] Falk & Salkind, *Alkaline Batteries*, pp. 18-19.

[9] British patent #9964 (1910), patented Sweden 1910 (A.T.K. Estelle).

technique as a means of producing iron active material, but it evolved later into a process for making anodes which were chiefly cadmium.[10] The Jungner-Estelle patents were commercialized throughout Europe from 1910 onward, although no attempt was ever made to manufacture in the United States. The Swedish company developed its own plant, as well as licensing its patents to companies in other countries. For example, in 1918 the firm of Batteries Ltd. was set up to work the Jungner-Estelle patents throughout the British empire, and a factory was established at Redditch, Worcestershire.[11] The Jungner company sold its batteries under the misleading name: NIFE; a remnant of the pre-cadmium days.

The mechanical design of the NIFE batteries was similar to that of the Edison cells, with a few variations in detail. The same perforated, nickel-plated steel pockets were used to hold the active material, although the Jungner construction was flat instead of tubular. Moreover, the NIFE cells were made with graphite conducting flakes instead of nickel. The similarities between these two technologies made them easy to blend. During the interwar period, a considerable amount of such blending went on in the various European firms created to make alkaline batteries—NIFE, DEAC, Jungner, Alkum, Britannia, to name a few.

One of the largest of these producers was SAFT. SAFT— Societé des Accumulateurs fixes et de traction—was created in France in 1920 to manufacture nickel-iron cells, although by the 1930s they were also making the nickel-cadmium model. The mutual influence of the Edison and Jungner models is evident in this company's activities. For example, in the 1920s, SAFT used the Jungner-preferred graphite flakes to add conductivity to the positive pockets, but used Edison-type nickel flake for the negatives.[12] The pockets used by the 1930s were also a compromise between the Jungner and Edison designs—Edison tubes for the positives and flatter pockets for the negatives.[13]

All alkaline batteries produced commercially before the Second World War were a combination of the Edison and Jungner designs. Nevertheless, the complexity and expense of these

---

[10] Falk & Salkind, *Alkaline Batteries*, p. 17.

[11] *Nife—The Original Alkaline Battery* (1939) (advert. pamphlet).

[12] Rodolphe Herold, *Étude theorique et pratique de l'Accumulateur Fer-Nickel* (Belgium, 1924) p. 8.

[13] J. Salauze, *Revue du Nickel*, 3/1936, pp. 2-3.

designs stimulated a number of attempts between the wars to develop simpler types of alkaline storage batteries. The electrochemical systems tried during this period were the same as those tried by Jungner, Edison, and their predecessors. For example, a number of attempts were made after 1908 to commercialize the silver oxide depolarized system, but none of these attempts advanced very far.[14]

Most of the interwar attempts to design mechanical grids other than the Jungner-Edison pocket pattern were also throwbacks to the design of Waddell-Entz, Desmazures, et al. For example, in 1927 Jirsa and Schneider in Germany suggested forming electrodes by subjecting silver or iron oxide to great pressure to force it into a metal screen framework.[15] They reported that plates so prepared did not break up under service—a remark which suggests that the researchers did not continue the tests long enough. As we have seen, this technique was proven impractical before the turn of the century. Nevertheless, it continued to be a popular theme with patentees between the wars. In 1920, for example, two French inventors filed an application for a United States patent describing various complicated methods of twisting nickel or nickel-plated wires into complex geometric shapes for the purpose of holding the nickel or iron active materials in place.[16] The active powders were to be pressed into the mass of filaments under intense pressure.

A number of attempts were also made to prepare electrodes by electrochemically depositing relatively thin deposits of active materials on thin nickel plates.[17] The most elaborate of these attempts was made in the 1930s by James Zimmerman, an electrical engineer in Madison, Wisconsin.[18] He experimented with various combinations of silver, nickel, iron, zinc, and cadmium deposits on thin nickel plates. Zimmerman reported that the results were encouraging, but as with most alkaline battery research in this period, no further developmental work was undertaken.

There was one exception, however; a spectacular attempt to produce a simpler, lighter, cheaper alkaline storage battery

---

[14] *Zeithshrift Für Electrochemistrie* 33 (1927): 129.

[15] *Zeit. f. Elektrochemie* 33 (1927): 129.

[16] U.S. patent #1,447,657—3/6/1923—filed 14 September 1920 (Paul Gouin & Edmond Roesel).

[17] *Trans. Amer. Electrochem. Soc.* #68 (1936): 231-249.

[18] Ibid.

was carried to commercial proportions in the interwar period.
This was the Irish government's Drumm battery project. The
Drumm battery fiasco is a good example of an inventor's dup-
lication of the mistakes of his predecessors because of his failure to study their work.

The Drumm project was begun in the late 1920s in the laboratory of Dr. James Drumm, a professor of chemistry at University College, Dublin.[19] Drumm, a part-time inventor, (he
had a new shaving soap and a method of preserving peas to his
credit) turned his attention in 1928 towards heavy electrical
engineering, with the intention of designing a storage battery
for a self-propelled train.

Although the late nineteenth-century vogue of engineering
interest in battery streetcars began to die out after the mid
1890s, the battery-powered railed vehicle continued to occupy
a serious, albeit small niche in the thinking of transportation
engineers well into the new century. The battery car seemed a
useful vehicle for low traffic feeder lines, where the service did
not justify the expense of overhead conductor electrification.
Thus, for example, the Edison storage battery was proposed
for use on a number of feeder lines and some cars were so
equipped in the years after 1910.[20]

No serious attempts, however, were made to run battery
cars on main lines, except for the Drumm project. In 1928
Drumm and a small group of Dublin businessmen formed
Celia Ltd; a small firm intended to finance Drumm's research
and the elaboration of his battery patents. These patents excited some enthusiasm in the electrical press in the late 1920s
and early 1930s. Drumm kept the construction details a secret,
at first, leading others to the belief that he had invented a new
form of alkaline cell. As one writer in the London *Electrical
Review* said in September 1930:[21] "The real reason why no one
can criticize the invention is that it involved a completely new
electro-chemical phenomenon. . . ."

In fact, however, if we examine Drumm's patent claims in
detail, there is nothing that had not appeared in many patents

---

[19] *Electrical World,* 25 October 1930, p. 778; 16 August 1930, p. 285; 19 September
1930, p. 447; 29 November 1930, p. 985; 12 December 1931, p. 1029; 12 March 1932, p.
478; 16 April 1932, p. 694; 29 October 1932, p. 584; 15 October 1932, p. 521; 2 September
1933, p. 294; 3 February 1934, p. 202; *Electrical Review* (London), 22 August 1930, pp.
286, 290-292; 5 December 1930, pp. 978-979; Illingworth, *Battery Traction,* pp. 20-22.
[20] *Electrical Review* (New York), 1/1910: 268; 13 January 1912: 68.
[21] *Electrical Review* (London) 19 September 1930: 447.

over several past decades: 1) Silver compound as a cathodic depolarizer; 2) graphite flakes mixed with the Ag_2O depolarizer; 3) nickel oxide mixed with silver oxide; 4) cerium oxide mixed with the silver oxide[22]; 5) zinc anode; 6) cadmium-iron anode; 7) container of nickel-plated steel or Monel metal.[23]

By 1930 Drumm had selected his desired system from among these possibilities. The positive electrodes were to use the same nickel hydroxide employed in the Jungner and Edison cells; in fact, the cathodes *were* Jungner's—they were manufactured for Drumm at the NIFE factory at Redditch and were an unmodified form of the standard Jungner NIFE positives. The negatives were constructed in the shape of flat plates made of nickel or Monel gauze to which zinc was electroplated. In essence, the Drumm battery was a hybrid Jungner-Desmazures battery. The mechanical construction was that of the NIFE.[24]

At about this point in the evolution of the project, the Irish government became interested. The government invested £5,000 in the development of the battery, and subsequently a further £30,000 in the construction of specially-designed trains to be run off the batteries. From this point on, Celia Ltd. essentially became a government enterprise. The scope of the project was expanded through the 1930s, with the trains being used as interurbans.[25] The main service for the cars was on the Dublin-Bray line—about 13 miles. As with the Waddell-Entz cars forty years before, the reports during the first years were enthusiastic. Professor Drumm claimed that a "huge leap forward" in electric traction has been made.[26] The trains were increased in size, so that by late 1930 they were capable of carrying 400 passengers. By 1934 plans were even afoot to use the Drumm cells in electric trucks; it was claimed that the life of such batteries was ten years as against only three years for lead-acids.[27]

This, of course, was nonsense; the problem of replating zinc from solution and preventing "trees" from forming has never yet been solved, although battery engineers continue today to struggle with the problem. It is quite possible that the batteries

[22] See U.S. patent #1,167,485—11 January 1916—filed 30 April 1912 (T. Edison).

[23] *Electrical Review* (London), 22 August 1930: 291.

[24] Ibid., 19 September 1930: 447.

[25] *Electrical World*, 16 August 1930: 285.

[26] *Electrical Review* (London) 19 September 1930: 448.

[27] *Electrical World*, 3 February 1934: 202.

lasted ten years, if only the life of the NIFE positives was considered. The replacement problem of the negatives, however, and the maintenance costs involved, made the project uneconomical, and it was abandoned in 1949. The trains were never tried outside Ireland.

Drumm believed that he was making an advance on Edison and Jungner by replacing their high resistance, complicated and expensive iron and cadmium negatives with his high energy, low resistance, cheaper zinc plates. What Drumm actually did, of course, was to resurrect a pre-Edison technology; a technology which both Edison and Jungner had tried and reluctantly rejected. It is worthy of note that the Drumm project was launched by men hitherto unconnected with the battery business. Moreover, none of the companies which had evolved out of the tradition of Jungner or Edison undertook projects like that of Drumm in the interwar period. These firms had a detailed knowledge of what had already been tried and what had failed, or at least what had not succeeded to anywhere near as great a degree as the Edison-Jungner approach. The pocket-type alkaline batteries, with their admixtures of nickel flake or graphite, were clearly an "inelegant" solution to the problem of making a durable battery—the pockets, tubes and flakes had a "brute force" character about them. Yet the existing companies realized that a new technological approach was needed to provide the more elegant engineering solution; the simple resurrection of nineteenth-century ideas and patents would not provide the solution.

This is not to suggest, however, that European alkaline battery manufacturers made no innovations during the interwar years. On the contrary, some of the improvements made were closely related to concurrent changes in lead-acid technology. For example, beginning in the mid 1930s, European alkaline battery makers began designing cells to capture part of the SLI market. These batteries were designed not for private automobiles, but for industrial and governmental vehicles— particularly motorbuses and diesel-powered trucks.[28] Despite the higher initial cost of these batteries over lead-acids, the European emphasis on durability of equipment led to a considerable use of these cells for SLI purposes in the later 1930s.

Cost, however, was not the only problem facing the use of nickel-cadmium or nickel-iron batteries for SLI applications.

---

[28] Trans. Am. Electrochem. Soc. 76 (1939): 435-453.

SLI batteries need to be able to produce high intensity discharges, and to do so even at low temperatures, at which sodium and potassium electrolytes grow considerably more viscous than the sulphuric acid of the lead-acids. This problem is compounded because electrolyte must penetrate the perforated pockets and the tightly compressed active powder of the alkaline cells.

Here we see another reason for the increased European emphasis upon nickel-cadmium cells in the 1930s and '40s, as compared with a decreasing emphasis on nickel-iron. In Edison and Jungner's day, the prime target market for the alkaline battery had been the electric vehicle—that is, a market for a deep-cycling battery. Moreover, maximum energy storage was of crucial importance during the first decade of the century, since the short range of the electric vehicle was a chief factor in the success of the electric's gasoline rival. Both Edison and (initially) Jungner emphasized nickel-iron over nickel-cadmium, since the former system has a considerable theoretical energy density advantage over both the Ni-Cd and the lead-acid:[29]

System	Watt Hours/Kilogram
Ni-Fe	267
Ni-Cd	209
Pb-Acid	175

This is fine, theoretically, but if we consider the practical problems of battery design, nickel-iron often stands at the bottom of the list of these three systems. Because of the high resistivity of the iron electrode and its tendency to become passive at low temperatures, for example, the Ni-Fe system is rated worst for low temperature performance, with Ni-Cd rated best.[30] The Ni-Cd system has a much lower electrical resistance than Ni-Fe; about the same as the lead-acid system. Moreover, its ability to hold a charge is superior to both Ni-Fe and lead-acid.

Consequently, as alkaline batteries were used more for brief discharges (SLI) rather than for cycling (electric vehicles) in the interwar years, the properties of the Ni-Cd system became of greater value than those of the Ni-Fe. The increased emphasis

[29] Falk & Salkind, *Alkaline Batteries*, p. 435.
[30] Ibid., p. 434.

on the nickel-cadmium system, therefore, can be seen as having been motivated by the same objectives as the concurrent reduction of antimony content in the grids and the increased use of expanders in lead-acid cells, but it should be pointed out that, although charge retention and preservation of capacity were the motivating factors in the abovementioned innovations in both the alkaline and acid storage batteries, the acid battery innovations were not quite concurrent with those in alkaline cells, but occurred somewhat later. The reduction in lead-acid SLI battery antimony content was begun only during World War II and continued afterwards, whereas large-scale European interest in shifting from nickel-iron to nickel-cadmium systems was already well underway in the 1930s.

This early European interest in nickel-cadmium was linked to the European market for military goods. Because of its charge-retaining properties, the Ni-Cd battery was widely used in all European armies by the later '30s.[31] The Ni-Cd cell's greater resistance to physical shock and resistance to damage by either over or undercharging also gave it a clear military advantage over conventional lead-acid batteries. The Ni-Fe batteries were no less favorable than the Ni-Cd's in these two latter qualities, but the ability of the Ni-Cd to hold its charge longer and to perform better at low temperatures gave it another military advantage over its alkaline rival. Given the interwar political situation, it is not surprising that much of Ni-Cd development occurred in Germany.

With regard to peace-time SLI battery innovation during the 1930s, however, European alkaline battery firms made another change which was also analogous to concurrent developments in lead-acids. In the early days of the alkaline storage battery it had been generally assumed that these cells were unsuitable for SLI use because of their high internal resistance, which would not permit the high current drains necessary for starting. In order to correct this, the thickness of the pocket-plates for alkaline SLI's was reduced by 100 percent in the 1930s from that of the pre-World War I alkalines. Also, the metallic pockets were provided with larger perforations to allow greater freedom of electrolyte penetration.[32]

The production of alkaline storage batteries at the AFA and

---

[31] *Trans. Am. Electrochem. Soc.* 76 (1939): 439.

[32] The decrease was from 2 to 4 mm thickness to 1 to 2 mm. Also, the interplate spacing was reduced from 4 to 2 mm to less than 1 mm. Ibid., p. 449.

Gottfried Hagen plants during the 1930s and early '40s is shown below:[33]

YEAR	AFA (Ni-Cd, in millions of amp. hours)	Hagen (in thousands of Marks)
1931		20
1935	5.5	200
1937	11.0	600
1940	21.0	840
1942	26.0	1050
1943	23.0	800
1944	26.0	

This spectacular growth stimulated German battery companies to develop the major technological change in alkaline batteries since the work of Edison and Jungner—the sintered plate. Unlike atomic energy and radar, sintered plate batteries are one of the lesser-known technologies which emerged out of the Second World War. Although this technology has not received much publicity, however, it has had a considerable impact economically and technologically on the post-war world. Before the war, all storage batteries fit the stereotype of the large, heavy, square black box.

The sintered plate, however, made it possible to miniaturize the storage battery to the size and shape of even the smallest dry cells. Moreover, sintered plate technology allowed manufacturers for the first time to alter the shape of the electrodes from the standard flat pattern. Theoretically, there is no limit to the thinness or shape of sintered electrodes. These properties, together with nickel-cadmium batteries, propensity to be totally sealed because they do not produce gas, permits present-day alkaline storage batteries to be made and used with the convenience of dry cells.

This has made possible the wide range of rechargeable tools and appliances developed during the past two decades—everything from rechargeable electronic calculators to cordless electric hedge trimmers.

The sintered plate is a variant on the mechanical techniques tried unsuccessfully by the earliest alkaline storage battery inventors, but combined with basic new technological methods. Waddell, Entz, Desmazures, et al had conceived the idea of depositing the active material on a fine metallic mesh plate. Edison had tried depositing his active materials in an electrode made of porous carbon, an idea that had its roots in the 1840s

---

[33] *FIAT Report #800; BIOS Report #708.*

telegraphic industry. The sintered plate is likewise a highly porous, metallic plate, whose random gridwork serves to hold large quantities of active material on its great surface area. The plate's structure is produced by heating a mass of finely-divided nickel crystals just until their points of contact melt and fuse; the interiors of the crystals do not melt. The result is a highly porous mass of great surface area, but one which also has coherence and good electrical conductivity. The porous masses form the electrode supports; the active nickel and cadmium salts are introduced into the porous mass as solutions and the plates dried.

The reason why the late nineteenth century porous grid concepts had all failed is that the porous structure of the plates must be truly microscopic in order for the plates to be durable. For example, in the sintered plate batteries, the size of the metallic crystals must be on the order of two to four microns for the process to be practical. Of equal if not greater importance is the shape of the crystals—they should be chain or filament-shaped, with small cross-sectional area. This will give the plates a low apparent density and provide a coherent, porous structure.[34]

The sinter powder crystals must also be chemically pure, particularly with regard to iron and copper impurities.[35] Such powders were not available in the nineteenth century, but beginning around the turn of the century a series of technological changes took place which made their production possible.

The first of these changes was the invention of the Mond process, which began as a method for the purification of nickel. Dr. Ludwig Mond, while working on improvements in the ammonia-soda process in the 1880s, noticed that valves containing nickel were attacked by carbon monoxide gas, causing the formation of a gaseous compound of the metal—nickel carbonyl: $Ni(CO)_4$.[36] This discovery, coming just at a time when world demand for nickel was rising rapidly, stimulated Mond to convert it into a commercial process. In 1892 Mond built an experimental plant to use the carbonyl gas to purify the metal. The first commercial Mond plant began operation in Wales in 1902.

The Mond process used chemical engineering techniques not available before the closing years of the nineteenth cen-

---

[34] Falk & Salkind, *Alkaline Batteries*, p. 114.

[35] Ibid., pp. 113-114.

[36] Charles Singer, *A History of Technology* (London, Oxford University Press, 1958) (Vol. V) pp. 88-89.

tury, but which rapidly became popular in the more advanced chemical process industries at the turn of the century. For example, the Mond process required the handling of a high temperature (200 °C), high pressure (200 atmospheres), highly toxic gas—nickel carbonyl. This process is carried out by passing a 200°C, 200 atmosphere stream of carbon monoxide over the finely-divided impure nickel. The resultant nickel carbonyl gas is then condensed and collected in high pressure vessels. It is then fractionally distilled to eliminate impurities, such as iron carbonyl—$Fe(CO)_5$. Finally, the purified nickel carbonyl is reduced to atmospheric pressure and the temperature raised to 250 °C, at which point it decomposes. The nickel precipitates as an extremely fine powder and the carbon monoxide gas is recycled to the start of the process.

At first, the Mond process was conceived only as a technique for producing pure nickel; the small size and thin, pointed, delicate shapes of the crystals were not of any importance. During the first decades of the twentieth century, however, a new form of metallurgy appeared which was to broaden the applicability of the Mond technique—powder metallurgy. This is based on the idea of the precision forming of pieces by pressing metallic powders into molds under high pressure and then sintering the pressing. Powder metallurgy demonstrated its value in making possible the production of new properties in metals. For example, the development of ductile tungsten for the tungsten filament incandescent lamp during the first decade of the century was an early triumph of powder metallurgy by the General Electric Research labs.[37]

In the 1920s, scientists at I.G. Farben conceived of using a modified form of powder metallurgy with Mond-prepared nickel powder to produce the sintered plate grid. In normal powder metallurgy, the intense pressure on the powder produces an object which has very little porosity after sintering. The Farben researchers found, however, that if the powder was gently pressed before sintering, a very delicate, open structure would be produced. In the basic patents for this process, filed in 1929, the Farben researchers used both nickel and iron carbonyl powders to produce sintered plates.[38]

Sintered plate technology is a synthetic one, that is, it was created from a combination of technological ideas and processes which were not created originally for battery manufac-

---

[37] Letter Edison to Bergman, 24 December 1910, *ENHSA*.

[38] British patents: #331,540—4 July 1930—filed 4 January 1929; #332,052—17 July 1930—filed 8 July 1929; #339,645—5 February 1930—filed 5 September 1929.

ture. The appearance of this method of battery manufacture was time dependent; it could not have arisen more than a decade or so earlier than it did, regardless of whatever need existed earlier for lighter alkaline storage battery construction. This is in contrast to the situation for organic technologies, whose components are evolved originally and directly for that technology. The technology of preparing pasted lead-acid battery grids is an example of such a technique; the practical physical-chemistry of the lead oxides, as developed empirically in the late nineteenth century battery industry, owed very little to prior technical processes.

But just as the sintered plate evolved as a synthetic technology, so also did the earlier plate-making technology developed by Jungner and Edison. The nickel plating of the tubes and pockets, as well as the outer cans, and the techniques used by Edison in preparing nickel flake and by Jungner in nickel-plating the graphite flakes, came directly out of the prior half century's evolution of commercial nickel plating. The Mond process and powder metallurgy engineering techniques upon which the I.G. Farben chemists based the sintered plate were innovative patterns of the new century, just as nickel plating was a tradition of the last century. Edison, of course, lived well into the twentieth century, but he never had anything to do with the sintered plate process. Brilliant in his synthesis of the technological traditions of the nineteenth century, he never took part in the technical traditions of the new century. Though it was chronologically possible for Edison and the other inventors of his generation to have invented the sintered plate, it was probably no accident that the invention of this process was undertaken by chemists and chemical engineers who grew up in the traditions of the new metallurgy and chemistry. I.G. Farben was not a battery manufacturer, nor was the Mond process a German invention. Yet it is appropriate that one of the leading companies to pioneer in the field of high temperature, high pressure chemical process equipment should have developed the sintered plate idea.

Although I.G. Farben was not a battery manufacturer itself, it had for many years been the raw materials supplier of the nickel and cadmium active materials for the alkaline battery industry. This was the motivation for Farben's sintered plate research in the late '20s. Nevertheless, for a number of years after the granting of the basic sintered patents, the German battery industry did not show any serious interest in the process. In 1936, however, Farben and AFA signed an agreement to produce batteries with sintered plates. The immediate stimulation for this commercialization was the national government's emphasis, starting in 1936, on decreasing Germany's reliance on foreign raw materials.

On 26 August 1936, Hitler gave Göring a directive to create a second Four Year Plan for industrial development. The basic goal of the plan was to give German industry 100 percent independence from foreign natural resources, particularly those resources most essential for military goods.[39] The details of the plan were published 16 October 1936. The plan created a pyramid-type structure, with Göring at the top exercising overall control, and a series of "plenipotentiaries" below him, each controlling a specific sector of industry. Göring exercised little of his power, however, and most of the plenipotentiaries exerted little influence either. The only plenipotentiary to exert any real influence and control over German industry in the early years before Albert Speer took control, was Karl Krauch.

Krauch who had been a leader of I.G. Farben was appointed by Göring plenipotentiary for "Special Problems of Chemical Production," with responsibility for research in raw materials and synthetics. Krauch, beginning in 1936, worked out a detailed plan (the Krauch Plan) for the utilization and development of all chemical and chemically-related resources. Göring approved the plan in July 1938 and gave Krauch his plenipotentiary role.[40]

By 1936 the German armaments industry was already facing shortages of lead. Lead and rubber, two of the most essential materials in lead-acid battery manufacture, were strictly rationed by the Four Year Plan. Of course, the provision of these materials for military batteries had to be sanctioned throughout the war. Nevertheless, a premium was placed on substitutes for the lead-acid system.[41] Moreover, the increasing importance of aircraft in war emphasized the importance of developing an alkaline cell without the "dead weight" present in the Edison-type pocket cell batteries.

The sintered plate development work began in earnest in 1938-1939.[42] AFA carried out the R&D work at its Hagen factory, but, since the Hagen plant was devoted to vehicle bat-

[39] Berenice A. Carroll, *Design for Total War—Arms & Economics in the Third Reich* (The Hague, Mouton, 1968) pp. 129, 144.

[40] Ibid., p. 136.

[41] Nadolny & Treue, *VARTA*, pp. 97-98.

[42] "The German Accumulator Industry," *BIOS Final Report #1129* (1947) p. 19.

teries, the actual manufacture of the sintered plate cells was carried out at AFA's new Hannover and Lethmathe plants which were created specifically for war industry needs. The raw materials were supplied by I.G. Farben as before.

The German sintered electrodes were called DURAC plates. The wartime production never reached very large proportions since the only plant designed specifically for DURAC manufacture—Lethmathe—only neared completion towards the end of the war. All DURAC cells made during the war were used in aircraft, and these were chiefly used in the new jet planes and rockets.[43] These batteries did not perform satisfactorily, however, and the Luftwaffe eventually had to readopt lead-acid batteries.

The problems with the early DURAC batteries were analogous to those facing the early lead-acids. It was difficult at first to produce a uniform sintering of the powder mass, and the plates were therefore often weak and nondurable. Also, the Germans found it hard to impregnate the sintered plates—a problem which may be imagined from the microscopic nature of the plate's pores—and therefore the batteries often suffered from low capacity. Moreover, the early DURAC batteries were rather large and ponderous, and they often buckled in use.[44] Storage battery engineering had once again shown itself to be a difficult and time consuming technology.

Nonetheless, AFA engineers did make considerable progress during the war in developing a practical sintered cell. Although the AFA research facilities at Hagen experimented with a number of new, high energy systems during the war, the only radically new system which impressed Allied intelligence teams after the war was the sintered plate process. For example, M. Barak and B.L. Davies, in a BIOS report published in 1947, recommended that the Lethmathe plant, which had escaped bombing, be shipped entire to Britain, along with the works manager of the Hannover plant, since it was Hannover which did about 80 percent of the actual DURAC manufacture.[45] The French, however, got to Lethmathe first, and it was they who took advantage of the opportunity to make the technology fully practical.

---

[43] The smaller DURAC's were made for the FW 190 and ME 162 and the larger cells for the ME 177; Ibid., p. 21.

[44] Private communication from Dr. M. Barak.

[45] BIOS Report #1129, p. 52.

During the late 1940s, SAFT engineers took the DURAC technology and converted it into a commercial process. The problems which the French engineers solved were several, but they were all based on the problem of quality control. SAFT converted what had essentially been a hand-craft, batch operation into a continuous, automated process. In the German plants the plates were prepared by cutting pieces of iron screening to the shape of the desired electrode. These pieces of screen were then edged with steel tape to give them rigidity, thinly nickel-plated and then gently pressed into a dry mixture of the sinter powder. All this was done by hand with a minimum of machinery. Some sort of internal screen or perforated sheet is always used in the plates to hold the sinter powder and to give the plate strength.

In the SAFT process, a slurry of carbonyl powder is mixed with an aqueous solution of a cellulose derivative. This gives the paste coherence and allows it to be used in this continuous feed process: [46]

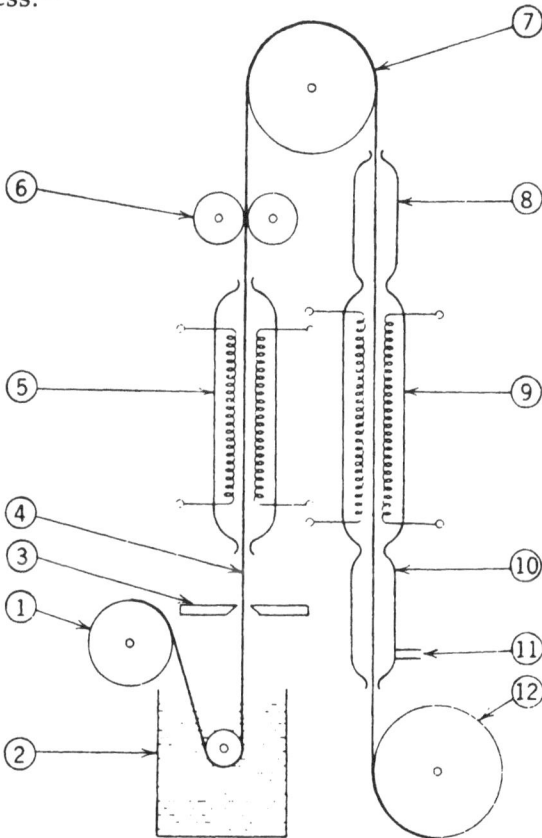

[46] Falk & Salkind, *Alkaline Batteries*, p. 121.

The perforated sheet is fed from a continuous roll (1) into the slurry (2), trimmed to a thickness between .4 and 2.5 mm (3); dried (5); passed through rollers (6&7); preheated (8); sintered (9); and cooled (10). During steps 8 through 10 the sheet is protected from oxidation by a reducing or inert gas (11). Finally, the sintered sheet is rolled up and prepared for impregnation with the active material solution. This is also carried out continuously and automatically. This is the basic technique used around the world today for manufacturing consumer oriented alkaline storage batteries.

VI. Conclusions: Environment and Evolution

It is a commonplace in the modern historiography of technology that scientific discoveries rarely if ever stimulate immediate technological response; the old Baconian idea that scientists discover new principles which engineers can swiftly apply to the solution of practical problems is attractive but unreal. Strong interactions between science and technology do exist, but these interactions are usually quite subtle and unpredictable. The evolution of the storage battery is a case in point.

Science stimulates technology by creating an intellectual climate or environment which directs the thinking of inventors and engineers into new paths. Thus, for example, the sixteenth-seventeenth century scientific image of the atmosphere as a great, ever-present weight led to the invention of the steam engine. In our own day, the technology of electronics emerged out of the scientific concepts of sub-atomic charged particles. So in the nineteenth century the dominant science of electrochemistry created a pervasive intellectual environment in which scientifically-trained inventors thought almost automatically in terms of electrochemical solutions to the practical problems facing them. This we saw in the work of Siemens, Sinsteden, Kirchhof, Planté, and others. This is not to imply that other intellectual environments did not coexist with that of electrochemistry; electromagnetisum was one of these other environments, and we can see that telegraphic inventors of the 1850s, '60s and '70s used both electrochemical and electromagnetic approaches to solve identical telegraphic problems. The work of Siemens is a case in point, although we have treated only one aspect of it here.

The storage battery emerged in this environment and evolved within it until the early 1890s. It is important to keep this relationship between environment and technological evolution in mind. If this is not done, the nature of many of the early proposed uses for storage batteries may seem illogical to the modern mind. This is because the modern electrical engineer uses batteries only when he cannot find a way to avoid them; he receives no stimulus either from his education or from the general scientific-technical intellectual environment to think in terms of electrochemical solution to problems. It was quite the opposite in the nineteenth century.

Moreover, when the technology of electric light and power began at the end of the 1870s, it was largely developed by men whose backgrounds were in telegraphy and electroplating. They therefore transferred the electrochemical orientation of the old to the new technologies. This is not to imply that a deterministic influence operated upon the older telegraphic engineers which forced them all to adopt the battery-approach to the solution of electric light and power problems. On the contrary, we saw that such an old telegraphic hand as Thomas Edison rejected the use of batteries in central stations. However, the influence of the electrochemical environment was strong nonetheless; one example of this is the scores of plans suggested in the 1880s and early '90s to operate domestic electric lighting plants from gargantuan primary batteries. Even Edison exaggerated the need for primary batteries with his development of the Edison-Lalande battery.

During the period 1850-1890 there were a number of examples of the proposed use of storage batteries which are only explicable in terms of the older mind-set. For example, the Planté-Niaudet scheme to increase the voltage of an electromechanical generator by passing its current through a bank of storage batteries was unnecessarily complicated compared to the much more direct method of redesigning the generator's windings. This would be obvious to a modern electrical engineer, and indeed would have been equally clear to most such engineers from the late 1880s on. The training of a nineteenth century telegraphic inventor, however, would not have led him to think in these terms. For the telegraphic expert, voltage and current changes were pictured in terms of changing the number of battery cells in series or parallel; indeed, it was not uncommon for telegraphic engineers to use the Daniell cell as their unit of voltage instead of the volt itself, just as lengths of standard commercial circuit wire were used often as units of resistance in place of the ohm.

Another example of the impress of the older intellectual environment was the early large-scale use of storage batteries as current and voltage dividers, or electrochemical transformers in electric lighting distribution systems. The large banks of primary batteries in major urban telegraphic offices had been used in this fashion for many years. and this led to the analog adaptation of storage batteries in light and power circuits. Such schemes were never widely adopted in practice since their economics were unfavorable. Nevertheless, that these schemes were widely discussed in the 1880s stimulated much of the early investment and innovation in the storage battery field.

One unfortunate result of the influence of the telegraphic technological tradition upon early storage battery manufacture was the tendency to design them like primary batteries. Primary batteries had always been made very simply, and they were usually sold as one of the large line of items stocked by telegraphic supply houses. In fact, in the sense that we today conceive of battery *manufacture*, it is inexact to speak of nineteenth century primary battery manufacture, since it was the consumer who actually manufactured the battery from components purchased in the supply store. The stores themselves obtained the components from several suppliers—the jars from bottle manufacturers; the copper and zinc electrodes from the smelters of these metals; the copper sulphate, nitric and sulphuric acids from chemical supply houses; the porous ceramic separators from pottery makers. Little quality control was possible, but fortunately, little was needed.

This pattern of manufacture was carried over into the field of storage batteries. Scientific instrument makers and telegraphic supply houses made storage batteries as a sideline to their other goods. This was a practice copied by several of the early power and lighting inventors, such as Charles Brush. This practice of methodological carry-over is typical whenever old industries spawn new ones. In the case of the electrical industrial revolution of the later 1870s and '80s, however, this pattern of carry-over has not yet been noted by historians of technology, and a worthwhile research project might be to seek such patterns in non-battery areas of electrical technology.

The popularity of storage batteries in electrical engineering practice during the half century following 1850 is a reflection of the phenomenon of scientific and technological pattern-copying, as discussed above. There was, however, a second major factor which also influenced this popularity. This was what I have chosen to call "technological buffering." A buffer technology is a technique which is introduced for the purpose of adapting a new technology to established industrial processes of which it is a part, but into which it does not yet fit in the most efficient or comfortable way. The function of the buffer is to permit an innovation to be introduced into a pre-existing system without disrupting the entire fabric of the system. Usually, the buffer serves to adapt existing manufacturing techniques to a new raw material, or vice versa.

Thus, for example, the making of coke emerged in the seventeenth century as a means of adapting a new industrial fuel to industrial processes for which wood had always been used, for example, brewing. Coke served to "buffer" the effects of the raw coal on the brewer's malt by removing undesirable elements such as sulphur. In a similar sense, the invention of "puddling" of coke-smelted iron at the end of the eighteenth century was a technique for converting the new iron into an ersatz charcoal iron so that it could be worked by contemporary iron fabrication techniques. Since contemporary fabrication methods (basically forging) could not have been used to work coke irons, the absence of the technological buffer would either have disrupted the entire structure of the iron-using industry, or would have prevented the use of the new irons. The latter of these two possibilities was essentially the situation which did exist from about 1710 until the invention of puddling seven decades later. Puddling disappeared in the late nineteenth century after new manufacturing techniques, such as machining, made the buffer unnecessary, and after newer and better buffering technologies, such as Bessemer and open-hearth steel, had been introduced.

In this same sense, the first proposed uses for the storage battery were for its function as a buffer. Indeed, in the case of storage batteries, the "buffer" metaphor takes on a more concrete reality—the substitution of generators in telegraphy for primary batteries necessitated the addition of some sort of device to "buff" or smooth the erratic generator output. In effect, the function of the electrochemical buffer was to convert the electromagnetically-generated current into an ersatz battery current, just as puddling made coke iron into an ersatz charcoal iron. In this sense, the electricity can be conceived as the raw material to be "finished" by the telegraphic instruments.

The development of "buffering" technologies occurs most commonly during periods of rapid technical change. It is not surprising, then, that during the early decades of electrical engineering practice, the storage battery should have been proposed for so many different applications, the great majority of which were abandoned once electrical engineering started to achieve a degree of maturity in the 1890s. During the infancy of the dynamo, when electrical engineers who were brought up in the telegraphic engineering tradition did not yet know how to use electromagnetic current directly in their circuits, the storage battery, like puddled iron, was an expensive, labor-intensive, but necessary intermediary.

One unexpected result of this study was the discovery of very different patterns of the early development of storage batteries in Europe and the United States. An even more unexpected result was the identification of the reason for these different patterns. My a priori assumption had been that the evolution of storage battery technology would have proceeded at about the same pace in Europe and the United States from 1880 onward, since the larger technologies which the batteries were intended to serve—electric lighting, power, and transportation—went through roughly identical patterns of development in both these areas at this time. And yet, for at least the first decade of the existence of a commercial storage battery industry, the American evolution of this device was far behind that of Europe.

Having determined this, I tried to find a rational, economically-determined reason for this divergence of innovative patterns. I assumed that there must have existed different economical or technological environments in these two areas which would have made the application of storage batteries less useful in the United States. However, all attempts to identify such different environments failed, at least as far as any connection with different relative usefulness of batteries is concerned. I was forced, therefore, to accept the statements of contemporary engineers, that the prejudice of American electrical engineers against storage batteries was the chief reason for the slow growth of the industry here.

The opposition of Thomas Edison to storage batteries set the pattern for early American electrical engineering attitudes towards their use. The only significant market for storage batteries during the first two decades of electric power and lighting was for powerhouse use, and the great majority of such stations built in the United States were designed by engineers trained in the Edison system. In Britain, no one man controlled the technology of electric distribution so extensively as Edison did in the United States; a number of those who favored D.C. distribution in Britain also favored the use of storage batteries as auxiliaries. Therefore, the British battery industry possessed a market for economic survival which the United States industry lacked.

Edison's opposition to batteries sprang from his knowledge of their poor performance during the first years of the 1880s; a fact which convinced him that their inclusion in an electric lighting system would be uneconomical. Yet the earliest British lead-acid cells were no better than those of the American industry. This did not stop such English engineers as R.E.B. Crompton from recommending battery use in what were basically the same kinds of D.C. systems which Edison was designing. Neither Edison's belief that storage batteries could never be used practically nor Crompton's belief that they soon would be was a purely rational decision, but upon these divergences in beliefs was based the different patterns of evolution of the industry in Britain and the United States.

The subsequent period of American catch-up during the 1890s is also surprising, although here the explanation can be attributed to clearly-defined economical-technical factors— chiefly to the American lead in trolley technology. It must have come as a surprise to the early promoters of storage battery streetcars, who saw the battery car as being in competition with the overhead-conductor car, that the very success of the latter system assured the storage battery of a large market in the 1890s and the early years of the new century. And yet, this pattern of technical evolution for battery transportation was analogous to the contemporary evolution of battery use in central stations.

The earliest central station battery schemes were designed to simplify and thereby to cheapen electric lighting service. Similarly, the self-propelled battery streetcar was conceived as simpler than either the trolley or conduit systems, with their complex and expensive conductors. Yet such simplicity, wherever it was sought in early electric lighting and power technology, whether with regard to battery usage or otherwise, proved to be illusory.

The survival of storage battery streetcar projects into the 1890s represents another example of the influence of noneconomic factors on the evolution of what was a technology which evolved chiefly through economic influence. The function of most streetcar lines was to move people from the suburbs into the central cities, or to move people inside the central cities themselves. Consequently, the environment of late nineteenth century central cities played a complex role in the evolution of urban electric transportation. On the economic side, central cities needed improved transport systems at the end of the century. On the other hand, such areas in those days were the locus of upper-class housing, or, in Europe at least, for very traditional and stable middle-class housing. Therefore, such areas often vigorously resisted the introduction of the overhead-conductor streetcar, since this was seen as a source of disfigurement of the community.

Although it would be an exaggeration to attribute all battery streetcar innovation to this big city anti-trolley environment, it is nonetheless true that all these systems which were more than experiments and which lasted for more than an insignificant period of time, were all built in areas where trolleys were either banned or strongly discouraged. It is therefore significant to note the crucial role of the central urban environment at the turn of the century in the evolution of storage battery technology. We have already seen how central cities, by their concentration of large numbers of well-to-do residents, were the ideal environment for the growth of the earliest central station lighting systems. These early systems were almost all D.C., and, since they usually remained D.C. through the 1890s and early years of the new century, they provided an excellent environment for one phase of storage battery evolution. For the reasons just discussed, central cities also provided a fertile environment for the development of the early transport storage battery in streetcars.

In addition, the post-1895 evolution of the electric automobile and its battery can also be attributed largely to the environment of the central cities. Just as the residents of these areas earlier opposed the trolley, they often chose the electric car over its gasoline rival not out of economic considerations, but from aesthetic considerations. Just as the relatively small size of these areas did not make D.C. current distribution (and thus battery use) impractically expensive, so this same relatively small size made the limited speed and range of the electric vehicles acceptable.

Unfortunately for the storage battery, the central urban environment which supported D.C. distribution, battery streetcars, and private electric automobiles did not long survive into the new century. The impoverishment of these areas and the growth of the suburbs began and continued to accelerate. Thus, A.C. steadily gained complete dominance over D.C. and the internal combustion vehicles triumphed over the battery vehicles and even over the once-triumphant trolleys.

But by one of the ironic turns of fate that enliven history, it was this very offshoot of the destruction of its original environment which led to the mushrooming growth of storage battery use during the past seven decades. Suburbs popularized the gasoline car and prepared the groundwork for the massive SLI battery industry of the twentieth century.

Select Bibliography

ABRAHAMS, HAROLD J. and SAVIN, MARION B. (ed). *Selections from the Scientific Correspondence of Elihu Thomson*. Cambridge: MIT Press, 1971.

ALGLAVE, Em. 7 Boulard, J., *La Lumiere Electrique*. Paris: Didot et Cie, 1882.

ALLSOP, F.C. *Practical Electric Bell Fitting*. N.Y.: E.&.F.N. Spon, 1890 Also, 1897 and 1905 Editions.

APPLEYARD, ROLLO. *The History of the Institution of Electrical Engineers, 1871-1931*. London: The Institution of Electrical Engineers, 1939.

ARNDT, K. "Das Leclanche-Element." *Elektrotechnische Zeitschrift* 22: 817.

BASCH, CARL. *Die Entwickkelung der Elecktrischen Beleuchtung und der Industrie Elektrischer Glühlampen in Deutschland*. Berlin: F. Siementroth, 1910.

BECQUEREL, A.C. & E., *Traite D'Electricite et de Magnetrome et des Applications de ces Sciences a la Chemie, a la Physiologie et Aux Arts*. Paris: Didot & Freres, 1856, 3 vols.

BELL, LOUIS. *Power Distribution for Electric Railroads*. N.Y.: McGraw-Hill, 1900.

BELLOC, ALEXIS. *La Telegraphie Historique Depuis les Temps. Les Plus Recules Jusqua nos jours*. Paris: Firmin-Didot, 1888.

BLAVIER, E.E. *Nouveau Traite de Telegraphie Electrique*. 2 vols. Paris: 1865.

————. *Considerations sur le Service Telegraphique*. Nancy: Sordoillet, 1872.

BOLEN, MARJORIE (ed). *Literature Search on Dry Cell Technology*. Atlanta: Georgia Institute of Technology, 1948.

BONEL, A. *Historie de La Telegraphie Description des Principaux Appareils Aeriens et Electriques*. Paris: Ballay et Conshon, 1857.

BOSTOCK, JOHN. *An Account of the History and Present State of Galvanism*. London: 1818.

BOYD, T.A. "The Self Starter." *Technology and Culture* 9: 585.

BRIGHT, ARTHUR A. *The Electric Lamp Industry: Technological Change and Economic Development from 1800 to 1947*. N.Y.: Macmillan Company, 1949.

BRYAN, GEORGE S. *Edison—The Man and His Work*. London: Alfred A. Knopf, 1926.

CAIRNS, E.J. *Development of High Energy Batteries for Electric Vehicles*. Argonne Nat. Lab.: 1971.

CALLAUD, ARMAND. *Essai Sur Les Piles*. Paris: Gauthier-Villars, 1875.

CAPRIOGLIO, GIOVANNI. "Review of Battery Systems for Electric Vehicles." SAE paper 690129, 13-17 January 1969.

CARDWELL, D.S.L. *Turning Points in Western Technology*. N.Y.: Neale Watson, 1972.

CARHART, HENRY S. *Primary Batteries*. Boston: Allyn & Bacon, 1891.

Carnegie Library of Pittsburgh. *Men of Science & Industry*. Pittsburgh: Carnegie Library, 1915.

CARROLL, BERNICE A. *Design for Total War—Arms and Economics in the Third Reich*. The Hague: Mouton, 1968.

CAZIN, A. *Traite Theorique et Pratique des Piles Electriques*. Paris: Gauthier-Villars, 1881. (One of the best nineteenth-century treatments.)

CLARK, WILLIAM J. "Electric Railways in America." *Cassier's Mag.* 16: 1899.

COHO, H.B. "An Historical Review of the Storage Battery." *Transactions of the American Electrochemical Society* 3: 159.

COOK, ARTHUR L. *Elements of Electrical Engineering*. N.Y.: John Wiley, 1935.

COOK, N.A. "Analysis of Fuel Cells for Vehicular Applications." SAE paper 680082, 8-12 January 1968.

COOPER, W.R. *Primary Batteries: Their Theory, Construction and Use*. London: *The Electrician*, 1901.

COTTON, HARRY. *Electrical Technology*. London: Isaac Pitman 1949.

CRENNELL, J.T. and LEA, F.M. *Alkaline Accumulators*. London: Longmans, Green and Company, 1928.

CROCKER, F.B. *Storage Batteries*. Chicago: American School of Correspondence, 1906.

CROMPTON, R.E.B. "Central Station Lighting: Transformers vs. Accumulators." *Journal of the Society of Telegraph Engineering and Electricians*. 17: 349.

————. *Reminiscences*. London, Constable & Co. 1930.

————. "The First Installation of House to House Supply in the United Kingdom." *Transactions of the Newcomen Society* 11: 90.

CROSBY, OSCAR, T. & BELL, LOUIS. *The Electric Railway in Theory and Practice*. N.Y.: W.J. Johnston Co., 1893.

CULLEY, R.S. *A Hand book of Practical Telegraphy*. London: Longmans, Green and Company, 1885. Eighth edition.

CUTHBERTSON, JOHN. *Practical Electricity and Galvanism*. London: J. Callow, 1821.

DALIN, GEORGE and KOBER, FREDERICK. "A Hybrid Battery System for Electric Vehicle Propulsion." SAE paper 690203, 13-17 January 1969.

DANIELS, FARRINGTON. *Physical—Chemical Aspects of the Leclanche Dry Cell*, 1928.

DAVIS, DANIEL. *Davis's Manual of Magnetism*. Boston: Daniel Davis, 1842.

————. *The Book of the Telegraph*. Boston: Daniel Davis, 1851.

DAWSON, PHILLIP. "Electric Railways and Tramways." *Engineering*, 1897.

DE CEW, GLASER. *Magneto and Dynamo—Electric Machines*. London: Symons and Company, 1884. Good historical review of the dynamo.

DICK, J.R. and FERNIE, F. *Electric Mains and Distributing Systems*. London: Benn Bros., 1919.

DOLEZALAK, FRIEDRICH. *The Theory of the Lead Accumulator*. N.Y.: John Wiley, 1904.

DONAVON, MICHAEL. *Essay on the Origin, Progress and Present*. Dublin: 1816.

DROTSCHMANN, C. "Uber eine Braunnstein-Salmiak-Reaktion." *Zeitschrift für Elektrochemie*. 35: 194.

DRUCKER, C. and FINKELSTEIN. *Galvanische Elemente und Akkumulatoren*. Leipzig: Akademische Verlagsgesellschaft, 1932.

DUB, JULIUS. *Die Anwendung des Elektromagnetismus*. Berlin: Julius Springer, 1873.

DU MONCEL, THEODORE. *Traite Theorique et Pratique de Telegraphie Electrique*. Paris: Gautheirs Villars, 1864.

————. *Electric Lighting*. London, George Routledge, 1883.

————. & GERALDY, FRANK. *Electricity as a Motive Power*. London: E. & F. Spon, 1883.

DUNSHEATH, PERCY. *History of Electrical Power Engineering*. Cambridge: MIT Press, 1962.

DURHAM, JOHN. *Telegraphs in Victorian London*. Cambridge: Golden Head Press, 1959.

DYER, FRANK L. & MARTIN, THOMAS C. *Edison—His Life and Inventions*. N.Y.: Harper & Bros., 1910. 2nd ed. 1924.

ELBS, KARL. *Die Akkumulatoren*. Leipzig: Johann Barth, 1893.

Electrical Storage Battery Company. *The Sale of Current for Electric Automobiles by Electric Lighting and Power Stations*. Bulletin #109, April 1908.

ELIESON, C.P. "Electric Locomotion." *Telegraphic Journal and Electrical Review* 23: 7.

FALK, S. UNO and SALKIND, ALVIN, J. *Alkaline Storage Batteries* N.Y.: John Wiley, 1969. Begins with good historical survey.

FARBER, EDUARD and SCOTT, WILSON L. *The History of the Electronic Theory of Oxidation and its Practical Application*. Actes du XIe Congres International d'Histoire des Sciences, Vol. 4. Warsaw: 1968.

FEDER, WALTER. "Electrode Polarization and Secondary Currents: Their Influence on Mutual Development in Electro-Chemistry During the First Half of the 19th Century." *Proceedings of the 10th International Congress on the History of Science*. Ithaca: 1962.

Federal Power Commission. *Development of Electrically Powered Vehicles*. Washington, D.C.: 1967.

FISKE, LT. BRADLEY A. *Electricity in Theory & Practice or The Elements of Electrical Engineering*. N.Y.: D. Van Nostrand, 1883.

FITZ-GERALD, DESMOND. *The Lead Storage Battery, Its History, Theory, Construction and Use*. London: Biggs & Co., 1916.

FLEMING, J.A. *Fifty Years of Electricity—The Memoirs of an Electrical Engineer*. London: Iliffe & Sons, 1921.

FRANKLIN, WM. *A Treatise on the Elements of Electrical Engineering* Vol. I—Direct and Alternating Current Machines & Systems. N.Y.: MacMillan, 1917.

FRIESS, RALPH. *The Reactions of Ammonia on the System $ZnCl_2$-NH_4Cl-H_2O*. 1930.

GAGET, MAURICE. *La Navigation Sous Marine*. Paris: 1901.

GILL, SYDNEY. "A Voltaic Enigma and a Possible Solution to It." *Annals of Science* 33: 351-370.

GIRAN, HENRI. "La Vie et L'Oeuvre de Gaston Plante." *Memoires deL'Academie des Sciences, Inscription et Belles Lettres de Toulouse* 12: 67.

GLADSTONE, J.H. & TRIBE, ALFRED. *The Chemistry of the Secondary Batteries of Plante and Faure*. London: MacMillan and Company. These were the two who first worked out the true chemistry of the cell.

GLASSTONE, SAMUEL. *Testbook of Physical Chemistry*, Second Edition Princeton: D. Van Nostrand, 1946.

GORE, G. *The Art of Electro-Metallurgy*. London: Longmans, Green and Company, 1877.

GORMAN, MEL. *The Early History of the Electrochemistry of Silver Chloride*. Vol. 4 Actes du XIe Congres International d'Histoire des Sciences. Warsaw: 1968.

GOSLING, WILLIAM. "The Genesis of Electrical Engineering." Univ. of Swansea, 24 January 1967.

GOTSHALL, W.C. *Notes on Electric Railway Economics and Preliminary Engineering*. N.Y.: McGraw-Hill, 1904.

GREER, HENRY. *Recent Wonders in Electricity, Electric Lighting, etc.* N.Y.: New York Agent College of Electrical Engineering, Co. 1885.

GRINDLE, G.A. "On the Use of Storage Batteries in Connection with Electric Tramways." *Journal of the Institution of Electrical Engineers* 30: 1098.

HADLEY, H.E. *A Class Book of Magnetism and Electricity*. London: MacMillan, 1936.

HALSALL, VINCENT. "The Lead-Acid Battery—New Horizons in Power through Use of Plastics." (SAE paper 680389, 20-24 May 1968).

HAMMOND, JOHN WINTHROP. *Men and Volts—The Story of General Electric*. N.Y.: J.B. Lippincott, 1941.

HAMPEL, CLIFFORD A. (ed.). *The Encyclopedia of Electrochemistry*. N.Y.: Reinhold Publishing Co., 1964.

HANSSON, SVEN A. "Waldemar Jungner och Jungnerackumulatorn." *Daedalus Tekniska Museets Arsbok* 1963: 77.

HARDING, FRANCIS. *Electric Railway Engineering*. N.Y.: McGraw-Hill, 1911.

HAUEL, ANNA P. "The Cadmium—Nickel Storage Battery." *Transactions of the Electrochemical Society*. 26.

HARTLEY, SIR HAROLD. *Humphry Davy*. London: Thomas Nelson and Sons, 1966.

HAUSMANN, ERICH. *Telegraph Engineering*. N.Y.: D. Van Nostrand, 1915.

Hawkins Electrical Guide #7. N.Y.: Theodore Audel & Company 1922.

HEAP, DAVID, P. *Electrical Appliances of the Present Day Being a Report on the Paris Exhibition of 1881*. N.Y., Van Nostrand, 1884.

HEDGES, KILLINGWORTH. *Central Station Electric Lighting*. London: E. &. F.N. Spon, 1888.

—————. *Continental Electric Light Central Stations*. London: E. &. F.N. Spon, 1892.

—————. *American Electric Street Railways*. London, E. & F.N. Spon, 1894.

HEISE, GEORGE W. "Pages from the History of the Primary Battery." *Proceedings of the Electrochemical Society* 93 20.

HEISE, GEORGE W. and CAHOON, N. COREY. *The Primary Battery*, Vol. 1 N.Y.: John Wiley, 1971. Begins with excellent historical survey.

HERING, CARL. *Recent Progress in Electric Railways*. N.Y., W. J. Johnston, 1892.

HEROLD, RODOLPHE. *Etude Theorique et Pratique de l'Accumulateur Fer-Nickel*. Belgium: 1924.

HERRICK, ALBERT B. & BOYNTON, EDWARD C. *American Electric Railway Practice*. N.Y.: McGraw-Hill, 1907.

HIRSCH, MARK D. *William C. Whitney—Modern Warwick*. N.Y.: Dodd, Mead, 1948.

HOFFMAN, E. *Die Entwicklung des Deutschen Telegraphwesens seit dem Jahre 1875*. Berlin: Herbig, 1880.

HOPKINS, GEORGE M. 20th ed. *Experimental Science*. N.Y.: Munn & Co., 1898.

HOWARD-WHITE, F.B. *Nickel; An Historical Review*. N.Y.: D. Van Nostrand, 1963.

HUGHES, THOMAS PARKE. *Elmer Sperry, Inventor and Engineer*. Baltimore: Johns Hopkins Press, 1971.

ICS Reference Library. *Storage Batteries*. Scranton: International Textbook Co., 1908.

IHDE, AARON J. *The Development of Modern Chemistry*. N.Y.: Harper and Row, 1964.

ILLINGWORTH, T. *Battery Traction on Tramways and Railways*, (Locomotion Paper #14). Surrey: The Oakwood Press, 1961. Good historical treatment by one of the developers of the Drumm Trains.

IRVINE, ANDREW CO. "The Promotion and First Twenty-two Year's History of a Corporation in the Electrical Manufacturing Industry." (unpublished master's thesis, Temple University, 1954) A history of the ESB Company.

JIRSA, FRANZ. "Studie über einen Silberakkumulator Ag/OH'/Fe." *Zeitschrift für Elektrochemie* 33: 129.

JONES, ALEXANDER. *Historical Sketch of the Electric Telegraph*. N.Y.: G. Putnam, 1852.

JONES, BERNARD. *Electric Primary Batteries*. London: Cassell and Company, 1913.

JONES, FRANCIS ARTHUR. *The Life Story of Thomas Alva Edison*. N.Y.: Grosset and Dunlap, 1907.

JONES, PAYSON. *A Power History of the Consolidated Edison System, 1878-1900*. N.Y.: Edison Company, 1940.

JOULE, JAMES PRESCOTT. *Joint Scientific Papers*. London, Dawsons of Pall Mall, 1963 (reprint of 1887).

JUNG, J. *Entwickelung des Deutschen Post und Telegraphwesens in den letzten 25 Jahren*. Leipzig: Duncker u. Humblot. 1893.

KARRASS, THEODOR. *Geschichte der Telegraphie*. Braunschweig: 1909.

KEMMANN, GUSTAV. *Die Berliner Elektrizitätswerke dis ende 1896*. Berlin: J. Springer, 1897.

KILGOUR, MARTIN H. & SWAN H. & BIGGS, C.H. *Electrical Distribution—Its Theory & Practice*. London: Biggs & Co.

KING, W. JAMES. *The Development of Electrical Technology in the 19th Century* (Bulletin #228). In three parts:
1) "The Electrochemical Cell and the Electromagnet"
   (Paper #28)
2) "The Telegraph and Telephone"
   (Paper #29)
3) "The Early Arc Light and Generator"
   (Paper #30)
Wash., D.C.: United States National Museum, 1962.

KLEIN, BURTON H. *Germany's Economic Preparations for War.* Cambridge: Harvard U. Press, 1959.

KOLLERT, J. and SIEG, E. *Stromquellen und Akkumulatoren.* Leipzig: S. Hirzel, 1901.

KRELLER, EMIL. *Die Entwicklung der Deutschen Elektrotechnischen Industrie.* Leipzig: van Duncker & Humblot, 1903.

LASZCZYNSKI, S. VON. "Neuere Arbeiten über Sammler aus anderen Metallen als Blei." *Zeitschrift für Elektrochemie* 1901: 821.

LAUMEISTER, BRUCE R. "The General Electric Vehicle." SAE paper 680430 20-24 May 1968.

LE BLANC, MAX. *The Elements of Electrochemistry.* London: MacMillan, 1896.

LEE, JOHN. *The Economics of Telegraphs and Telephones.* London: 1913, Isaac Pitman.

LINES, ROBERT R. *Report on Telegraphs and on Telegraphic Administration.* Washington: Government Printing Office, 1876.

LLOYD, HERBERT. "Storage Batteries and Electric Railways." *Cassier's Magazine* 16.

LOEB, LEON. "The MIT Electric Car." SAE paper 690118, 13-17 January 1969.

LORING, A.E. *A Handbook of the Electro-Magnetic Telegraph.* N.Y.: Van Nostrand, 1878.

LUDEWIG, JULIUS. *Der Reichstelegraphist; Ein Handbuch zum Selbstunterricht.* Berlin: W. Baensch, 1877.

LÜPKE, ROBERT. *The Elements of Electrochemistry.* London: H. Grevel, 1897.

LYNDON, LAMAR. *Storage Battery Engineering.* London: McGraw-Hill, 1911.

MACKENZIE, J.K. "The Distribution of Electricity by Means of Secondary Generator or Transformer." *Journal of the Society of Telegraph Engineers and Electricians* 17: 120.

MANTELL, C.L. *Batteries and Energy Systems.* N.Y.: McGraw-Hill, 1970.

MARLAND, E.A. *Early Electrical Communication.* London: Abelard-Schuman, 1964. Good technical treatment of early telegraphy.

MARTIN, THOMAS and COLES, STEPHEN. *The Story of Electricity.* N.Y.: Story of Electricity Company, 1919.

————. & SACHS, JOSEPH. *Electrical Boats and Navigation.* N.Y.: C.C. Shelley, 1894.

————. WETZLER, JOSEPH. *The Electric Motor and Its Applications.* N.Y.: W.J. Johnston, 1887.

MAVER, WILLIAM. *Practical Systems of Electrical Telegraphy.* Newport: Newport Torpedo Station, 1888.

————. *American Telegraphy: Systems, Apparatus and Operation.* New York: 1892.

————. & DAVIS, MINOR M. *The Quadruplex.* N.Y.: W.J. Johnston, 1885.

McQUEEN, ALEXANDER. *A Romance in Research—The Life of Charles F. Burgess.* Pittsburgh: Instruments Publishing Company, 1951.

McNICOL, DONALD. *American Telegraph Practice.* N.Y.: McGraw-Hill, 1913. An excellent guide to the fundamental technology of the telegraph.

MEYER, HUGO R. *The British State Telegraphs.* N.Y.: MacMillan 1907. Good statistics but must be taken with a grain of salt because of polemical nature.

MILWARD, ALAN S. *The German Economy at War.* London: The Athlone Press, 1965.

MIYAKE, YOSHIZO. "Development of Electric Vehicles in Japan." SAE Paper 690072, 13-17 January 1969.

MOERMAN, THEOPHILE. *Notice sur L'Electro-Metallurgie.* Paris: Baudry, 1882.

MOORE, F.J. *A History of Chemistry.* N.Y.: McGraw-Hill, 1931.

NATIONAL BUREAU OF STANDARDS. *Specification for Dry Cells and Batteries.* Washington, D.C.: U.S. Govt. Printing Office, 1959. Contains history of the specifications.

NEBEL, B. "Über die an einem die Lalande-Element gemachten Beobachtungen." *Centralblatt für Elektrotechnik* 9: 75.

————. "Neuerungen an Nichtblei-Akkumulatoren." *Zeitschrift für Elektrochemie* 1902: 265.

NERNST, WALTER. *Theoretical Chemistry.* London: 1923. Translation by L.W. Codd.

NIAUDET, ALFRED. *Elementary Treatise on Electric Batteries*, 1890-6th ed. N.Y.: John Wiley, 1882.

NIBLETT, J.T. *Secondary Batteries*. London: Biggs and Co. 1895.

NIVEN, JOHN; CANBY, COURTLANDT; and WELSH, VERNON. *Dynamic American, A History of General Dynamics Corporation*. N.Y.: Doubleday.

NOAD, HENRY. *A Manual of Electricity*. London: Lockwood and Company, 1859.

O'DEA, W.T. "Electrical Invention and Reinvention." *Transactions of the Newcomen Society* 20: 75.

OETTEL, FELIX. *Introduction to Electrochemical Experiments*. Phila.: P. Blakiston, 1897.

OPPERMANN, G. "Eine einfache Methode zur herstellung von Kupferoxyd-Elektroden für Kupfer-Oxyd-Zinc-Alkali Elemente." *Elektrochemische Zeitschrift* 10: 191.

OSTWALD, WILHELM. *Elektrochemie—Ihre Geschichte und Lehre*. Leipzig: von Veit, 1895-1896.

PARSONS, R.H. *The Early Days of the Power Station Industry*. Cambridge U. Press, 1940.

PARTINGTON, J.R. *A History of Chemistry*, vol 4. London: MacMillan and Company, 1964.

PELTIER, J.C.A. *Notice sur la vie et les Travaux Scientifiques de J.C.A. Peltier*. Paris: Edouard Bautruche, 1847.

PENDER, HAROLD. *Direct Current Machinery*. N.Y.: John Wiley, 1928.

Philadelphia Electrical Handbook. N.Y.: American Institure of Electrical Engineers. 1904.

PLANTÉ, GASTON. *The Storage of Electrical Energy*. London: Whittaker and Company, 1887. Trans. Paul B. Elwell.

POLE, WILLIAM. *The Life of Sir William Siemens*. London: John Murray, 1888.

POPE, FRANK L. *Modern Practice of the Electric Telegraph*, 5th Ed. N.Y.: Van Nostrand, 1871.

POTTER, EDMUND C. *Electrochemistry—Principles & Applications*. London: Cleaver-Hume.

PREECE, W.H. "The Leclanche Battery." *Journal of the Society of Telegraph Engineers and of Electricians* 11: 150.

————. "Use of Secondary Batteries in Telegraph." *The Telegraphic Journal and Electrical Review*. 1884: 410.

PREECE, W.H. and SIVEWRIGHT, J. *Telegraphy*, 6th ed. London: Longmans, Green and Co., 1887.

PRESCOTT, GEORGE B. *History, Theory and Practice of the Electric Telegraph*. Boston: Ticknor and Fields, 1860.

————. *Electricity and the Electric Telegraph*. N.Y.: D. Appleton and Co., 1877.

Primary Battery Division—Thomas A. Edison Industries. *Seventy-Five years of Packaged Power*. Bloomfield: McGraw-Edison, 1964.

REYNIER, EMILE. *The Voltaic Accumulator*. London: E. & F.N. Spon, 1889. Trans. by J.A. Berly. Good review of French batteries with dates and patents.

RICHARDSON, H.K. "The Modern Dry Cell, Its Development, Construction and Manufacture." *Metallurgical and Chemical Engineering* 1912: 531.

RIDER, JOHN HALL. *Electrical Traction* (Specialists Series) London: Whittaker & Co. 1903.

ROBERTS, MARTIN F. "On Batteries." *Journal of the Society of Telegraph Engineers* 6: 257.

ROLOFF, MAX & SIEDE, ERICH. "Neuerungen auf dem Gebiete Der Akkumulatorentechnik im Jahre 1905." *Zeitschrift für Elektrochemie* 1906: 321.

ROLT, L.T.C. *A Short History of Machine Tools*. Cambridge: MIT Press, 1965.

ROLPH, S. WYMAN. *Exide—The Development of an Engineering Idea—A Brief History of the Electric Storage Battery Company*. Phila.: The Newcomen Society of North America, 1951.

ROLT-WHEELER, FRANCIS. *Thomas Alva Edison*. N.Y.: MacMillan, 1922.

ROSENBERG, NATHAN. "Economic Development and the Transfer of Technology: Some Historical Perspectives." *Technology and Culture* 11: 550.

————. "Factors Affecting the diffusion of Technology." *Explorations in Economic History* 10: 3.

ROUSCH, G.A. "The Manufacture of Carbon Electrodes." *The Journal of Industrial and Engineering Chemistry* 1909: 286.

ROWSOME, FRANK. *Trolley Car Treasure.* N.Y.: Bonanza Books, 1956.

SABINE, ROBERT. *The Electric Telegraph.* London: Virtue Bros. and Company, 1867.

SALOMONS, SIR DAVID. *Management of Accumulators and Private Electric Light Installations,* 4th Ed. London: Whittaker and Company, 1888.

SCHMOOKLER, JACOB. *Invention and Economic Growth.* Boston: Harvard U. Press, 1966.

SCHNEER, CECIL J. *Mind and Matter.* N.Y.: Grove Press, 1969.

SEIGER, H.N. & LYALL, A.E. "The High Energy Lithium Battery System in Electric Vehicle Propulsion." SAE Paper 680454, 20-24 May 1968.

SEITH, WOLFGANG. "Über den Kreislauf des Bleis im Bleiakkumulator." *Zeitschrift für Elektrochemie* 34: 362.

SHARLIN, HAROLD I. *The Making of the Electrical Age.* London: Abelard Schuman, 1963.

————. *The Convergent Century.* London: Abelard-Schuman, 1966.

SIEMENS, C.W. "A Contribution to the History of Secondary Batteries." *The Telegraphic Journal* 1881: 376.

SMEE, ALFRED. *Elements of Electro-Metallurgy.* London: Longmans, Brown, Green and Longmans, 1843. (1st ed. 1840).

SMEE, ELIZABETH MARY. *Memoir of the late Alfred Smee.* London: George Bell and Sons, 1878.

SMITH, E.F. *Life of Robert Hare—An American Chemist.* Phila.: J.B. Lippincott, 1917.

SPRAGUE, JOHN T. *Electricity: Its Theory, Sources & Applications* London: E. &. F.N. Spon, 1884.

STANLEY, ROBERT C. *Nickel, Past and Present.* Toronto: 1927.

STARKMAN, E.S. "Prospects of Electric Power for Vehicles." SAE Paper #680541, 12-15 August 1968.

STURGEON, WILLIAM. *A Course of Twelve Elementary Lectures on Galvanism.* London: 1843.

TATON, RENE. *Science in the Nineteenth Century.* N.Y.: Basic Books, 1965.

TEGG, WILLIAM. *Posts & Telegraphs, Past & Present.* London: William Tegg, 1878.

THOMPSON, SILVANUS P. *Storage of Electricity.* London: E. &. F.N. Spon, 1881.

TOMMASI, DONATO. *Traite des Piles Electriques.* Paris: George Carre, 1889.

TOULMIN, H.A. *Patent Law for the Inventor and Executive.* N.Y.: Harper and Bros., 1928.

TREADWELL, AUGUSTUS. *The Storage Battery.* N.Y.: MacMillan, 1898.

TROMMSDORFF, JOHANN B. *Geschichte des Galvanismus, oder Der Galvanischen Elektricität Vorzüglich in Chemischer Hinsicht.* Erfurt: 1803.

TURNBULL, LAURENCE. *The Electro-Magnetic Telegraph With an Historical Account of its Rise, Progress & Present Condition.* Phila.: A. Hart, Carey & Hart, 1853.

TURNOCK, L.C. *Electrochemistry for Engineers.* Ann Arbor: Edwards Bros., 1923.

UNDERHILL, CHARLES R. *Power Factor Wastes.* N.Y.: McGraw-Hill, 1926.

UNITED GAS IMPROVEMENT COMPANY, *Forty Five Years of Service 1882-1927.* Phila.: UGIC.

URQUHART, J.W. *Electro-Plating—A Practical Handbook.* London: Crosby Lockwood, 1880.

VANDERBILT, BYRON M. *Thomas Edison, Chemist.* Wash., D.C.: American Chemical Society, 1971.

"Die Verwendung von Accumulatoren im Telegraphenbetrieb." *Centralblatt für Elektrotechnik* 11: 358.

VINAL, GEORGE W. *Storage Batteries.* N.Y.: John Wiley, 1930.

————. *Primary Batteries.* N.Y.: John Wiley, 1950.

————. "Manuscript Biography of George Wood Vinal" (typescript) American Institute of Physics Library, 1963.

WADE, E.J. *Secondary Batteries: Their Theory, Construction and Use*. London: The Electrician Printing and Publishing Company, 1902.

WADSWORTH, C. *Primary Battery Ignition*. N.Y.: Van Nostrand, 1912.

WAHL, WILLIAM H. *Galvanoplastic Manipulations: A Practical Guide for the Gold & Silver Electroplater*. Phila.: Henry Carey Baird, 1883.

WALKER, CHARLES V. *Electric Telegraph Manipulation*. London: George Knight, 1850.

"Where The Alkaline Battery Leads. *Electrical Review* 110: 157.

WETZELS, W. "Johann Wilhelm Ritter: Physik im Wirkungsfeld der Duetschen Romantik." Princeton University Dissertation, 1969.

WHITTAKER, EDMUND. *A History of the Theories of Aether and Electricity*. Vol 1. N.Y.: Harper Torchbooks, 1960.

WILKINSON, C.H. *Elements of Galvanism* 2 vols London: John Murray, 1804.

WILLIAMS, L. PEARCE. *Michael Faraday*. N.Y.: Basic Books, 1964.

WOODBURY, DAVID O. *A Measure of Greatness*. N.Y.: McGraw-Hill, 1949.

WOUK, VICTOR & SEIGER, HARVEY. "Design of an Electronic Automobile Employing Nickel-Cadmium Batteries." SAE Paper #690454, 19-23 May 1969.

WRIGHT, AUGUSTINE M. *American Street Railways: Their Construction, Equipment & Maintenance*. Chicago: Rand McNally 1888.

ZETSCHE, KARL EDMUND. *Geschichte der Elektrischen Telegraphie*. Berlin: Julius Springer, 1877.

Index

I